THE BOOK ®

BMW R850 & 1100
4-valve Twins
Service and Repair Manual

by Matthew Coombs

Models covered

(3466 -248)

R850R. 848cc. UK 1995-on, US 1996-on
R1100R. 1085cc. UK and US 1995-on
R1100RS. 1085cc. UK 1993-on, US 1994-on
R1100RT. 1085cc. UK and US 1996-on
R1100GS. 1085cc. UK 1994-on, US 1995-on

© Haynes Publishing 1998

A book in the **Haynes Service and Repair Manual Series**

ABCDE
FGHIJ
KLMNO
PQRST

Printed by **J. H. Haynes & Co. Ltd., Sparkford, Nr Yeovil, Somerset, BA22 7JJ, England**

Haynes Publishing
Sparkford, Nr Yeovil, Somerset, BA22 7JJ, England

Haynes North America, Inc
861 Lawrence Drive, Newbury Park, California 91320, USA

Editions Haynes SA
147/149, rue Saint Honoré, 75001 PARIS, France

Haynes Publishing Nordiska AB
Box 1504, 751 45 UPPSALA, Sweden

ISBN 1 85960 466 8

British Library Cataloguing in Publication Data
A catalogue record for this book is available from the British Library

Library of Congress Catalog Card Number 97-77261

Contents

LIVING WITH YOUR BMW TWIN

Introduction

Daily (pre-ride checks)

MAINTENANCE

Routine maintenance and servicing

Contents

BMW – They did it their way

by Julian Ryder

BMW - Bayerische Motoren Werke

If you were looking for a theme tune for BMW's engineering philosophy you'd have to look no further than Francis Albert Sinatra's best known ditty: 'I did it my way.' The Bayerische Motoren Werke, like their countrymen at Porsche, takes precious little notice of the way anyone else does it, point this out to a factory representative and you will get a reply starting: 'We at BMW... '. The implication is clear.

It was always like that. The first BMW motorcycle, the R32, was, according to motoring sage L J K Setright: 'the first really outstanding post-War design, argued from first principles and uncorrupted by established practice. It founded a new German school of design, it established a BMW tradition destined to survive unbroken from 1923 to the present day.' That tradition was, of course, the boxer twin. The nickname 'boxer' for an opposed twin is thought to derive from the fact that the pistons travel horizontally towards and away from each other like the fists of boxers.

Before this first complete motorcycle, BMW had built a horizontally-opposed fore-and-aft boxer engine for the Victoria company of Nuremburg. It was a close copy of the British

BMW's well known 2-valve Boxer engine seen here fitted to a 1000 cc GS model

Douglas motor which the company's chief engineer Max Friz admired, a fact the company's official history confirms despite what some current devotees of the marque

A later GS model fitted with the 4-valve engine

will claim. In fact BMW didn't really want to make motorcycles at all, originally it was an aero-engine company - a fact celebrated in the blue-and-white tank badge that is symbolic of a propeller. But in Germany after the Treaty of Versailles such potentially warlike work was forbidden to domestic companies and BMW had to diversify, albeit reluctantly.

Friz was known to have a very low opinion of motorcycles and chose the Douglas to copy simply because he saw it as fundamentally a good solution to the engineering problem of powering a two-wheeler. In the R32 the engine was arranged with the crankshaft in-line with the axis of the bike and the cylinders sticking out into the cooling airflow, giving a very low centre-of-gravity and perfect vibration-free primary balance. It wasn't just the motor's layout that departed from normal practice, the clutch was a single-plate type as used in cars, final drive was by shaft and the rear wheel could be removed quickly. The frame and suspension were equally sophisticated, but the bike was quite heavy. Most of that description could be equally well applied to any of the boxer-engined bikes BMW made in the next 70-plus years.

Development within the surprisingly flexible confines of the boxer concept was quick.

The second BMW, the R37 of 1925, retained the 68 x 68 mm Douglas bore and stroke but had overhead valves in place of the side valves. In 1928 two major milestones were passed. First, BMW acquired the car manufacturer Dixi and started manufacturing a left-hand-drive version of the Austin 7 under license. Secondly, the larger engined R62 and R63 appeared, the latter being an OHV sportster that would be the basis of BMW's sporting and record-breaking exploits before the Second World War.

The 1930s was the era of speed records on land, on sea and in the air, and the name of Ernst Henne is in the record books no fewer than ten times: eight for two-wheeled exploits, twice for wheel-on-a-stick 'sidecar' world records. At first he was on the R63 with supercharging, but in 1936 he switched to the 500 cc R5, high-pushrod design reminiscent of the latest generation of BMW twins. Chain-driven camshafts operated short pushrods which opened valves with hairpin - not coil - springs. A pure racing version of this motor also appeared, this time with shaft and bevel-gear driven overhead camshafts, but with short rockers operating the valves so the engine can't be called a true DOHC design. Again with the aid of a blower, this was the motor that powered the GP 500s of the late '30s to many wins including the 1939 Senior TT. After the War, this layout would re-emerge in the immortal Rennsport.

From 1939 to 1945 BMW were fully occupied making military machinery, notably the R75 sidecar for the army. The factory didn't restart production until 1948, and then only with a lightweight single. There was a false start in 1950 and a slump in sales in 1953 that endangered the whole company,

The R850R is the smallest 4-valve engined model . . .

before the situation was rescued by one of the truly classic boxers. Their first post-War twin had been the R51/2, and naturally it was very close to the pre-War model although simplified to a single-camshaft layout. Nevertheless, it was still a relatively advanced OHV design not a sidevalve sidecar tug which enabled a face-lift for the 1955 models to do the marketing trick.

The 1955 models got a swinging arm - at both ends. The old plunger rear suspension was replaced by a swinging arm while leading-link Earles forks adorned the front. Thus were born the R50, the R60 and the R69. The European market found these new bikes far too expensive compared to British twins

but America saved the day, buying most of the company's output. The car side of the company also found a product the market wanted, a small sports-car powered by a modified bike engine, thus BMW's last crisis was averted

In 1960 the Earles fork models were updated and the R69S was launched with more power, closer transmission ratios and those funny little indicators on the ends of the handlebars. Very little changed during the '60s, apart from US export models getting telescopic forks, but in 1970 everything changed...

The move to Spandau and a new line of Boxers

BMW's bike side had outgrown its site in Munich at the company's head-quarters, so, taking advantage of government subsidies for enterprises that located to what was then West Berlin, surrounded by the still Communist DDR, BMW built a new motorcycle assembly plant at Spandau in Berlin. It opened in 1969, producing a completely new range of boxers, the 5-series, which begat the 6-series, which begat the 7-series.

In 1976, at the same time as the launch of the 7-series the first RS boxer appeared. It's hard to believe now, but it was the only fully faired motorcycle, and it set the pattern for all BMWs, not just boxers, to come. The RS suffix came to mean a wonderfully efficient fairing that didn't spoil a sporty riding position. More sedate types could buy the RT version with a massive but no less efficient fairing that protected a more upright rider. Both bikes could carry luggage in a civilised

. . . and is produced with either cast alloy or spoked wheels

The R1100RT touring styled model

The R1100RS sport styled model

fashion, too, thanks to purpose-built Krauser panniers. Both the RT and RS were uncommonly civilised motorcycles for their time.

When the boxer got its next major makeover in late 1980 BMW did something no-one thought possible, they made a boxer trail bike, the R80G/S. This wasn't without precedent as various supermen had wrestled 750 cc boxers to honours in the ISDT and in '81 Hubert Auriol won the Paris-Dakar on a factory boxer. Some heretics even dared to suggest the roadgoing G/S was the best boxer ever.

The K-series

By the end of the '70s the boxer was looking more and more dated alongside the opposition, and when the motorcycle division's management was shaken up at the beginning of 1979 the team working on the boxer replacement was doubled in size. The first new bike wasn't launched until late '83, but when it was it was clear that BMW had got as far away from the boxer concept as possible.

The powerplant was an in-line water-cooled DOHC four just like all the Japanese

opposition, but typically BMW did it their way by aligning the motor so its crank was parallel to the axis of the bike and lying the motor on it side. In line with their normal practice, there was a car-type clutch, shaft drive and a single-sided swinging arm. It was totally novel yet oddly familiar. And when RT and RS version were introduced to supplement the basic naked bike, the new K-series 'flying bricks' felt even more familiar.

It was clear that BMW wanted the new four, and the three-cylinder 750 that followed it, to be the mainstay of the company's production - but in a further analogy with Porsche the customers simply wouldn't let go of the old boxer. Just as Porsche were forced to keep the 911 in production so BMW had to keep the old air-cooled boxer going by pressure from their customers. It kept going until 1995, during which time the K-bikes had debuted four-valve heads and ABS. And when the latest generation of BMWs appeared in 1993 what were they? Boxers. Granted they were four-valve, water-cooled and equipped with non-telescopic fork front ends, but they were still boxers. And that high camshaft, short pushrod layout looked remarkably similar to something that had gone before...

BMW's flat triple engine fitted to a K75S model

The New 4-valve Boxers

Even by BMW's standards, the new-generation Boxers were a shock. Maybe we shouldn't have been surprised after the lateral thinking that gave us the K-series, but the way in which the men from Munich took the old opposed-twin Boxer concept that launched the company and projected it into the 21st-Century was nothing short of breath-taking in its audacity. About the only design features the old and new Boxers have in common is that they both have two wheels and two cylinders.

The new bike uses fuel injection and four-valve heads; nothing unusual there for a design of the 1990s, various specials builders and racers had done that to old Boxers. The potential problem of drastically decreased ground clearance was averted by using an overhead-valve cylinder head – overhead cams would have made the heads stick out too far. Instead the valves and their rockers are operated by short pushrods activated by chain driven camshafts, a layout used by, among others, Riley in the 1950s. Dammit, the thing is even still air cooled!

So far very predictable, but the chassis is something else. For a start there isn't much that you can call a frame, the massive engine and gearbox castings are used as load-bearing members and have the front and rear suspension bolted directly to them. At the back there's the Paralever single-sided swingarm and its parallelogram linkage that's been around for years, but look at the proportions of the swingarm tube that hides the driveshaft; it's enormous, no lack of stiffness there. But the revolutionary bit is at the front, the Telelever. From a distance, it looks like a strangely-styled but conventional set of telescopic forks but when you get closer you notice the wishbone under the steering head and finally the centrally-mounted single shock absorber.

What BMW have achieved is the theoretically valuable result of separating the steering, braking and suspension forces.

What the rider notices is a very un-BMW lack of front-end dive under braking. Just before that the rider will also have noticed a very un-BMW tendency for the bike to charge up to the next braking point with the sort of enthusiasm very few of the old Boxers were noted for. And when you get to the corner you find out that this newfangled chassis/ suspension concept works, and works well. The new bikes are far less quirky than the old 'uns.

Unusually, the first variant to arrive was the RS in 1993, the sports touring model complete with sleek yet efficient faring and ABS as standard. It was impressive enough to become Bike magazine's Bike of the Year, something that no previous BMW would have been considered for. Next along was the GS, a bike that looks like it came straight off the set of the latest Aliens movie, where it would probably have been playing the monster. Like previous GS models, this was the go anywhere bike but also a surprisingly useful everyday bike and two-up tourer. We had to wait until 1995 for what has become the definitive new-generation Boxer, the R1100. Previously, the basic naked version of any of BMW's ranges has been the one to avoid. Not this time. The R is that most unusual of modern motorbikes, a great all-rounder. And don't be put off the smaller 850 cc (87 mm stroke as opposed to 99 mm) version by BMW referring to it as an entry-level machine, it's much more than that. Above all, the R models, large and small, are

The R1100RT

fun, in fact they're great fun. And you haven't been able to say that about too many BMW Boxers.

The full touring RT version arrived in 1996 with all the home comforts we've come to expect, and you might have thought that would have completed the range. But no, they've got a custom version too, the R1200C, although it is very different from the rest of the new Boxers. A custom BMW? Who'd have thought it? Certainly not all those critics who dismissed Boxers as museum pieces. You can't do that any more and in many ways their chassis design – or lack of it – puts the new generation Boxers at the very forefront of motorcycle design.

Acknowledgements

Our thanks are due to V & J Superbikes of Taunton who supplied the machines featured in the photographs throughout this manual. We would also like to thank NGK Spark Plugs (UK) Ltd for supplying the colour spark plug condition photos and the Avon Rubber Company for supplying information on tyre fitting. Certain model photographs used in the manual were kindly supplied by BMW (GB) Ltd.

Thanks are also due to Paul Tanswell for carrying out the front cover model photography and to Kel Edge for supplying the rear cover photograph of the R1100RS. The introduction, "BMW – They did it their way" was written by Julian Ryder.

About this Manual

The aim of this manual is to help you get the best value from your motorcycle. It can do so in several ways. It can help you decide what work must be done, even if you choose to have it done by a dealer; it provides information and procedures for routine maintenance and servicing; and it offers diagnostic and repair procedures to follow when trouble occurs.

We hope you use the manual to tackle the work yourself. For many simpler jobs, doing it yourself may be quicker than arranging an appointment to get the motorcycle into a dealer and making the trips to leave it and pick it up. More importantly, a lot of money can be saved by avoiding the expense the

shop must pass on to you to cover its labour and overhead costs. An added benefit is the sense of satisfaction and accomplishment that you feel after doing the job yourself.

References to the left or right side of the motorcycle assume you are sitting on the seat, facing forward.

We take great pride in the accuracy of information given in this manual, but motorcycle manufacturers make alterations and design changes during the production run of a particular motorcycle of which they do not inform us. No liability can be accepted by the authors or publishers for loss, damage or injury caused by any errors in, or omissions from, the information given.

Frame and engine numbers

The frame serial number is stamped into the right-hand side of the frame behind the steering head, above the manufacturer's vehicle identification plate. The engine number is stamped into the crankcase below the right-hand cylinder. Both of these numbers should be recorded and kept in a safe place so they can be furnished to law enforcement officials in the event of a theft.

The frame serial number, engine serial number and colour code should also be kept in a handy place (such as with your driving licence) so they are always available when purchasing or ordering parts for your machine. The colour code label is on the top of the fuse/relay holder cover.

The procedures in this manual identify the bikes by model code and, if necessary, also by production year. The model code (eg R1100RS) is used by itself if the information applies to all bikes produced over the life of the model. The production year is added (eg 1995 and 1996 R1100RS) where the information applies only to bikes produced in certain years of the model's life, usually when a component has been changed or upgraded. Note that the code R259 may be encountered on BMW material – this refers to the entire 4-valve engine series, as distinct from their 2-valve engined predecessors.

Colour code label can be found under the seat, on the fuse/relay holder cover

Buying spare parts

When ordering replacement parts, it is essential to identify exactly the machine for which the parts are required. While in some cases it is sufficient to identify the machine by its title eg 'R1100RS', any modifications made to components mean that it is usually essential to identify the machine by its BMW **production** or model year eg '1995 R1100RS'. The BMW production year starts in September of the previous calendar year, after the annual holiday, and continues until the following August. Therefore a 1995 R1100RS was **produced** at some time between September 1994 and August 1995; it may have been **sold** (to its first owner) at any time from September 1994 onwards. Models are referred to **at all times** in this Manual by their BMW **production** or model year.

To identify your own machine, record its full engine and frame numbers and take them to any BMW dealer who should have the necessary information to identify it exactly. Finally, in some cases modifications can be identified only by reference to the machine's engine or frame number; these should be noted and taken with you whenever replacement parts are required.

To be absolutely certain of receiving the correct part, not only is it essential to have the machine's identifying title and engine and frame numbers, but it is also useful to take the old part for comparison (where possible). Note that where a modified component has superseded the original, a careful check must be made that there are no related parts which have also been modified and must be used to enable the replacement to be correctly refitted; where such a situation is found, purchase all the necessary parts and fit them, even if this means replacing apparently unworn items.

Always purchase replacement parts from an authorised BMW dealer who will either have the parts in stock or can order them quickly from the importer, and always use genuine parts to ensure the machine's performance and reliability. Pattern parts are available for certain components (ie disc brake pads, oil and air filters); if used, ensure these are of recognised quality brands which will perform as well as the original.

Expendable items such as lubricants, spark plugs, some electrical components, bearings, bulbs and tyres can usually be obtained at lower prices from accessory shops, motor factors or from specialists advertising in the national motorcycle press.

The engine number is stamped into the crankcase below the right-hand cylinder.

The frame number is stamped into the right-hand side of the frame behind the steering head.

Professional mechanics are trained in safe working procedures. However enthusiastic you may be about getting on with the job at hand, take the time to ensure that your safety is not put at risk. A moment's lack of attention can result in an accident, as can failure to observe simple precautions.

There will always be new ways of having accidents, and the following is not a comprehensive list of all dangers; it is intended rather to make you aware of the risks and to encourage a safe approach to all work you carry out on your bike.

Asbestos

● Certain friction, insulating, sealing and other products - such as brake pads, clutch linings, gaskets, etc. - contain asbestos. Extreme care must be taken to avoid inhalation of dust from such products since it is hazardous to health. If in doubt, assume that they do contain asbestos.

Fire

● Remember at all times that petrol is highly flammable. Never smoke or have any kind of naked flame around, when working on the vehicle. But the risk does not end there - a spark caused by an electrical short-circuit, by two metal surfaces contacting each other, by careless use of tools, or even by static electricity built up in your body under certain conditions, can ignite petrol vapour, which in a confined space is highly explosive. Never use petrol as a cleaning solvent. Use an approved safety solvent.

● Always disconnect the battery earth terminal before working on any part of the fuel or electrical system, and never risk spilling fuel on to a hot engine or exhaust.
● It is recommended that a fire extinguisher of a type suitable for fuel and electrical fires is kept handy in the garage or workplace at all times. Never try to extinguish a fuel or electrical fire with water.

Fumes

● Certain fumes are highly toxic and can quickly cause unconsciousness and even death if inhaled to any extent. Petrol vapour comes into this category, as do the vapours from certain solvents such as trichloro-ethylene. Any draining or pouring of such volatile fluids should be done in a well ventilated area.
● When using cleaning fluids and solvents, read the instructions carefully. Never use materials from unmarked containers - they may give off poisonous vapours.
● Never run the engine of a motor vehicle in an enclosed space such as a garage. Exhaust fumes contain carbon monoxide which is extremely poisonous; if you need to run the engine, always do so in the open air or at least have the rear of the vehicle outside the workplace.

The battery

● Never cause a spark, or allow a naked light near the vehicle's battery. It will normally be giving off a certain amount of hydrogen gas, which is highly explosive.

● Always disconnect the battery ground (earth) terminal before working on the fuel or electrical systems (except where noted).
● If possible, loosen the filler plugs or cover when charging the battery from an external source. Do not charge at an excessive rate or the battery may burst.
● Take care when topping up, cleaning or carrying the battery. The acid electrolyte, evenwhen diluted, is very corrosive and should not be allowed to contact the eyes or skin. Always wear rubber gloves and goggles or a face shield. If you ever need to prepare electrolyte yourself, always add the acid slowly to the water; never add the water to the acid.

Electricity

● When using an electric power tool, inspection light etc., always ensure that the appliance is correctly connected to its plug and that, where necessary, it is properly grounded (earthed). Do not use such appliances in damp conditions and, again, beware of creating a spark or applying excessive heat in the vicinity of fuel or fuel vapour. Also ensure that the appliances meet national safety standards.
● A severe electric shock can result from touching certain parts of the electrical system, such as the spark plug wires (HT leads), when the engine is running or being cranked, particularly if components are damp or the insulation is defective. Where an electronic ignition system is used, the secondary (HT) voltage is much higher and could prove fatal.

Remember...

✗ **Don't** start the engine without first ascertaining that the transmission is in neutral.

✗ **Don't** suddenly remove the pressure cap from a hot cooling system - cover it with a cloth and release the pressure gradually first, or you may get scalded by escaping coolant.

✗ **Don't** attempt to drain oil until you are sure it has cooled sufficiently to avoid scalding you.

✗ **Don't** grasp any part of the engine or exhaust system without first ascertaining that it is cool enough not to burn you.

✗ **Don't** allow brake fluid or antifreeze to contact the machine's paintwork or plastic components.

✗ **Don't** siphon toxic liquids such as fuel, hydraulic fluid or antifreeze by mouth, or allow them to remain on your skin.

✗ **Don't** inhale dust - it may be injurious to health (see Asbestos heading).

✗ **Don't** allow any spilled oil or grease to remain on the floor - wipe it up right away, before someone slips on it.

✗ **Don't** use ill-fitting spanners or other tools which may slip and cause injury.

✗ **Don't** lift a heavy component which may be beyond your capability - get assistance.

✗ **Don't** rush to finish a job or take unverified short cuts.

✗ **Don't** allow children or animals in or around an unattended vehicle.

✗ **Don't** inflate a tyre above the recommended pressure. Apart from overstressing the carcass, in extreme cases the tyre may blow off forcibly.

✔ **Do** ensure that the machine is supported securely at all times. This is especially important when the machine is blocked up to aid wheel or fork removal.

✔ **Do** take care when attempting to loosen a stubborn nut or bolt. It is generally better to pull on a spanner, rather than push, so that if you slip, you fall away from the machine rather than onto it.

✔ **Do** wear eye protection when using power tools such as drill, sander, bench grinder etc.

✔ **Do** use a barrier cream on your hands prior to undertaking dirty jobs - it will protect your skin from infection as well as making the dirt easier to remove afterwards; but make sure your hands aren't left slippery. Note that long-term contact with used engine oil can be a health hazard.

✔ **Do** keep loose clothing (cuffs, ties etc. and long hair) well out of the way of moving mechanical parts.

✔ **Do** remove rings, wristwatch etc., before working on the vehicle - especially the electrical system.

✔ **Do** keep your work area tidy - it is only too easy to fall over articles left lying around.

✔ **Do** exercise caution when compressing springs for removal or installation. Ensure that the tension is applied and released in a controlled manner, using suitable tools which preclude the possibility of the spring escaping violently.

✔ **Do** ensure that any lifting tackle used has a safe working load rating adequate for the job.

✔ **Do** get someone to check periodically that all is well, when working alone on the vehicle.

✔ **Do** carry out work in a logical sequence and check that everything is correctly assembled and tightened afterwards.

✔ **Do** remember that your vehicle's safety affects that of yourself and others. If in doubt on any point, get professional advice.

● If in spite of following these precautions, you are unfortunate enough to injure yourself, seek medical attention as soon as possible.

Note: *The daily (pre-ride) checks outlined in your owners manual covers those items which should be inspected on a daily basis.*

1 Engine oil level check

Before you start:

✔ Take the motorcycle on a short run to allow it to reach normal operating temperature. *Caution: Do not run the engine in an enclosed space such as a garage or workshop.*

✔ Stop the engine and support the motorcycle on its centre stand, making sure it is in an upright position on level ground. Allow it to stand undisturbed for at least ten minutes to allow the oil level to stabilise. This time period

is important and must be adhered to, otherwise the oil from the cooler(s) will not have drained down, giving an incorrect level in the window. Note that the rate at which the oil returns from the oil cooler(s) can be increased by fitting a vent valve in the oil pipe under the engine's front cover – refer to a BMW dealer for details.

✔ The oil level is viewed through the window in the left-hand side of the engine, where applicable via a cut-out in the fairing. Wipe the glass clean before inspection to make the check easier.

Bike care:

● If you have to add oil frequently, you should check whether you have any oil leaks. If there is no sign of oil leakage from the joints and gaskets the engine could be burning oil (see *Fault Finding*).

● Never run the engine with the oil level below the minimum level, and do not fill it above the maximum level.

The correct oil

● Modern, high-revving engines place great demands on their oil. It is very important that the correct oil for your bike is used.
● Always top up with a good quality oil of the specified type and viscosity and do not overfill the engine. **Note:** *BMW advise that to ensure smooth running in of a new engine, synthetic oil is not used until at least 6000 miles (10,000 km) has been covered.*
● To fill the engine from MIN to MAX marks on the oil level window requires 0.5 litre of oil.

Oil type	HD oil, API grade SF, SG or SH
Oil viscosity	see accompanying chart and select the best grade to suit prevailing temperatures

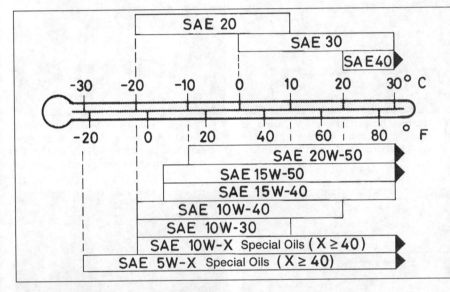

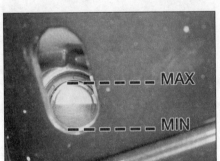

1 Check the oil level through the window in the left-hand side of the engine, where applicable via the cut-out in the fairing. Wipe the oil level window so that it is clean. With the motorcycle vertical, the oil level should lie between the top and bottom of the graduation circle.

If the level is below the bottom of the circle, remove the filler cap from the top of the left-hand cylinder valve cover. **2**

3 Top the engine up with the recommended type and chosen grade of oil (see accompanying viscosity chart), to bring the level up to the top of the circle on the window.

On completion, fit the filler cap, making sure it is secure. If there are any signs of oil leakage from around the cap, remove the sealing O-ring and replace it with a new one. **4**

2 Brake fluid level checks

Warning: Brake hydraulic fluid can harm your eyes and damage painted surfaces, so use extreme caution when handling and pouring it and cover surrounding surfaces with rag. Do not use fluid that has been standing open for some time, as it absorbs moisture from the air which can cause a dangerous loss of braking effectiveness.

Before you start:

When checking the front brake fluid level, position the handlebars until the top of the master cylinder is as level as possible.

When checking the rear brake fluid level on RS and RT models, the reservoir level can be seen via the cut-out in the right-hand fairing side panel. If topping up is required, the seat (all models) and side panel (R models) or fairing side panel (RS and RT models) must be removed – see Chapter 7.

Make sure you have the correct hydraulic fluid, **DOT 4 (ATE SL)** is recommended. Wrap a rag around the reservoir being worked on to ensure that any spillage does not come into contact with painted surfaces.

Bike care:

● The fluid in the front and rear brake master cylinder reservoirs will drop slightly as the brake pads wear down, especially during the bedding-in period of new pads.

● If any fluid reservoir requires repeated topping-up this is an indication of an hydraulic leak somewhere in the system, which should be investigated immediately.

● Check for signs of fluid leakage from the hydraulic hoses and components – if found, rectify immediately.

● Check the operation of both brakes before taking the machine on the road; if there is evidence of air in the system (spongy feel to lever or pedal), it must be bled as described in Chapter 6.

1 The front brake fluid level is checked via the sightglass in the reservoir – the entire sightglass must be covered by fluid.

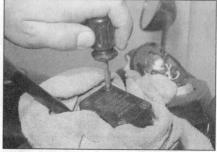

2 To top up, remove the four screws to free the fluid reservoir cover and diaphragm. Note how much the fluid level drops once the diaphragm has been removed.

3 Top up the fluid level with new hydraulic fluid, until the level is sufficient to cover the sightglass once the diaphragm and cover have been replaced.

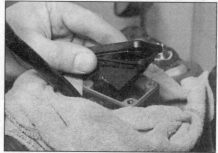

4 Ensure that the diaphragm is correctly seated before installing the cover. Recheck the level after installing the cover.

5 On R and GS models, the rear brake fluid reservoir is located below the seat on the right-hand side. The fluid must lie between the MAX and MIN level lines (arrowed).

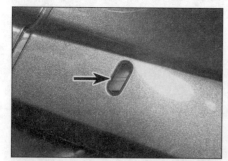

6 On RS and RT models, the fluid level can be seen via the cut-out in the right-hand fairing side panel. The fluid must lie between the MAX and MIN level lines, though only the MIN level line (arrowed) is readily visible. If topping up is required, the fairing side panel must be removed (see Chapter 7).

7 To top up, free the reservoir from its clip, then remove the cap and diaphragm. Note how much the fluid level drops once the diaphragm has been removed. Top up with new hydraulic fluid, until the level is sufficient to reach the MAX level line once the diaphragm and cover have been replaced.

8 Ensure that the diaphragm is correctly seated before installing the cover. Recheck the level after installing the cover. Fit the reservoir into its retaining clip. On RS and RT models, install the fairing side panel (see Chapter 7).

3 Tyre checks

The correct pressures:

● The tyres must be checked when **cold**, not immediately after riding. Note that low tyre pressures may cause the tyre to slip on the rim or come off. High tyre pressures will cause abnormal tread wear and unsafe handling.

● Use an accurate pressure gauge.

● Proper air pressure will increase tyre life and provide maximum stability and ride comfort. Ensure that the pressures are suited to the load the machine is carrying.

Tyre care:

● The need for frequent topping-up indicates a leak, which should be investigated immediately.
● Check the tyres carefully for cuts, tears, embedded nails or other sharp objects and excessive wear. Operation of the motorcycle with excessively worn tyres is extremely hazardous, as traction and handling are directly affected.
● Check the condition of the tyre valve and ensure the dust cap is in place.
● Pick out any stones or nails which may have become embedded in the tyre tread. If left, they will eventually penetrate through the casing and cause a puncture.
● If tyre damage is apparent, or unexplained loss of pressure is experienced, seek the advice of a tyre fitting specialist without delay.

Tyre tread depth:

● At the time of writing UK law requires that tread depth must be at least 1 mm over ¾ of the tread breadth all the way around the tyre, with no bald patches. Many riders, however, consider 2 mm tread depth minimum to be a safer limit. BMW recommend a minimum tread depth of 2 mm on the front tyre and 3 mm on the rear tyre, measured at the centre of the tread.

● Many tyres now incorporate wear indicators in the tread. Identify the arrow, triangular pointer or TWI marking on the tyre sidewall to locate the indicator bars and replace the tyre if the tread has worn down level with the bars.

Loading – all models	Front	Rear
Rider only	32 psi (2.2 Bar)	36 psi (2.5 Bar)
Rider and passenger	36 psi (2.5 Bar)	39 psi (2.7 Bar)
Rider and passenger with luggage	36 psi (2.5 Bar)	42 psi (2.9 Bar)

1 Check the tyre pressures when the tyres are **cold** and keep them properly inflated.

2 Measure tread depth at the centre of the tyre using a tread depth gauge.

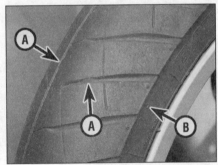

3 Tyre tread wear indicator bars (A) and their location marking (B) – usually either an arrow, a triangle or the letters TWI on the sidewall.

4 Suspension, steering and drive checks

Suspension and Steering:

● Check that the front and rear suspension operates smoothly without binding. Note that due to the design of the front suspension, very little movement can actually be obtained using this method, and riding the bike is the best way to determine whether the suspension is working correctly.
● Check that the rear suspension is adjusted as required.
● Check that the steering moves smoothly from lock-to-lock.

Gearbox and final drive:

● Check for signs of oil leakage around the gearbox and final drive housing. If any is evident, check their respective oil levels (Chapter 1). Note that signs of oil may be seen around the final drive housing breather after the machine has been standing for a while. The breather bore was increased on late 1996 models as a preventative measure. If oil traces are noticed, check the final drive oil level and refer to a BMW dealer for advice.

5 Legal and safety checks

Lighting and signalling:
● Take a minute to check that the headlight, taillight, brake light, instrument lights and turn signals all work correctly.
● Check that the horn sounds when the switch is operated.
● A working speedometer is a statutory requirement in the UK.

Safety:
● Check that the throttle grip rotates smoothly and snaps shut when released, in all steering positions. Also check for the correct amount of freeplay (see Chapter 1).

● Check that the clutch lever operates smoothly, and that the cable is secure and not worn or frayed. Also check for the correct amount of freeplay (see Chapter 1).

● Check that the clutch lever, front brake lever, gearchange lever and rear brake pedal function correctly, are secure and correctly adjusted.

● Check that the engine shuts off when the kill switch is operated.

● Check that side stand return spring holds the stand securely up when retracted. The same applies to the centre stand.

● Check that the maximum weight limit is not exceeded (see *Dimensions and Weights* in the Reference section).

Fuel:
● This may seem obvious, but check that you have enough fuel to complete your journey. If you notice signs of fuel leakage – rectify the cause immediately.

● Ensure you use the correct grade unleaded fuel, minimum 95 RON (Research Octane Number). The use of unleaded fuel is essential where a catalytic converter is fitted.

Chapter 1
Routine maintenance and Servicing

Contents

1

Degrees of difficulty

Easy, suitable for novice with little experience		**Fairly easy,** suitable for beginner with some experience		**Fairly difficult,** suitable for competent DIY mechanic		**Difficult,** suitable for experienced DIY mechanic		**Very difficult,** suitable for expert DIY or professional

Specifications

Engine

Spark plugs
  Type . . . . Bosch FR6 DDC or NGK BCPR7ET
  Electrode gap
    Standard . . . . 0.8 mm
    Service limit (see text) . . . . 1.0 mm
Valve clearances (COLD engine)
  Intake valves . . . . 0.15 mm
  Exhaust valves . . . . 0.30 mm
Engine idle speed . . . . 1000 to 1150 rpm
Cylinder compression
  Good . . . . above 145 psi (above 10 Bar)
  Normal . . . . 125 to 145 psi (8.5 to 10 Bar)
  Poor . . . . less than 125 psi (8.5 Bar)
Oil pressure (with engine warm) . . . . 50 to 87 psi (3.5 to 6.0 Bar)
Exhaust CO content (models without catalytic converter) . . . . 1.5 ± 0.5%

Cycle parts

Clutch lever freeplay at handlebar	5 mm
Throttle cable freeplay – 1993 to 1995 models	
Main cable	0.5 mm
Joining cable	zero
Choke cable freeplay – 1993 to 1995 models	1 mm
Throttle and choke cable freeplay – 1996-on models	see text
Brake pad minimum friction material thickness	1.5 mm

Tyre pressures (cold)*	Front	Rear
Rider only	32 psi (2.2 Bar)	36 psi (2.5 Bar)
Rider and passenger	36 psi (2.5 Bar)	39 psi (2.7 Bar)
Rider and passenger with luggage	36 psi (2.5 Bar)	42 psi (2.9 Bar)
Tyre tread depth (minimum)	2 mm	3 mm

Tyre pressures are also stated on the tyre information label under the seat.

Torque settings

Engine oil drain plug	32 Nm
Engine oil filter	11 Nm
Spark plugs	20 Nm
Valve clearance adjuster screw locknut	8 Nm
Front shock absorber lower mounting bolt (RS models to 1995)	43 Nm
Gearbox oil filler cap	23 Nm
Gearbox oil drain plug	23 Nm (30 Nm on late 1996-on models)
Final drive oil filler cap	23 Nm
Final drive oil drain plug	23 Nm
Intake manifold clamp screws	2 Nm
Fuel tank mounting bolt	22 Nm
Brake caliper mounting bolts	40 Nm
Alternator drive belt pre-load (on adjusting bolt)	8 Nm
Alternator mounting bolt and nuts	20 Nm

Recommended lubricants and fluids

Engine oil type	HD oil, API grade SF, SG or SH motor oil
Engine oil viscosity	see chart in *Daily (pre-ride)* checks
Engine oil capacity	
Oil change	3.5 litres
Oil and filter change	3.75 litres
Difference between max. and min. levels	0.5 litre
Gearbox oil type	Hypoid gear oil, API class GL5
Gearbox oil viscosity	
Above 5°C	SAE 90
Below 5°C	SAE 80
All conditions	SAE 80W 90
Gearbox oil capacity	
Oil change	0.8 litre
Following overhaul	1.0 litre
Final drive oil type	Hypoid gear oil, API class GL5
Final drive oil viscosity	
Above 5°C	SAE 90
Below 5°C	SAE 80
All conditions	SAE 80W 90
Final drive oil capacity	
Oil change	230 ml
Following overhaul	250 ml
Brake fluid	DOT 4
Miscellaneous	
Wheel bearings	High melting point lithium grease, e.g. Shell Retinax A
Rear suspension bearings	High melting point lithium grease, e.g. Shell Retinax A
Cables, lever and stand pivot points	Motor oil
Throttle grip	Multi-purpose grease or dry film lubricant

Note: *The daily (pre-ride) checks outlined in your owners manual covers those items which should be inspected on a daily basis. Always perform the pre-ride inspection at every* maintenance interval (in addition to the procedures listed). The intervals listed below are the intervals recommended by the manufacturer for each particular operation during the model years covered in this manual. Your owners manual may have different intervals for your model.

Daily (pre-ride)
See 'Daily (pre-ride) checks' at the beginning of this manual.

After the initial 600 miles (1000 km)
Note: *This check is usually performed by a BMW dealer after the first 600 miles (1000 km) from new. Thereafter, maintenance is carried out according to the following intervals of the schedule.*

Minor service – after the first 6000 miles (10,000 km), and every 12,000 miles (20,000 km) thereafter
Carry out all the items under the Daily (pre-ride) checks, plus the following:
- [ ] Change the engine oil and oil filter (Section 1) – see Note 1.
- [ ] Check the spark plug condition and gaps (Section 2).
- [ ] Check the valve clearances (Section 3).
- [ ] Check clutch lever freeplay (Section 4).
- [ ] Check the brake pads and discs for wear (Section 5).
- [ ] Check the front and rear suspension and, on R models, the steering damper (Section 6).
- [ ] Check and lubricate the stands, lever pivots and cables (Section 7) – see Note 2.
- [ ] Check the engine idle speed and throttle synchronisation (Section 8) – see Note 3.
- [ ] Check throttle/choke cable operation and freeplay (Section 9).
- [ ] Check the gearbox oil level (Section 10).
- [ ] Check the final drive oil level (Section 11).
- [ ] Check the tightness of all nuts and bolts (Section 12).

Major service – after the first 12,000 miles (20,000 km), and every 12,000 miles (20,000 km) thereafter
Carry out all the items under the Minor service (except the spark plug check), plus the following:
- [ ] Change the gearbox oil (Section 13) – see Note 4.
- [ ] Change the final drive oil (Section 14) – see Note 5.
- [ ] Replace the spark plugs (Section 15).
- [ ] Check tightness of intake manifold clamps (Section 16).
- [ ] Replace the air filter (Section 17) – see Note 6.
- [ ] Adjust the alternator drive belt (non maintenance-free belts) (Section 18).
- [ ] Change the brake fluid (Section 19) – see Note 7.
- [ ] Check the operation of the brakes, and for fluid leakage (Section 20).
- [ ] Check the battery (Section 21) – see Note 8.
- [ ] Check the wheels and wheel bearings (Section 22).
- [ ] Grease the windshield height adjuster – RS models (Section 23).

Every 18,000 miles (30,000 km)
- [ ] Replace the throttle cables (Section 24).

Every 24,000 miles (40,000 km)
- [ ] Replace the fuel filter (Section 25) – see Note 9.
- [ ] Replace the alternator drive belt (non maintenance-free belts) (Section 26).

Every 36,000 miles (60,000 km)
- [ ] Replace the alternator drive belt (maintenance-free belts) (Section 27).

Non-scheduled maintenance
- [ ] Re-grease the swingarm bearings (Section 28).
- [ ] Replace the brake master cylinder and caliper seals (Section 29).
- [ ] Replace the brake hoses (Section 30).
- [ ] Check the cylinder compression (Section 31).
- [ ] Check the engine oil pressure (Section 32).
- [ ] Check the headlight beam alignment (Section 33).

Notes – Additional recommendations:
1 Engine oil – in normal use the engine oil should be changed at least once annually. If the machine is used in temperatures below 0°C (32°F), or for short, local journeys only, the oil should be changed every 2000 miles (3000 km) or three months at the latest.
2 Side stand pivot and clutch cable nipple – should be greased annually.
3 Engine idle speed, throttle synchronisation and CO level – should be checked annually.
4 Gearbox oil – must be changed at least once annually.
5 Rear bevel drive oil – must be changed at least once annually.
6 Air filter – if the machine is used in very dusty or dirty conditions, the air filter should be renewed every 6000 miles (10,000 km) or more frequently if conditions require.
7 Hydraulic brake fluid – must be changed annually.
8 Battery – should be checked annually.
9 Fuel filter – renew every 12,000 miles (20,000 km) if poor quality fuel has been used.

Component locations on left-hand side

1 Alternator drive belt
2 Clutch cable adjuster at handlebar
3 Choke cable adjuster – 1996-on
4 Throttle cable adjuster – 1993 to 1995
5 Choke cable adjuster – 1993 to 1995
6 Air filter
7 Rear shock absorber
8 Swingarm (Paralever) pivot bolt
9 Throttle body
10 Engine oil drain plug
11 Spark plugs and valves
12 Engine oil filter
13 Engine oil sightglass
14 Front brake pads
15 ABS sensor

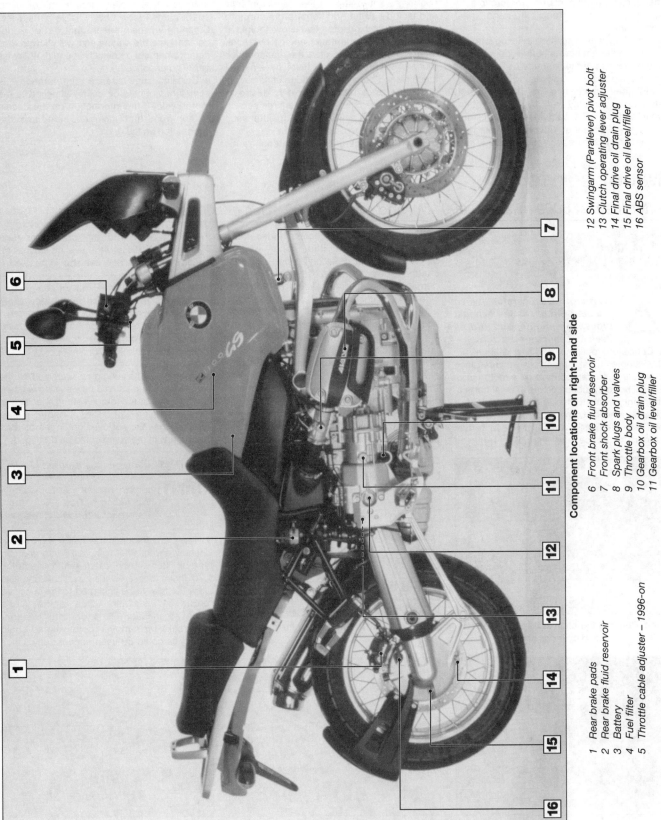

Component locations on right-hand side

1 Rear brake pads
2 Rear brake fluid reservoir
3 Battery
4 Fuel filter
5 Throttle cable adjuster – 1996-on

6 Front brake fluid reservoir
7 Front shock absorber
8 Spark plugs and valves
9 Throttle body
10 Gearbox oil drain plug
11 Gearbox oil level/filler

12 Swingarm (Paralever) pivot bolt
13 Clutch operating lever adjuster
14 Final drive oil drain plug
15 Final drive oil level/filler
16 ABS sensor

1

1 This Chapter is designed to help the home mechanic maintain his/her motorcycle for safety, economy, long life and peak performance.
2 Deciding where to start or plug into the routine maintenance schedule depends on several factors. If the warranty period on your motorcycle has just expired, and if it has been maintained according to the warranty standards, you may want to pick up routine maintenance as it coincides with the next mileage or calendar interval. If you have owned the machine for some time but have

never performed any maintenance on it, then you may want to start at the nearest interval and include some additional procedures to ensure that nothing important is overlooked. If you have just had a major engine overhaul, then you may want to start the maintenance routine from the beginning. If you have a used machine and have no knowledge of its history or maintenance record, you may desire to combine all the checks into one large service initially and then settle into the maintenance schedule prescribed.
3 Before beginning any maintenance or

repair, the machine should be cleaned thoroughly, especially around the oil filter, spark plugs, valve covers, side panels etc. Cleaning will help ensure that dirt does not contaminate the engine and will allow you to detect wear and damage that could otherwise easily go unnoticed.
4 Certain maintenance information is sometimes printed on decals attached to the motorcycle. If the information on the decals differs from that included here, use the information on the decal.

Minor service

After the first 6000 miles (10,000 km), and every 12,000 miles (20,000 km) thereafter

1 Engine – oil and oil filter change

Warning: Be careful when draining the oil, as the exhaust pipes, the engine, and the oil itself can cause severe burns.

1 Consistent routine oil and filter changes are the single most important maintenance procedure you can perform on a motorcycle. The oil not only lubricates the internal parts of the engine, but it also acts as a coolant, a cleaner, a sealant, and a protectant. Because of these demands, the oil takes a terrific amount of abuse and should be replaced

often with new oil of the recommended grade and type. Saving a little money on the difference in cost between a good oil and a cheap oil won't pay off if the engine is damaged. The oil filter should be changed with every oil change.
2 Before changing the oil, warm up the engine so the oil will drain easily.
3 On GS models, remove the engine sump guard (see Chapter 7).
4 Put the motorcycle on its centre stand, and position a clean drain tray below the engine. Unscrew the oil filler cap on the left-hand cylinder head to vent the crankcase and to act as a reminder that there is no oil in the engine **(see illustration)**. Next, unscrew the oil drain plug from the bottom of the engine and allow

the oil to flow into the drain tray **(see illustrations)**. Discard the sealing washer on the drain plug as a new one should be used.

HAYNES HINT *An oil drain tray can be easily made by cutting away the front or back of an old five litre oil container, or any other such container that is of an adequate size.*

5 When the oil has completely drained, fit the plug to the sump using a new sealing washer, and tighten it to the torque setting specified at the beginning of the Chapter **(see illustrations)**. Do not overtighten the plug as the threads in the sump could be damaged.

HAYNES HINT *Damaged sump threads can be repaired with a Heli-Coil thread insert – see Tools and Workshop Tips in the Reference section for details.*

6 Now place the drain tray below the oil filter. Unscrew the oil filter using a 76 mm end cap type oil filter wrench, as shown in illustration 1.7c; these tools are widely available and also allow the filter to be tightened to the specified torque on installation. Due to the filter's recessed location, it is doubtful whether universal filter removal tools or strap/chain wrenches could be used successfully. Once

1.4a Remove the oil filler cap from the left-hand cylinder head

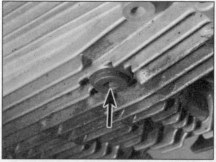
1.4b Unscrew the drain plug (arrowed) ...

1.4c ... and allow the oil to drain completely

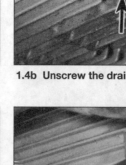
1.5a Install the drain plug ...

1.5b ... and tighten it to the specified torque setting

1.7a Smear the sealing ring with clean oil . . .

1.7b . . . then install the filter

1.7c Fit the filter wrench . . .

1.7d . . . and tighten it to the specified torque setting

1.8a Fill the engine to the correct level using a funnel to avoid spillage

— — MAX

— — MIN

1.8b Check the oil level via the window

the filter has been unscrewed, tip any residue oil into the drain tray.

7 Smear clean engine oil onto the rubber seal on the new filter, then screw the filter onto the engine and tighten it to the specified torque setting using the filter wrench **(see illustrations)**.

8 Refill the engine to the proper level using the recommended type and amount of oil (see *Daily (pre-ride) checks*) **(see illustration)**. With the motorcycle vertical, the oil level should lie between the top and bottom of the graduation circle on the inspection window **(see illustration)**. Install the filler cap **(see illustration 1.4a)**. Start the engine and let it run for two or three minutes (make sure that the oil pressure light extinguishes after a few seconds). Shut it off, wait ten minutes, then check the oil level. If necessary, add more oil to bring the level up to the top of the circle on the inspection window. Check around the

drain plug and the oil filter for leaks. On GS models install the engine sump guard.

9 The old oil drained from the engine cannot be re-used and should be disposed of properly. Check with your local refuse disposal company, disposal facility or environmental agency to see whether they will accept the used oil for recycling. Don't pour used oil into drains or onto the ground.

> **HAYNES HINT** *Check the old oil carefully – if it is very metallic coloured, then the engine is experiencing wear from break-in (new engine) or from insufficient lubrication. If there are flakes or chips of metal in the oil, then something is drastically wrong internally and the engine will have to be disassembled for inspection and repair.*

2 Spark plugs – check

1 Make sure your spark plug socket is the correct size before attempting to remove the plugs – a suitable one is supplied in the motorcycle's tool kit (which is stored under the seat) along with a spark plug cap removing tool.

2 Clean the area around the plug caps to prevent any dirt falling into the spark plug channels.

3 Remove the spark plug cover from the valve cover **(see illustration)**. Pull the spark plug cap off each spark plug, using the tool provided in the toolkit **(see illustrations)**. Clean the area around the base of the plugs to prevent any dirt falling into the engine. Using

1

2.3a Remove the spark plug cover . . .

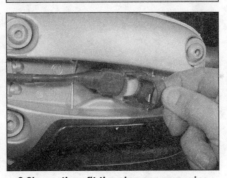

2.3b . . . then fit the plug cap removing tool . . .

2.3c . . . and pull off the cap

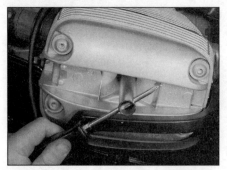

2.3d Unscrew the spark plug using the tool provided or a suitable alternative

2.7 Measuring the spark plug gap using a wire type gauge

2.8 Thread the plugs into the head taking care not to cross the threads

either the plug removing tool supplied in the bike's toolkit or a deep socket type wrench, unscrew the plug from the cylinder head **(see illustration)**.

4 Inspect the electrodes for wear. Both the centre and side electrodes should have square edges and the side electrodes should be of uniform thickness. Look for excessive deposits and evidence of a cracked or chipped insulator around the centre electrode. Compare your spark plugs to the colour spark plug reading chart at the end of this manual. Check the threads, the washer and the ceramic insulator body for cracks and other damage.

5 If the electrodes are not excessively worn, and the deposits can be easily removed with a wire brush, and no cracks or chips are visible in the insulator the plugs can be re-used. If in doubt concerning the condition of the plugs, replace them with new ones, as the expense is minimal.

6 Cleaning spark plugs by sandblasting is permitted, provided you clean the plugs with a high flash-point solvent afterwards.

7 Before installing the plugs, make sure they are the correct type and heat range and check the gap between the electrodes. Due to the curved face of the three earth electrodes, this is best achieved using a wire type plug measuring tool **(see illustration)**. In the absence of a suitable tool, you may be able to lay a ruler on the top of the central electrode and measure the gap between it and the centre of each earth electrode. Note that BMW advise against bending the earth

electrodes to make adjustment in case this causes them to fracture and break off when the engine is running. If a plug exceeds the service limit specified it should be replaced – new plugs are pre-set to the correct gap. Make sure the washer is in place before installing each plug.

8 Since the cylinder head is made of aluminium, which is soft and easily damaged, thread the plugs into the heads turning the tool by hand **(see illustration)**. Once the plugs are finger-tight, tighten them to the torque setting specified at the beginning of the Chapter, or, if a torque wrench is not available, according to the manufacturers instructions – do not over-tighten them **(see illustration 2.3d)**.

> **HAYNES HINT** *A short length of hose can be fitted over the end of the plug to use as a tool to thread it into place. The hose will grip the plug well enough to turn it, but will start to slip if the plug begins to cross-thread in the hole – this will prevent damaged threads.*

9 Reconnect the spark plug caps and install the spark plug covers.

> **HAYNES HINT** *Stripped plug threads in the cylinder head can be repaired with a Heli-Coil insert – see 'Tools and Workshop Tips' in the Reference section.*

3 Valve clearances – check and adjustment

1 The engine must be completely cool for this maintenance procedure, so let the machine sit overnight before beginning.

2 Place the motorcycle on its centre stand. Remove the spark plugs (see Section 2).

3 Remove the valve covers (see Chapter 2). Each cylinder is referred to according to its side, i.e. right or left. On R1100RS and RT models, remove the right-hand fairing side panel if not already done (see Chapter 8).

4 Make a chart or sketch of all valve positions so that a note of each clearance can be made against the relevant valve.

5 Remove the timing inspection plug from behind the right-hand cylinder **(see illustration)**. The engine can be turned by selecting a high gear and rotating the rear wheel by hand in its normal direction of rotation. Alternatively, remove the engine front cover and turn the engine using a 17 mm spanner or socket on the belt drive pulley bolt, turning it in a clockwise direction only **(see illustrations)**.
Caution: Be sure to turn the engine in its normal direction of rotation only.

6 Turn the engine until the "OT" mark on the flywheel, visible via the timing inspection hole, aligns with the middle of the hole, and the valves of the cylinder being worked on are all closed (i.e. there is freeplay in the rocker arm in the form of a clearance between the rocker

3.5a Remove the inspection hole blanking plug (arrowed)

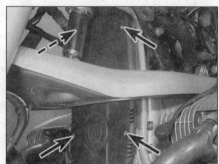

3.5b Remove the four bolts (arrowed) securing the front cover . . .

3.5c . . . and turn the engine using a spanner or socket as shown

3.6 Align the "OT" mark with the middle of the hole

3.7a Measure the clearance with a feeler gauge

3.7b If adjustment is required, slacken the locknut and turn the adjuster with an Allen key

3.7c Tighten the locknut to the specified torque setting

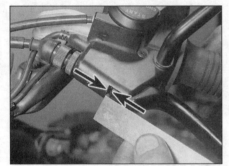

4.1 Measure the amount of freeplay at the clutch lever

and the valve stem) **(see illustration)**. The piston for that cylinder is now at TDC (top dead centre) on the compression stroke. If any of the valves on that cylinder are open when the mark aligns, rotate the engine clockwise one full turn until the "OT" mark again aligns with the middle of the hole. The valves will now be closed.

> **HAYNES HiNT**
> *As there is no static timing mark on the timing inspection hole, mark your own using a dab of paint or Tippex to make alignment easier. Use a ruler to find the exact middle of the hole and mark the side of it accordingly.*

7 Insert a feeler gauge of the correct thickness (see Specifications) between each rocker arm adjuster screw and valve and check that it is a firm sliding fit **(see illustration)**. If it is not, slacken the locknut and turn the adjuster using a suitable Allen key until a firm sliding fit is obtained, then tighten the locknut to the torque setting specified at the beginning of the Chapter, making sure the adjuster does not turn as you do so **(see illustrations)**. Re-check the clearances, not forgetting that there is a difference between the intake valve clearance and the exhaust valve clearance.
8 Moving to the other cylinder, rotate the engine until it is at TDC (top dead centre) on the compression stroke. At this point the "OT"

mark on the flywheel should be centred in the timing hole, and the valves for that cylinder should all be closed. Check and adjust the valve clearance as described in Step 7 above.
9 Install all disturbed components in a reverse of the removal sequence, referring to the relevant Chapters where necessary, and not forgetting the plug for the inspection cover. Apply engine oil to the valve assemblies, rockers and camshafts before installing the valve covers.

4 Clutch – check

Freeplay measurement

1 Clutch lever freeplay is measured at the handlebar lever. Gently pull in the clutch lever to take up cable slack, and measure the gap between the lever blade and stock **(see illustration)**. Freeplay should measure 5 mm. If freeplay is less than specified, carry out full adjustment of the clutch as described below.

Freeplay adjustment

2 Slacken off the lockring on the adjuster at the handlebar **(see illustration)** and turn the adjuster out so that when the lockring is threaded down against the lever stock, the distance between the adjuster head and lockring measures 10 mm **(see illustration)**.

1

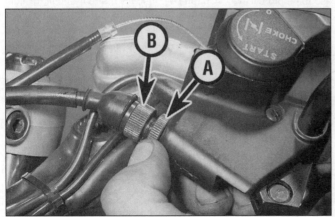

4.2a Clutch handlebar lever lockring (A) and adjuster (B)

4.2b Set distance between adjuster head and lockring to 10 mm

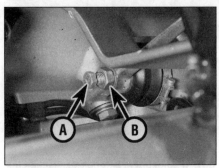

4.3 Clutch operating lever locknut (B) and adjuster bolt (A)

5.2 Brake pad wear indicator cutout location – R, RT and GS models

5.3 Measuring the amount of friction material on the brake pads – RS models

3 Moving to the lower end of the clutch cable, locate the operating lever adjuster mechanism on the back of the gearbox. The adjuster mechanism is accessed from the right-hand side of the bike. Slacken the adjuster locknut and turn the adjuster bolt **(see illustration)** until the distance between the clutch lever blade and stock measures 5 mm **(see illustration 4.1)**. Hold the adjuster whilst the locknut is tightened. Make sure that the lockring at the clutch lever is tightened against the lever stock.

5 Brake pads and discs – wear check

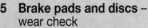

Brake pads – R, RT and GS models

1 Each brake pad has a wear indicator, in the form of a cut-out in the friction material, which can be viewed without removing the pads from the caliper, although unbolting the calipers will allow close inspection of the cut-outs.
2 The pad wear indicator cut-outs are positioned at the leading or trailing edge of the pads. If the friction material has worn down level with the base of the cut-outs or the specified minimum thickness has been reached, the pads must be renewed **(see illustration)**. **Note:** *Some after-market pads may use different indicators to BMW parts. If this is the case, the indicator used should still be obvious.* If the pads are dirty or if you are in

doubt as to the amount of friction material remaining, remove them for closer inspection (see Chapter 6). Always renew both pads in the caliper at the same time, and in the case of the front brake renew both sets of pads at the same time.

Brake pads – RS models

3 On RS models, remove the pads (see Chapter 6), and measure the thickness of friction material remaining and compare it to the minimum thickness specified at the beginning of the Chapter **(see illustration)**.
4 If the pads are worn to or beyond the specified minimum thickness, they must be replaced. Always renew both pads in the caliper at the same time, and in the case of the front brake renew both sets of pads at the same time.

Brake discs – all models

5 Check the front and rear brake discs for wear, damage and distortion (warpage). Refer to Chapter 6, Sections 4 and 7 for information.

6 Suspension and steering damper (R models) – check

1 The suspension components must be maintained in top operating condition to ensure rider safety. Loose, worn or damaged suspension parts decrease the motorcycle's stability and control.

Front suspension

2 Check that the front suspension operates smoothly and without binding of the forks. Note that due to its unconventional design, the front suspension cannot be checked by pushing and pulling on the handlebars. If problems have been noted whilst riding the bike, the front forks or Telelever pivots should be dismantled and checked thoroughly.
3 Inspect the area above the dust seal on the fork tubes for signs of oil leakage, then carefully lever off the dust seal using a flat-bladed screwdriver and inspect the area around the fork seal **(see illustrations)**. If leakage is evident, the seals must be replaced (see Chapter 5).
4 Inspect the shock absorber for fluid leakage and tightness of its mountings. If leakage is found, the shock should be replaced (see Chapter 5).
5 On RS models, remove the shock absorber lower mounting bolt and apply Never Seez or copper-based grease to the pivot section of the bolt **(see illustration)**. Counter-hold the bolt and tighten the nut to the torque setting specified at the beginning of the Chapter **(see illustration)**.
6 Check the tightness of all suspension nuts and bolts to be sure none have worked loose.

Rear suspension

7 Where fitted, remove the panniers (see Chapter 7). Inspect the rear shock for fluid leakage and tightness of its mountings. If

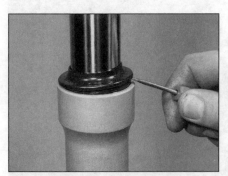

6.3a Lever up the dust seal . . .

6.3b . . . and check for signs of leakage past the oil seal

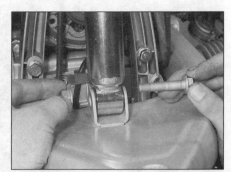

6.5a Remove the bolt and grease the pivot (unthreaded) section . . .

6.5b . . . then counter-hold the bolt and tighten the nut to the specified torque setting

6.9 Checking for play in the rear shock absorber mountings

6.10 Checking for play in the swingarm bearings

leakage is found, the shock should be replaced (see Chapter 6).

8 With the aid of an assistant to support the bike, compress the rear suspension several times by pressing down on the passenger grab-rail. It should move up and down freely without binding. If any binding is felt, the worn or faulty component must be identified and replaced. The problem could be due to either the shock absorber or the swingarm components.

9 Grasp the top of the rear wheel and pull it upwards – there should be no discernible freeplay before the shock absorber begins to compress **(see illustration)**. If freeplay is felt, check that the shock absorber mountings are tight. Check also for wear in the shock absorber mountings.

10 Position the motorcycle on its centre stand so that the rear wheel is off the ground. Grab the swingarm and rock it from side to side to check for freeplay in the swingarm pivot **(see illustration)**. If freeplay is detected in the swingarm pivot, this could be due to worn swingarm bearings or loose pivot bolts; remove the rear wheel (see Chapter 6) and shock absorber (see Chapter 5) to make an accurate assessment. The swingarm should move smoothly about its pivots, without any binding or rough spots. There should be no discernible freeplay felt front-to-back or side-to-side. If bearing damage or freeplay is evident, and this is not due to loose pivot bolts, the swingarm should be removed and the bearings inspected (see Chapter 5).

11 Note that there have been instances of the swingarm pivot bolts and locknut working loose and as a result a higher strength locking

compound is recommended on their threads. If the swingarm pivots bolts or locknut are found to be loose, they should be removed and all thread-locking compound cleaned off before being reassembled using the specified strength thread-lock and correct torque settings as described in Chapter 5, Section 12. *Caution: It is imperative that the correct removal and installation procedure is followed if damage to the pivot bolts is to be avoided.*

Steering damper (R models)

12 On R models, a steering damper is fitted between the lower fork bridge and the Telelever arm **(see illustration)**. Check that there is no play at the damper mountings, and that the damper rod moves smoothly in the damper as the handlebars are turned. Also check for signs of pitting on the rod and for fluid leakage past the seal in the body.

7 Stands, lever pivots and cables – lubrication

1 Since the controls, cables and various other components of a motorcycle are exposed to the elements, they should be lubricated periodically to ensure safe and trouble-free operation.
2 The following pivot points should be lubricated frequently.
Footrest pivots
Clutch lever pivot
Clutch cable nipples at each end of cable
Brake lever pivot
Brake pedal pivot

Gearshift lever linkage pivots
Side stand and centre stand pivots
In order for the lubricant to be applied where it will do the most good, the component should be disassembled. However, if chain and cable lubricant is being used, it can be applied to the pivot joint gaps and will usually work its way into the areas where friction occurs. If motor oil or light grease is being used, apply it sparingly as it may attract dirt (which could cause the controls to bind or wear at an accelerated rate). The centre stand pivots are equipped with grease nipples, however we found that the grease tends to squirt out of the ends of the pivots rather than working its way into them, so it is advisable to remove the centre stand every now and then to properly lubricate it. **Note:** *One of the best lubricants for the control lever pivots is a dry-film lubricant (available from many sources by different names).*
3 To lubricate the cables, disconnect the relevant cable at its upper end, then lubricate the cable with a pressure adapter, or if one is not available, using the set-up shown **(see illustrations)**. See Chapter 2 for the clutch cable and Chapter 3 for the choke and throttle cable removal procedures.

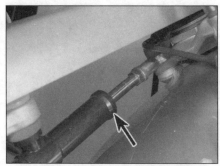

6.12 On R models, check the steering damper (arrowed) as described

7.3a Lubricating a cable with a pressure lubricator. Make sure the tool seals around the inner cable

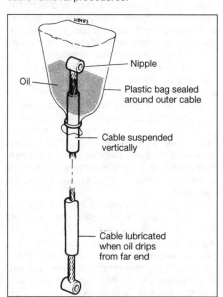

Nipple
Oil
Plastic bag sealed around outer cable
Cable suspended vertically
Cable lubricated when oil drips from far end

7.3b Lubricating a cable with a makeshift funnel and motor oil

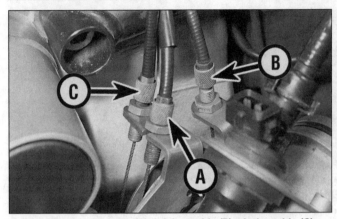

8.5 Main throttle cable (A), joining cable (B), choke cable (C) –
left-hand throttle body shown

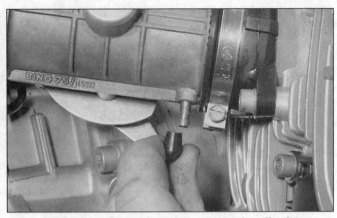

8.6 Remove the cap from the vacuum take-off point

4 The speedometer cable should be removed (see Chapter 8) and the inner cable withdrawn from the outer cable and lubricated with motor oil or cable lubricant. Do not lubricate the upper few inches of the cable as the lubricant may travel up into the instrument head.

8 Idle speed and throttle synchronisation – check and adjustment

Warning: Do not allow exhaust gases to build up in the work area; either perform the check outside or use an exhaust gas extraction system. When carrying out synchronisation, but careful not to allow the engine to overheat.

1 The idle speed is checked and adjusted in conjunction with throttle synchronisation. Before adjustment, make sure the valve clearances and spark plug gaps are correct. Also, turn the handlebars back-and-forth and see if the idle speed changes as this is done. If it does, the throttle cable may not be adjusted or routed correctly, or may be worn out. This is a dangerous condition that can cause loss of control of the bike. Be sure to correct this problem before proceeding.

2 Throttle synchronisation is simply the process of adjusting the throttle bodies so they pass the same amount of fuel/air mixture to each cylinder. This is done by measuring the vacuum produced in each cylinder. If out of synchronisation decreased fuel mileage, increased engine temperature, less than ideal throttle response and higher vibration levels will result.

3 To properly synchronise the throttles, you will need a pair of vacuum gauges or calibrated tubes to indicate engine vacuum. The equipment used should be suitable for a two cylinder engine and come complete with the hoses to fit the take-off points. **Note:** *Because of the nature of the synchronisation procedure and the need for special instruments, most owners leave the task to a BMW dealer equipped with the Bosch Synchrotester.*

4 The engine should be at normal operating temperature, which is usually reached after 10 to 15 minutes of stop-and-go riding. Place the motorcycle on its centre stand, and make sure the transmission is in neutral. On R1100RT and R1100RS models with a full fairing, remove the fairing side panels (see Chapter 7). Refer to the following procedure according to model year; note that 1993 to 1995 models have the throttle and choke cable adjusters located on the left-hand throttle body bracket, whereas 1996-on models have adjusters at the handlebar ends of the cables.

1993 to 1995 models

5 Using the adjusters on the left-hand throttle body, set choke cable freeplay to 1 mm, throttle cable freeplay to 0.5 mm and joining cable freeplay to zero **(see illustration)**. In all cases, freeplay is measured between the outer cable and its seat in the adjuster.

Warning: Take great care not to burn your hands on the hot engine unit when accessing the gauge take-off points on the intake manifolds.

6 Remove the vacuum take-off point blanking plug from the bottom of each intake manifold **(see illustration)**. Connect the vacuum gauge hoses, making sure they are a good fit – air leaks will result in false readings.

7 Start the engine and note the vacuum difference between the cylinders; if using vacuum gauges, set their damping so that needle flutter is just eliminated. Using the cable adjuster at each end of the joining cable, adjust so that both throttle bodies produce the same reading, indicating that they are synchronised. Slowly open the throttle whilst carefully observing the reading on each vacuum gauge. The readings for both cylinders should be equal at idle speed, change at the same time when the throttle is opened and remain equal with the throttle held steady, indicating that the throttles are opening at the same time. Tighten the cable adjuster locknuts, taking care not to move the adjuster positions whilst this is done.

8 With the throttles synchronised and the engine running, set the engine idle speed to 1000 – 1150 rpm. Idle speed can be adjusted independently on each cylinder via the adjuster screw **(see illustration)**. Snap the throttle open and shut a few times, then recheck synchronisation and idle speed. If necessary, repeat the adjustment procedure. *Caution: Do not adjust the position of the TPS (Throttle Position Sensor) or the two limiter screws on the inner face of the throttle pulley – each has been carefully set up by the manufacturer.*

9 Remove the vacuum gauges and fit the cap over the take-off point.

1996-on models

10 Pull back the rubber covers to expose the choke and throttle cable adjusters at the handlebar. Slacken their lockwheels and screw the adjusters in until the gap between the lockwheel and adjuster head is 1 mm with the lockwheel tightened **(see illustrations)**.

11 Moving to the throttle bodies, slacken the cable adjuster locknuts and rotate each adjuster so that there is 2 mm freeplay in the cables, measured between the outer cable and its seat in the adjuster **(see illustrations)**.

12 Connect the vacuum gauges and adjust throttle synchronisation and idle speed as described above in Steps 6 to 9.

13 Reset throttle cable freeplay at the handlebar adjuster to 0.5 mm, and choke cable freeplay to zero. Slip the rubber caps back into place over the adjusters.

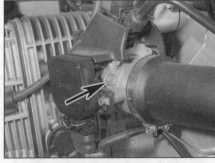

8.8 The idle speed is adjusted using this screw – left-hand body shown

8.10a Choke cable adjuster at handlebar

8.10b Throttle cable adjuster at handlebar

8.11a Left-hand throttle body cable adjuster

8.11b Right-hand throttle body cable adjuster

All models

14 On R1100RT and R1100RS models with a full fairing, install the fairing (see Chapter 7).
15 If a smooth, steady idle can't be achieved, check for any possible air leaks around the throttle body and intake manifold assemblies and replace any worn components (see Chapter 3). Otherwise there could be a problem with either the valves (see Chapter 2) or the Motronic system (see Chapter 3).

9 Throttle and choke cables – check

Throttle cable

1 Make sure the throttle grip rotates easily from fully closed to fully open with the front wheel turned at various angles. The grip should return automatically from fully open to fully closed when released.

10.2 The gearbox oil should be level with the lower edge of the filler hole

2 If the throttle sticks, this is probably due to a cable fault. Remove the cable (see Chapter 3) and lubricate it (see Section 7). Install the cable, making sure it is correctly routed. If this fails to improve the operation of the throttle, the cable must be replaced. Note that in very rare cases, poor throttle action could be due to a sticking throttle pulley or joining cable between the two throttle bodies; on 1996-on models, the cable distributor box could be at fault (see Chapter 3).

Warning: Turn the handlebars all the way through their travel with the engine idling. Idle speed should not change. If it does, the cable may be routed incorrectly. Correct this condition before riding the bike.

3 On 1993 to 1995 models, check for a small amount of freeplay in the main cable between the handlebar and the left-hand throttle body, measured in terms of the amount of slack between the outer cable and its seat in the cable adjuster, and compare the amount to that listed in this Chapter's Specifications. If

10.3a Install the filler cap . . .

it's incorrect, adjust the cable to correct it. On 1996-on models cable freeplay should be 0.5 mm (see illustration 8.10b).
Caution: Adjustment of the joining cable linking the two throttle bodies can only be made in conjunction with vacuum testing equipment to preserve throttle body synchronisation (see Section 8).

Choke cable

4 If the choke does not operate smoothly this is probably due to a cable fault. Remove the cable (see Chapter 3) and lubricate it (see Section 7). Install the cable, routing it so it takes the smoothest route possible. If this fails to improve the operation of the choke, the cable must be replaced.
5 Check that the amount of freeplay is as specified in the Specifications section of this chapter. On 1993 to 1995 models, freeplay is measured in terms of slack between the outer cable and its seat in the cable adjuster (see illustration 8.5). On 1996-on models cable freeplay should be zero (see illustration 8.10a).

10 Gearbox – oil level check

Note: *The gearbox oil level is unlikely to fall unless there is leakage from the oil seals or the oil drain plug.*
1 Place the motorcycle on its centre stand, making sure it is on level ground.
2 The check should be made after the machine has been standing for a few hours. Unscrew the oil filler cap and check that the oil is up to the lower edge of the filler hole (see illustration). If the level is below this, look for signs of leakage, such as oil staining on the underside of the gearbox. If leakage is evident, the problem must be rectified to avoid the possibility of damage to the gearbox and oil contaminating the rear tyre (see Chapter 2).
3 Replenish the oil if necessary to the correct level using the type and grade specified at the beginning of the Chapter (see illustration 13.4), then install the filler cap, using a new sealing washer, and tighten it to the torque setting specified at the beginning of the Chapter (see illustrations).

10.3b . . . and tighten it to the specified torque setting

1

11.2a Remove the final drive oil level plug (arrowed)

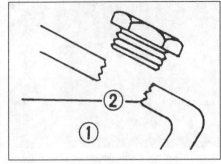

11.2b Oil (1) must be level with bottom of threads (2)

11 Final drive – oil level check

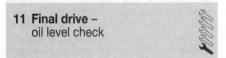

Note: *The final drive oil level is unlikely to fall unless there is leakage from the oil seals or the oil drain plug.*

1 Place the motorcycle on its centre stand, making sure it is on level ground.
2 The check should be made after the machine has been standing for a few hours. Unscrew the oil filler cap and check that the oil is up to the lower edge of the filler hole threads **(see illustrations)**. If the level is below this, look for signs of leakage, such as oil staining on the underside of the casing. If leakage is evident, the problem must be rectified to avoid the possibility of damage to

the final drive and oil contaminating the rear tyre (see Chapter 5).
3 Replenish the oil if necessary to the correct level using the type and grade specified at the beginning of the Chapter, then install the filler cap, using a new sealing washer, and tighten it to the torque setting specified at the beginning of the Chapter.

12 Nuts and bolts – tightness check

1 Since vibration of the machine tends to loosen fasteners, all nuts, bolts, screws, etc. should be periodically checked for proper tightness.
2 Pay particular attention to the following:

Spark plugs
Engine oil, gearbox oil and final drive oil filler and drain plugs
Gearshift lever, front brake lever and rear brake pedal bolts
Footrest and stand bolts
Engine mounting bolts
Shock absorber mounting bolts and Telelever/swingarm pivot bolts
Handlebar clamp bolts
Front axle bolt and axle clamp bolts
Front fork yoke bolts
Rear wheel bolts
Brake caliper mounting bolts
Brake hose banjo bolts and caliper bleed valves
Brake disc bolts
Exhaust system bolts/nuts

3 If a torque wrench is available, use it along with the torque specifications at the beginning of this and other Chapters.

> **HAYNES HINT** *Complete the Minor service with a test ride, paying attention to all freshly-adjusted or serviced components while riding. The ride must be of sufficient length to warm up all components to normal operating temperature and a careful check must be made for any fuel, oil or brake fluid leaks that may have developed. If necessary repeat any service operation.*

Major service

After the first 12,000 miles (20,000 km), and every 12,000 miles (20,000 km) thereafter

Carry out all the items under the minor service (except the spark plug check), plus the following:

13 Gearbox – oil change

1 Before changing the oil, take the bike for a ride to warm up the oil so it will drain easily. Place the motorcycle on its centre stand,

making sure it is on level ground.
2 Place an oil drain tray below the gearbox. To avoid the oil running over the hot exhaust system, make a chute using a piece of card and shape it to fit under the oil drain passage, so that it will channel the oil into the tray. Unscrew the oil filler plug to act as a vent, then unscrew the drain plug and allow the oil to flow into the drain tray **(see illustrations)**. Discard the sealing washer (where fitted) on the drain plug as a new one should be used.

> **HAYNES HINT** *An oil drain tray can be easily made by cutting away the front or back of an old five litre oil container, or any other such container that is of an adequate size.*

3 When the oil has completely drained, fit the drain plug using a new sealing washer, and tighten it to the torque setting specified at the

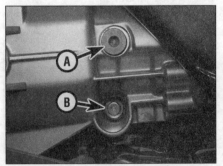

13.2a Unscrew the filler plug (A), followed by the drain plug (B) . . .

13.2b . . . and allow the oil to drain completely – note the cardboard chute

13.4 Fill the gearbox with the specified oil

14.2a Unscrew the final drive oil drain plug (arrowed) . . .

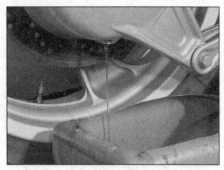
14.2b . . . and allow the oil to drain completely

14.3 Tighten the drain bolt to the specified torque setting

14.4 Fill the housing with the specified oil

14.5a Install the filler cap . . .

14.5b . . . and tighten it to the specified torque setting

beginning of the Chapter. Do not overtighten the plug as the threads in the sump could be damaged. **Note:** *On models from late 1996 no sealing washer is fitted to the drain plug.*

4 Fill the gearbox using the amount and type of oil specified at the beginning of the Chapter until it just starts to run out of the filler hole **(see illustration)**. The oil level is correct when it is up to the lower edge of the filler hole.

5 Install the filler cap, using a new sealing washer, and tighten it to the specified torque setting.

14 Final drive – oil change

1 Before changing the oil, take the bike for a ride to warm up the oil so it will drain easily. Place the motorcycle on its centre stand, making sure it is on level ground.

2 Place an oil drain tray below the final drive housing. Unscrew the oil filler plug to act as a vent **(see illustration 11.2a)**, then unscrew the drain plug and allow the oil to flow into the drain tray **(see illustrations)**. Discard the sealing washer on the drain plug as a new one should be used.

3 When the oil has completely drained, fit the drain plug using a new sealing washer, and tighten it to the torque setting specified at the beginning of the Chapter **(see illustration)**.

Do not overtighten the plug as the threads in the sump could be damaged.

4 Fill the gearbox using the amount and type of oil specified at the beginning of the Chapter **(see illustration)** until it is up to the lower threads of the filler hole **(see illustration 11.2b)**.

5 Install the filler cap, using a new sealing washer, and tighten it to the specified torque setting **(see illustrations)**.

15 Spark plugs – replacement

1 Remove the old spark plugs as described in Section 2 and install new ones.

16.2 Check that the clamp screws (arrowed) are tightened as described

16 Intake manifold clamps – tightness check

1 Check that the clamps securing the throttle bodies to the intake manifolds and the air ducts, and the air ducts to the air box are secure, otherwise air leaks will result.

2 Tighten them if necessary to take up any slack, but take care not to overtighten them causing distortion **(see illustration)**. Note that BMW specify a torque setting for the clamp screws.

17 Air filter – replacement

Caution: If the machine is continually ridden in continuously dusty or dirty conditions, the filter should be replaced more frequently – see Service Schedule.

1 Remove the seat (see Chapter 7).

2 On GS models, remove the bolt securing the rear of the fuel tank to the right-hand side of the frame, then raise the tank and carefully support it in that position, making sure there is no strain on the fuel pipes.

3 Press in the clip on the air temperature sensor wiring connector and detach the

17.3a Press in the clip and detach the temperature sensor wiring connector

17.3b Release the clips . . .

17.3c . . . and remove the cover

17.3d Remove the old filter from the housing

17.4a Ensure the new element seats correctly

17.4b Secure the cover with the clips

connector **(see illustration)**. Release the clips securing the filter cover, then remove the cover **(see illustrations)**. Withdraw the filter from the housing and discard it **(see illustration)**.

4 Install the new filter by reversing the removal procedure **(see illustration)**. Make sure the filter cover locates on its tabs at the front of the housing and that the cover is properly seated before securing with the clips **(see illustration)**. On GS models, remove the support and tighten the fuel tank mounting bolt to the torque setting specified at the beginning of the Chapter.

5 Replacement, rather than cleaning, is recommended by BMW. If cleaning is attempted, tap the element on a hard surface to dislodge any dirt, then use compressed air directed from the inside outwards to clear any particles of dust and dirt.

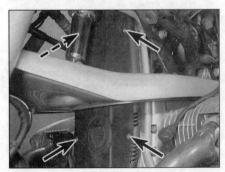

18.2 Unscrew the four bolts (arrowed) and remove the cover

18 Alternator drive belt – adjustment (R1100RS and GS to engine no. 38946129 only)

Note 1: *A new type maintenance-free belt was fitted to all R1100GS and R1100RS models from engine no. 38946130 onwards and to all R850/1100R and R1100RT models. Modified pulleys were fitted at the same time as the new belt and can be identified as described in Step 3. Note that the new type belt and pulley arrangement requires no adjustment of belt tension. The new belt and pulleys can be fitted retrospectively to older RS and GS models.*
Note 2: *The maintenance-free type belt may produce slight squealing noises when the engine is cold. This should stop when engine temperature rises – no adjustment is necessary.*

1 On R1100RS models, remove the left-hand fairing side panel (see Chapter 7).
2 Unscrew the four bolts securing the engine front cover and remove the cover **(see illustration)**.
3 At this stage, check the markings on the belt and the belt pulleys to determine whether the old belt and pulleys have already been replaced by the new maintenance-free type. The new belt is marked either 1231 1342059 Fa. Dayco or 1231 1341779 Fa. Conti. The old top pulley is made of aluminium, the new one is made of steel; the old bottom pulley has three spot welds on the front, the new bottom pulley does not. If new components have

been fitted, there is no need for adjustment. If the old components are fitted adjust the belt tension as follows; note that if the machine is approaching its next 24,000 mile (40,000 km) service, install the new belt and pulleys as fitted to later models.

4 Slacken the two nuts and the bolt securing the alternator to the engine casing, then re-tighten the nut onto the adjusting bolt finger-tight **(see illustration)**. Check the drive belt along its entire length for splits, cracks, worn or damaged teeth, fraying and any other damage or deterioration. Be careful not to bend the belt excessively or to get oil or grease on it. Replace the belt if it is in any way worn, damaged or deteriorated (see Section 25).

5 Tighten the adjusting bolt using a torque wrench to pre-load the belt to the torque setting specified at the beginning of the

18.4 Slacken the nuts (A) and (B) and the bolt (C), then tighten the nut (B) finger-tight

Chapter **(see illustration)**. Keeping the torque wrench held on the adjuster to prevent it from slackening, tighten the top retaining nut and the bolt to the specified torque setting, then tighten the adjuster bolt nut to the specified torque setting **(see illustration)**.

6 Turn the engine over once on the starter motor to settle the belt.

7 Install the engine front cover and the fairing panels in a reverse of the removal procedure.

19 Brakes – fluid change

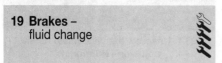

1 The brake fluid should be replaced at the prescribed interval or once every year regardless of mileage. Refer to the brake bleeding section in Chapter 6, noting that all old fluid must be pumped from the fluid reservoir and hydraulic line before filling with new fluid.

 HAYNES HINT *Old brake fluid is invariably much darker in colour than new fluid, making it easy to see when all old fluid has been expelled from the system.*

21.2 The electrolyte level must be between the level lines (arrowed)

21.3a Unscrew the cell caps . . .

18.5a Apply specified torque to the adjuster bolt to pre-load the belt (arrowed) . . .

20 Brake system – check

General checks

1 Make sure all brake fasteners are tight. Check the brake pads and discs for wear (see Section 5) and make sure the fluid level in the reservoirs is correct (see *Daily (pre-ride) checks*).

2 Look for leaks at the hose connections and check for cracks and bulges in the hoses, particularly where the hose joins the banjo union and where it passes through hose guides. If the lever or pedal is spongy, bleed the brakes (see Chapter 6).

Front brake lever

3 Check the brake lever for loose connections, improper or rough action, excessive play, bends, and other damage. Replace any damaged parts with new ones (see Chapter 6).

4 The front brake lever has a span adjuster which alters the distance of the lever from the handlebar. Each setting is identified by a number on the adjuster which aligns with the arrow on the lever. Pull the lever away from the handlebar and turn the adjuster ring until the setting which best suits the rider is obtained. There are four settings.

Rear brake pedal

5 Check the brake pedal for loose

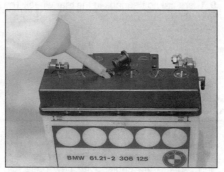

21.3b . . . and top up the level as required

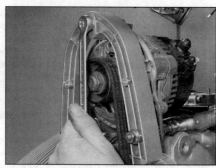

18.5b . . . and tighten the mounting nuts and bolt as described

connections, improper or rough action, excessive play, bends, and other damage. Replace any damaged parts with new ones (see Chapter 6).

Brake light

6 Make sure the brake light operates when the front brake lever is pulled in and also when the rear brake pedal is depressed. The brake light switches are not adjustable. If they fail to operate properly, check them (see Chapter 8).

21 Battery – checks

Caution: Be extremely careful when handling or working around the battery. The electrolyte is very caustic and an explosive gas (hydrogen) is given off when the battery is charging.

1 Remove the battery (see Chapter 8).

2 The electrolyte level is visible through the translucent battery case – it should be between the MIN and MAX level marks **(see illustration)**.

3 If the electrolyte is low, remove the cell caps and fill each cell to the upper level mark with distilled water **(see illustrations)**. Do not use tap water (except in an emergency), and do not overfill. The cell holes are quite small, so it may help to use a clean plastic squeeze bottle with a small spout to add the water. Install the battery cell caps, tightening them securely **(see illustration)**.

21.3c Make sure the caps are secure

4 The battery case should be kept clean to prevent current leakage, which can discharge the battery over a period of time (especially when it sits unused). Wash the outside of the case with a solution of baking soda and water. Rinse the battery thoroughly, then dry it.

5 Look for cracks in the case and replace the battery if any are found. If acid has been spilled on the frame or battery box, neutralise it with a baking soda and water solution, dry it thoroughly, then touch up any damaged paint.

6 If the motorcycle sits unused for long periods of time, remove the battery and charge it once every month to six weeks (see Chapter 8).

7 The condition of the battery can be assessed by measuring its specific gravity and open-circuit voltage (refer to *Fault Finding Equipment* in the Reference section for details).

8 Check the battery terminals and leads for tightness and corrosion. If corrosion is evident, unscrew the terminal screws and disconnect the leads from the battery, disconnecting the negative (-ve) terminal first, and clean the terminals and lead ends with a wire brush or knife and emery paper. Reconnect the leads, connecting the negative (-ve) terminal last, and apply a thin coat of petroleum jelly (Vaseline) to the connections to slow further corrosion.

22 Wheels and wheel bearings – check

Cast wheels

1 Cast wheels are virtually maintenance free, but they should be kept clean and checked periodically for cracks and other damage. Never attempt to repair damaged cast wheels; they must be replaced with new ones. Also check the wheel runout and front/rear wheel alignment as described in Chapter 6.

2 Check the valve rubber for signs of damage or deterioration and have it replaced if necessary. Also, make sure the valve stem cap is in place and tight.

Wire spoke wheels

3 Visually check the spokes for damage, breakage or corrosion. A broken or bent spoke must be renewed immediately because

22.7 Checking the front wheel bearings

the load taken by it will be transferred to adjacent spokes which may in turn fail. If you suspect that any of the spokes are incorrectly tensioned, tap each one lightly with a screwdriver and note the sound produced. Properly tensioned spokes will make a sharp pinging sound, loose ones will produce a lower pitch and tight ones will be higher pitched.

4 Unevenly tensioned spokes will promote rim misalignment – refer to information on wheel runout in Chapter 6 and seek the advice of a BMW dealer or wheel building specialist if the wheel needs realigning. Check front and rear wheel alignment as described in Chapter 6.

5 Check the valve rubber for signs of damage or deterioration and have it replaced if necessary. Also, make sure the valve stem cap is in place and tight.

Front wheel bearings

6 The bearings in the front wheel will wear over a period of time and result in handling problems.

7 Place the motorcycle on its centre stand and take the weight off the front wheel. Check for any play in the bearings by pushing and pulling the wheel against the hub **(see illustration)**. Also rotate the wheel and check that it rotates smoothly.

8 If any play is detected in the hub, or if the wheel does not rotate smoothly (and this is not due to brake drag), the wheel bearings must be removed and inspected for wear or damage (see Chapter 6).

Rear wheel bearings

9 There are no bearings in the rear wheel. The rear wheel and brake disc are bolted to the

23.1 Unscrew the windshield height adjuster – RS models

final drive unit and pivot on the bearings contained within the drive unit. Before checking for play, first check that the rear wheel bolts are tight – refer to the torque setting in the Specifications section of this Chapter.

10 Place the motorcycle on its centre stand and take the weight off the rear wheel. Grasp the wheel and check for any play between the rear wheel and final drive unit. If excessive play is evident the final drive bearings or shims may need attention – seek the advice of a BMW dealer.

> **HAYNES HiNT**
>
> *Complete the Major service with a test ride, paying attention to all freshly-adjusted or serviced components while riding. The ride must be of sufficient length to warm up all components to normal operating temperature and a careful check must be made for any fuel, oil or brake fluid leaks that may have developed. If necessary repeat any service operation.*

23 Windshield height adjuster – lubrication (RS models)

1 Unscrew and remove the windshield height adjuster **(see illustration)**. Clean off all traces of old and hardened grease, then smear the shaft with fresh grease – BMW recommend Shell Retinax A.

2 Screw the adjuster back into place and set the windshield height as required.

Every 18,000 miles (30,000 km)

24 Throttle cables – replacement

1 The throttle cables stretch and wear over time, making throttle action heavy and synchronisation inaccurate. Refer to Chapter 3 and replace the cables with new ones. On completion, the throttles must be synchronised and the idle speed reset (see Section 8).

25.2 Release the clamps (arrowed), detach the hoses and remove the fuel filter

26.4 Fit the belt around the pulleys, making sure it seats correctly

Every 24,000 miles (40,000 km)

25 Fuel filter –
replacement

 Warning: Petrol (gasoline) is extremely flammable, so take extra precautions when you work on any part of the fuel system. Don't smoke or allow open flames or bare light bulbs near the work area, and don't work in a garage where a natural gas-type appliance is present. If you spill any fuel on your skin, rinse it off immediately with soap and water. When you perform any kind of work on the fuel system, wear safety glasses and have a fire extinguisher suitable for a Class B type fire (flammable liquids) on hand.

1 Remove the fuel tank (see Chapter 3) and the fuel pump (see Chapter 8). The fuel filter is mounted adjacent to the fuel pump.
2 Release the clamps securing the fuel hoses to the filter and detach the filter, noting which hose fits where and which way round the filter is **(see illustration)**. Install the new filter, making sure the arrow on its body points in the direction of fuel flow, and secure the hoses with the clamps.

3 Check all the internal and external fuel hoses for signs of leakage, deterioration or damage; in particular check that there is no leakage from the external fuel hoses, particularly around the clamps. It is best to check for leakage with the ignition switched on, as the fuel pump will have pressurised the system. Replace any hoses which are cracked or deteriorated, and any clamps that are weak or deformed.
4 Install the fuel pump (see Chapter 8) and the fuel tank (see Chapter 3).

26 Alternator drive belt –
replacement (R1100RS and GS to engine no. 38946129 only)

Note: *A new type maintenance-free belt was fitted to all R1100GS and R1100RS models from engine no. 38946130 onwards and to all R850/1100R and R1100RT models. Modified pulleys were fitted at the same time as the new belt and can be identified as described in Section 18, Step 3. The new belt and pulleys can be fitted retrospectively to older RS and GS models and may have already been implemented at an earlier service interval. The new belt is maintenance-free, but must be replaced every 36,000 miles (60,000 km).*

1 On R1100RS and RT models, remove the left-hand fairing side panel (see Chapter 7).
2 On R850R and R1100R models, remove the left-hand fuel tank cover (see Chapter 7).
3 Unscrew the four bolts securing the engine front cover and remove the cover **(see illustration 18.2)**.
4 Slacken the two nuts and the bolt securing the alternator to the engine casing, then push the alternator down in its mountings to release the tension in the belt and tighten the nut onto the adjusting bolt finger-tight **(see illustration 18.4)**. Remove the old belt and install a new one, making sure its ribs sit correctly in the channels in the pulleys **(see illustration)**.
5 Tighten the adjusting bolt using a torque wrench to pre-load the belt to the torque setting specified at the beginning of the Chapter **(see illustration 18.5a)**. Keeping the torque wrench held on the adjuster to prevent it from slackening, tighten the top retaining nut and the bolt to the specified torque setting, then tighten the adjuster bolt nut to the specified torque setting **(see illustration 18.5b)**.
6 Turn the engine over once on the starter motor to settle the belt.
7 Install the engine front cover and the fairing panels in a reverse of the removal procedure.

1

Every 36,000 miles (60,000 km)

27 Alternator drive belt –
replacement (maintenance - free type)

1 The procedure for changing and tensioning the maintenance-free belt is the same as that for the earlier type belt (see Section 25). Refer to Section 18, Step 3 for details of belt type identification.

Non-scheduled maintenance

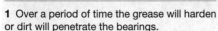

28 Swingarm (Paralever) bearings – lubrication

1 Over a period of time the grease will harden or dirt will penetrate the bearings.
2 The swingarm is not equipped with grease nipples. Remove the swingarm as described in Chapter 5 for greasing of the bearings.

29 Brake calipers and master cylinders – seal replacement

1 Brake seals will deteriorate over a period of time, particularly if the bike has been in long-term storage, and lose their effectiveness, leading to sticking operation or fluid loss, or allowing the ingress of air and dirt. Refer to Chapter 6 and dismantle the components for seal replacement, noting that it is worth checking first on the availability of replacement seals with a BMW dealer.
2 Always change the brake fluid when new seals are installed.

30 Brake hoses – replacement

1 The flexible brake hoses will in time deteriorate with age and should be replaced with new ones. Regular checks should be made for cracks and bulges in the hose material, particularly where the hose joins the banjo union and where it passes through hose guides.
2 Refer to Chapter 6 and disconnect the brake hoses from the master cylinders and calipers. Always replace the banjo union sealing washers with new ones.
3 Always change the brake fluid when new hoses are installed.

31 Cylinder compression – check

Note: *Refer to the general information on compression tests in the Fault Finding Equipment section of Reference, together with the following specific instructions for BMW twins.*
1 Among other things, poor engine performance may be caused by leaking valves, incorrect valve clearances, a leaking head gasket, or worn pistons, rings and/or cylinder walls. A cylinder compression check will help pinpoint these conditions and can also indicate the presence of excessive carbon deposits in the cylinder heads.

2 The only tools required are a compression gauge with a threaded adapter for the spark plug hole and a spark plug wrench **(see illustration)**.
3 Note that during a normal compression test one would go on to temporarily seal the piston rings by pouring a quantity of oil into the barrel and then take a second set of readings. If the pressure increased noticeably it could then be assumed that the piston rings were worn rather than the valves. Since it would be very difficult to get a full seal from such a method in a warm flat-twin engine there is little point in doing this; check the pistons and rings as well as the head gasket and valves when looking for the cause of compression loss.

32 Engine – oil pressure check

1 The oil pressure warning light should come on when the ignition (main) switch is turned ON and extinguish a few seconds after the engine is started – this serves as a check that the warning light bulb is sound. If the oil pressure light comes on whilst the engine is running, low oil pressure is indicated – stop the engine immediately and carry out an oil level check *(see Daily (pre-ride) checks)*.
2 An oil pressure check must be carried out if the warning light comes on when the engine is running and the oil level is good (Step 1). It can also provide useful information about the condition of the engine's lubrication system.
3 To check the oil pressure, a suitable gauge and adapter piece (which screws into the crankcase) will be needed (see *Tools and Workshop Tips*).
4 Warm the engine up to normal operating temperature then stop it.
5 Remove the oil pressure switch (see Chapter 8) and swiftly screw the adapter into the crankcase threads. Connect the gauge to the adapter.
6 Start the engine and increase the engine speed whilst watching the gauge reading. The oil pressure should be similar to that given in

31.2 Cylinder compression tester and adapter

the Specifications at the start of this Chapter.
7 If the pressure is significantly lower than the standard, either the pressure relief valve is stuck open, the oil pump is faulty, the oil strainer or filter is blocked, or there is other engine damage. Begin diagnosis by checking (or replacing) the oil filter, then the oil pump, followed by the strainer and relief valve (both of which require engine removal and crankcase separation) (see Chapter 2). If those items check out okay, chances are the bearing oil clearances are excessive and the engine needs to be overhauled.
8 If the pressure is too high, either an oil passage is clogged, relief valve is stuck closed or the wrong grade of oil is being used.
9 Stop the engine and unscrew the gauge and adapter from the crankcase.
10 Install the oil pressure switch (see Chapter 8). Check the oil level (see *Daily (pre-ride) checks*).

33 Headlight aim – check and adjustment

Note: *An improperly adjusted headlight may cause problems for oncoming traffic or provide poor, unsafe illumination of the road ahead. Before adjusting the headlight aim, be sure to consult with local traffic laws and regulations – for UK models refer to MOT Test Checks in the Reference section.*
1 The headlight beam can adjusted both horizontally and vertically. Before making any adjustment, check that the tyre pressures are correct and the suspension is adjusted as required. Make any adjustments to the headlight aim with the machine on level ground, with the fuel tank half full and with an assistant sitting on the seat. If the bike is usually ridden with a passenger on the back, have a second assistant to do this.

RS and RT models

2 Vertical adjustment is made by turning the adjuster knob on the right-hand side of the headlight **(see illustration)**. Turn it anti-clockwise to move the beam up, and

33.2a Vertical beam adjuster – RS and RT models

33.2b Vertical beam pre-set adjuster – RS and RT models

33.3 Horizontal beam adjuster – RS and RT models

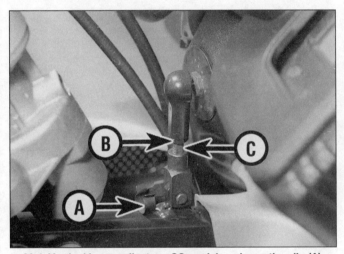

33.4 Vertical beam adjuster – GS models: release the clip (A), slacken the locknut (B) and turn the adjuster (C) as required

33.5 Horizontal beam adjuster – GS models: slacken the bolt (arrowed) on each side and move the headlight in the slots as required

1

clockwise to move it down. The adjuster lever mounted around the knob can be used as an instant adjustment to compensate for the effect on the headlight beam of carrying a passenger and/or luggage. Moving the lever from one position to the other tilts the headlight by a pre-set amount **(see illustration)**. Fine adjustments can be made using the knob.

3 Horizontal adjustment is made by turning the adjuster screw on the left-hand side of the headlight **(see illustration)**.

GS models

4 Vertical adjustment is made by releasing the clip on the linkage on the back of the headlight, then loosening the locknut and turning the linkage adjuster **(see illustration)**.

5 Remove the windshield (see Chapter 7) to gain access to the horizontal adjusters. Adjustment is made by slackening the

mounting bolts and sliding the headlight as required in the slots in the headlight brackets **(see illustration)**.

R models

6 On models without a headlight cowl, beam adjustment is made by slackening the mounting bolts on each side of the headlight shell and tilting the headlight up or down (vertical) or side-to-side (horizontal) until the position is correct **(see illustration)**. Hold the headlight steady while the bolts are tightened.

7 On models with a headlight cowl, horizontal adjustment is made by slackening the headlight shell mounting bolts and moving the shell to the left or right within the bracket slots. To adjust vertically, slacken the headlight shell mounting bolts and rotate the adjuster bar at the left-hand underside of the shell to make adjustment. When adjustment is complete, tighten the shell mounting bolts.

33.6 Slacken headlight bolt (arrowed) on each side of shell to adjust beam on R models

Chapter 2
Engine, clutch and transmission

Contents

2

Degrees of difficulty

Easy, suitable for novice with little experience	Fairly easy, suitable for beginner with some experience	Fairly difficult, suitable for competent DIY mechanic	Difficult, suitable for experienced DIY mechanic	Very difficult, suitable for expert DIY or professional

Specifications

General

Type	Four-stroke, horizontally-opposed twin with 4 valves per cylinder
Capacity	
850 model	848 cc
1100 models	1085 cc
Bore	
850 model	87.5 mm
1100 models	99.0 mm
Stroke	70.5 mm
Compression ratio	
R850R, R1100R and R1100GS	10.3 to 1
R1100RS and R1100RT	10.7 to 1
Cooling system	Air/oil
Clutch	Dry single plate with diaphragm spring
Transmission	Five-speed constant mesh
Final drive	Shaft

Rocker arms and shafts
Rocker shaft diameter . . . . 15.973 to 15.984 mm
Rocker arm bore diameter . . . . 16.016 to 16.027 mm
Clearance
  Standard . . . . 0.032 to 0.054 mm
  Service limit (max) . . . . 0.10 mm
Rocker shaft end-float
  Minimum . . . . 0.05 mm
  Maximum . . . . 0.40 mm

Camshafts and followers
Camshafts
  Intake valve lift . . . . 9.68 mm
  Exhaust valve lift
    R850R, R1100R and R1100GS . . . . 8.55 mm
    R1100RS and R1100RT . . . . 9.26 mm
  Journal diameter . . . . 20.97 to 21.00 mm
  Journal holder diameter . . . . 21.02 to 21.04 mm
  Clearance
    Standard . . . . 0.02 to 0.07 mm
    Service limit (max) . . . . 0.15 mm
  Guide bearing width . . . . 15.92 to 15.95 mm
  Camshaft bearing width . . . . 16.0 to 16.5 mm
  Axial play
    Standard . . . . 0.08 to 0.13 mm
    Service limit (max) . . . . 0.25 mm
Followers
  Follower diameter . . . . 23.947 to 23.960 mm
  Follower bore diameter . . . . 24.000 to 24.021 mm
  Clearance
    Standard . . . . 0.040 to 0.074 mm
    Service limit (max) . . . . 0.18 mm

Valves, guides and springs
Valve clearances . . . . See Chapter 1
Intake valve
  Seat width
    Standard . . . . 0.95 to 1.25 mm
    Service limit (max) . . . . 2.5 mm
  Head diameter
    1993 to 1995 models . . . . 36 mm
    1996-on models . . . . 34 mm
  Head edge thickness
    Standard . . . . 0.8 to 1.2 mm
    Service limit (min) . . . . 0.5 mm
  Stem diameter
    All 1100 models up to 1995
      Standard . . . . 5.960 to 5.975 mm
      Service limit (min) . . . . 5.940 mm
    All 1100 models 1996-on, and 850 model
      Standard . . . . 4.966 to 4.980 mm
      Service limit (min) . . . . 4.946 mm
  Guide internal diameter
    All 1100 models up to 1995 . . . . 6.000 to 6.015 mm
    All 1100 models 1996-on, and 850 model . . . . 5.000 to 5.012 mm
  Stem-to-guide clearance
    All 1100 models up to 1995
      Standard . . . . 0.025 to 0.055 mm
      Service limit (min) . . . . 0.15 mm
    All 1100 models 1996-on, and 850 model
      Standard . . . . 0.020 to 0.046 mm
      Service limit (min) . . . . 0.15 mm
  Guide external diameter . . . . 12.533 to 12.544 mm
  Guide bore in cylinder head . . . . 12.500 to 12.518 mm
  Guide to bore overlap . . . . 0.015 to 0.044 mm
  Replacement guide external diameter . . . . 12.550 to 12.561 mm
  Oversize replacement guide external diameter . . . . 12.733 to 12.744 mm

Valves, guides and springs (continued)

Exhaust valve
  Seat width
    Standard . . . . 1.25 to 1.55 mm
    Service limit (max) . . . . 3.0 mm
  Head diameter
    1993 to 1995 models . . . . 31 mm
    1996-on models . . . . 29 mm
  Head edge thickness
    R850R, R1100R and R1100GS
      Standard . . . . 0.8 to 1.2 mm
      Service limit (min) . . . . 0.5 mm
    R1100RS and RT models to 1995
      Standard . . . . 1.45 to 1.85 mm
      Service limit (min) . . . . 1.0 mm
    R1100RS and RT models 1996-on
      Standard . . . . 0.8 to 1.2 mm
      Service limit (min) . . . . 0.5 mm
  Stem diameter
    All 1100 models up to 1995
      Standard . . . . 5.945 to 5.960 mm
      Service limit (min) . . . . 5.925 mm
    All 1100 models 1996-on, and 850 model
      Standard . . . . 4.956 to 4.970 mm
      Service limit (min) . . . . 4.936 mm
  Guide internal diameter
    All 1100 models up to 1995 . . . . 6.000 to 6.015 mm
    All 1100 models 1996-on, and 850 model . . . . 5.000 to 5.012 mm
  Stem-to-guide clearance
    All 1100 models up to 1995
      Standard . . . . 0.040 to 0.070 mm
      Service limit (max) . . . . 0.17 mm
    All 1100 models 1996-on, and 850 model
      Standard . . . . 0.030 to 0.056 mm
      Service limit (max) . . . . 0.17 mm
  Guide external diameter . . . . 12.533 to 12.544 mm
  Guide bore in cylinder head . . . . 12.500 to 12.518 mm
  Guide to bore overlap . . . . 0.015 to 0.044 mm
  Replacement guide external diameter . . . . 12.550 to 12.561 mm
  Oversize replacement guide external diameter . . . . 12.733 to 12.744 mm
  Valve guide height above cylinder head . . . . 15.3 to 15.5 mm
Valve spring free length (intake and exhaust)
  Standard . . . . 41.1 mm
  Service limit (min) . . . . 39.0 mm

Cylinders

Bore
  850 model
    Type A
      Standard . . . . 87.492 to 87.500 mm
      Service limit (max) . . . . 87.550 mm
    Type B
      Standard . . . . 87.500 to 87.508 mm
      Service limit (max) . . . . 87.558 mm
  All 1100 models
    Type A
      Standard . . . . 98.992 to 99.000 mm
      Service limit (max) . . . . 99.050 mm
    Type B
      Standard . . . . 99.000 to 99.008 mm
      Service limit (max) . . . . 99.058 mm
Ovality (out-of-round) (max)
  20 mm from top . . . . 0.03 mm
  100 mm from bottom . . . . 0.04 mm
Cylinder compression . . . . see Chapter 1

2

Pistons

Piston diameter (measured 6.0 mm up from skirt, at 90° to piston pin axis)

850 model
  Type A
    Standard .... 87.465 to 87.477 mm
    Service limit (min) .... 87.390 mm
  Type B
    Standard .... 87.477 to 87.485 mm
    Service limit (min) .... 87.400 mm
  Type AB
    Standard .... 87.473 to 87.481 mm
    Service limit (min) .... 87.395 mm

1100 models
  Type A
    Standard .... 98.965 to 98.977 mm
    Service limit (min) .... 98.890 mm
  Type B
    Standard .... 98.977 98.989 mm
    Service limit (min) .... 98.900 mm
  Type AB
    Standard .... 98.973 to 98.981 mm
    Service limit (min) .... 98.895 mm

Piston-to-bore clearance
  Standard .... 0.011 to 0.035 mm
  Service limit (max) .... 0.12 mm
Piston pin diameter
  Standard .... 21.995 to 22.000 mm
  Service limit (min) .... 21.960 mm
Piston pin bore diameter in piston .... 22.005 to 22.011 mm
Piston pin-to-bore clearance
  Standard .... 0.005 to 0.016 mm
  Service limit (max) .... 0.070 mm

Piston rings

850 model
  Ring height
    Top and second rings
      Standard .... 1.170 to 1.190 mm
      Service limit (min) .... 1.100 mm
    Oil ring
      Standard .... 2.470 to 2.490 mm
      Service limit (min) .... 2.400 mm
  Ring end gap (installed)
    Top ring
      Standard .... 0.10 to 0.30 mm
      Service limit (max) .... 0.80 mm
    2nd ring
      Standard .... 0.30 to 0.50 mm
      Service limit (max) .... 1.00 mm
    Oil ring side-rail
      Standard .... 0.30 to 0.60 mm
      Service limit (max) .... 1.20 mm
  Ring-to-groove clearance
    Top and second rings
      Standard .... 0.030 to 0.070 mm
      Service limit (max) .... 0.150 mm
    Oil ring
      Standard .... 0.020 to 0.060 mm
      Service limit (max) .... 0.150 mm
1100 models
  Ring height
    Top and second rings
      Standard .... 1.175 to 1.190 mm
      Service limit (min) .... 1.100 mm
    Oil ring
      Standard .... 2.475 to 2.490 mm
      Service limit (min) .... 2.400 mm

Ring end gap (installed)

Top and second rings

Standard . . . . 0.10 to 0.30 mm

Service limit (max) . . . . 0.80 mm

Oil ring side-rail

Standard . . . . 0.30 to 0.60 mm

Service limit (max) . . . . 1.20 mm

Ring-to-groove clearance

Top ring

Standard . . . . 0.040 to 0.075 mm

Service limit (max) . . . . 0.150 mm

2nd ring

Standard . . . . 0.030 to 0.065 mm

Service limit (max) . . . . 0.150 mm

Oil ring side-rail

Standard . . . . 0.020 to 0.055 mm

Service limit (max) . . . . 0.150 mm

Ring installation . . . . "TOP" mark facing up

Connecting rods

Small-end internal diameter . . . . 22.015 to 22.025 mm

Small-end to piston pin clearance

Standard . . . . 0.015 to 0.030 mm

Service limit (max) . . . . 0.06 mm

Big-end bore diameter (without bearing shells) . . . . 51.000 to 51.013 mm

Big-end bore diameter (with bearing shells) . . . . 48.016 to 48.050 mm

Crankpin diameter . . . . 47.975 to 47.991 mm

Big-end oil clearance

Standard . . . . 0.025 to 0.075 mm

Service limit (max) . . . . 0.13 mm

Big-end side clearance

Standard . . . . 0.130 to 0.312 mm

Service limit (max) . . . . 0.5 mm

Maximum twist at 150 mm spacing . . . . 0.07 mm

Lubrication system

Oil pressure . . . . see Chapter 1

Pressure relief valve opens . . . . 80 psi (5.5 Bar)

Pump housing depth

Pump 1 . . . . 12.02 to 12.05 mm

Pump 2 . . . . 10.02 to 10.05 mm

Pump rotor thickness

Pump 1 . . . . 11.95 to 11.98 mm

Pump 2 . . . . 9.95 to 9.98 mm

Clearance (end-float)

Standard . . . . 0.04 to 0.1 mm

Service limit (max) . . . . 0.25 mm

Crankshaft and bearings

Crankshaft identification

Un-ground, stage 0 . . . . no paint mark

Re-ground +0.25 mm, stage 1 . . . . paint mark (subtract 0.25 mm from all specifications)

Main bearing (front)

Bearing shell bore diameter in crankcase . . . . 60.000 to 60.019 mm

Bearing shell inside diameter

Green . . . . 54.998 to 55.039 mm

Yellow . . . . 55.008 to 55.049 mm

Crankshaft journal diameter

Green . . . . 54.971 to 54.980 mm

Yellow . . . . 54.981 to 54.990 mm

Oil clearance

Standard . . . . 0.018 to 0.163 mm

Service limit (max) . . . . 0.13 mm

Guide bearing (rear)

Bearing shell bore diameter in crankcase . . . . 64.949 to 64.969 mm

Bearing shell inside diameter

Green . . . . 59.964 to 60.003 mm

Yellow . . . . 59.974 to 60.013 mm

White . . . . Not available

Crankshaft and bearings (continued)

Crankshaft journal diameter
Green	59.939 to 59.948 mm
Yellow	59.949 to 59.958 mm
White	Not available

Oil clearance
Standard	0.016 to 0.064 mm
Service limit (max)	0.100 mm
Bearing shell width	24.890 to 24.940 mm
Crankshaft journal width (bearing surface)	25.020 to 25.053 mm

Clearance (end-float)
Standard	0.080 to 0.163 mm
Service limit (max)	0.200 mm

Clutch

Friction plate diameter	180 mm
Friction plate minimum thickness (wear limit)	4.5 mm

Transmission

Gear ratios

R1100RS to 1993
1st gear	4.030 to 1
2nd gear	2.576 to 1
3rd gear	1.886 to 1
4th gear	1.538 to 1
5th gear	1.318 to 1

R1100RS 1994-on and all other models
1st gear	4.163 to 1
2nd gear	2.914 to 1
3rd gear	2.133 to 1
4th gear	1.740 to 1
5th gear	1.450 to 1
Input shaft pre-load (with tapered-roller bearings)	0.2 mm
Input shaft friction (with tapered-roller bearings)	0.5 to 1.0 Nm
Input shaft endfloat (with ball bearings)	0.05 to 0.15 mm
Intermediate shaft and output shaft end-float	0.05 to 0.15 mm

Selector drum

Selector drum end float (max)	0.1 mm

Auxiliary shaft

Shaft diameter	24.959 to 24.980 mm
Shaft bore diameter (in crankcase)	25.02 to 25.041 mm

Clearance
Standard	0.040 to 0.082 mm
Service limit (max)	0.17 mm

Torque settings

Front frame strut-to-engine through-bolt	82 Nm

Front frame strut side rails-to-strut bolts
8.8 bolt (number marked on bolt head)	47 Nm
10.9 bolt (number marked on bolt head)	58 Nm

Front frame strut side rails-to-engine through-bolt
RS and RT models	47 Nm
R and GS models	58 Nm
Rear sub-frame bolts	47 Nm
Oil cooler hose banjo bolt	25 Nm
Oil cooler mounting bolts	9 Nm
Valve cover bolts	8 Nm
Cam chain tensioners	32 Nm
Camshaft journal bolts	15 Nm
Rocker shaft cap bolts	15 Nm
Rocker/camshaft assembly holder 6 mm bolts	9 Nm
Camshaft sprocket bolt	65 Nm
Camshaft sprocket cover bolts	9 Nm

Cylinder head stud nuts (see text)
1st stage setting	20 Nm
2nd stage setting	angle-tighten 90°
Final setting	angle-tighten 90°

Torque settings (continued)

Cylinder head 10 mm bolt	40 Nm
Cylinder head 6 mm bolts	9 Nm
Cylinder 8 mm bolt	20 Nm
Cylinder 6 mm bolts	9 Nm
Cam chain guide blade pivot bolts	18 Nm
Connecting rod cap bolts (see text)	
Initial setting	20 Nm
Final setting	angle-tighten 80°
Engine inner front cover	
6 mm bolts	9 Nm
8 mm bolts	20 Nm
Alternator belt drive pulley bolt	50 Nm
Breather pipe – all except RS models to 5/94	
8 mm bolt	20 Nm
Banjo bolt	25 Nm
Auxiliary shaft drive chain upper sprocket (crankshaft sprocket) bolts	10 Nm
Auxiliary shaft drive chain lower sprocket bolt	70 Nm
Auxiliary shaft drive chain tensioner bolts	10 Nm
Oil pipe bracket bolts	10 Nm
Oil pipe banjo bolt	25 Nm
Oil pump cover bolts	9 Nm
Oil pressure relief valve cap	35 Nm
Crankcase 10 mm bolts (oiled)	45 Nm
Crankcase 8 mm bolts (oiled)	20 Nm
Crankcase 6 mm bolts	9 Nm
Oil strainer bolt	9 Nm
Right-hand cylinder camchain tensioner blade pivot bolt	18 Nm
Flywheel bolts	
Initial setting	40 Nm
Final setting	angle-tighten 32°
Clutch cover plate bolts	18 Nm
Clutch operating lever pivot bolt	18 Nm
Gearbox mounting bolts	22 Nm
Gearchange lever pinch bolt	9 Nm
Gearbox cover bolts	10 Nm
Neutral detent ball bolt	13 Nm
Gearbox oil deflector plate bolts	9 Nm

2

1 General information

The engine/transmission unit is an air/oil-cooled horizontally-opposed twin. The four valves per cylinder are operated by camshafts via followers, pushrods and rockers. The camshafts are driven by chain off the auxiliary shaft. The auxiliary shaft is driven by chain off the crankshaft and runs at half crankshaft speed. The engine/transmission assembly is constructed from aluminium alloy. The crankcase is divided vertically.

The crankcase incorporates a wet sump, pressure-fed lubrication system which uses a dual-rotor oil pump driven by the auxiliary shaft, an oil filter and by-pass valve assembly, a relief valve and an oil pressure switch. The oil pump front rotor feeds the cooling oil circuit while the rear rotor feeds the lubrication circuit.

Power from the crankshaft is routed to the transmission via the clutch and input shaft which has a sprung damper. The clutch is of the dry, single-plate type and is driven off the crankshaft. The transmission is a five-speed constant-mesh unit, housed in a separate casing. Final drive to the rear wheel is by shaft.

The alternator is belt driven off a pulley on the end of the crankshaft.

2 Operations possible with the engine in the frame

1 The components and assemblies listed below can be removed without having to remove the engine assembly from the frame.
Oil filter and oil cooler(s)
Valve covers
Camchain tensioners
Rocker/camshaft assemblies
Cylinder heads and valves
Cylinders, pistons and piston rings
Connecting rods and bearings
Gearbox
Clutch
Alternator drive belt
Alternator
Starter motor

3 Operations requiring engine removal

It is necessary to remove the engine assembly from the frame to gain access to the following components.
Auxiliary shaft
Camchains
Camchain tensioner blades and guide blades
Oil pump, oil strainer and oil pressure relief valve
Crankshaft and bearings

4 Major engine repair – general note

1 It is not always easy to determine when or if an engine should be completely overhauled, as a number of factors must be considered.
2 High mileage is not necessarily an indication that an overhaul is needed, while low mileage, on the other hand, does not preclude the need for an overhaul. Frequency of servicing is probably the single most important consideration. An engine that has regular and frequent oil and filter changes, as well as other required maintenance, will most likely give many miles of reliable service. Conversely, a neglected engine, or one which has not been run in properly, may require an overhaul very early in its life.
3 Exhaust smoke and excessive oil consumption are both indications that piston rings and/or valve guides are in need of attention, although make sure that the fault is not due to oil leakage.
4 If the engine is making obvious knocking or rumbling noises, the connecting rod and/or main bearings are probably at fault.
5 Loss of power, rough running, excessive valve train noise and high fuel consumption rates may also point to the need for an overhaul, especially if they are all present at the same time. If a complete tune-up does not remedy the situation (including checking of the Motronic system by a BMW dealer), major mechanical work is the only solution.
6 An engine overhaul generally involves restoring the internal parts to the specifications of a new engine. The piston rings and main and connecting rod bearings are usually replaced and the valve seats are re-ground. The end result should be a like new engine that will give as many trouble-free miles as the original.
7 Before beginning the engine overhaul, read through the related procedures to familiarise yourself with the scope and requirements of the job. Overhauling an engine is not all that difficult, but it is time consuming. Plan on the motorcycle being tied up for a minimum of two weeks. Check on the availability of parts

5.6 Unscrew the bolt (arrowed) and remove the air duct

and make sure that any necessary special tools, equipment and supplies are obtained in advance.
8 Most work can be done with typical workshop hand tools, although a number of precision measuring tools are required for inspecting parts to determine if they must be replaced. Often a dealer will handle the inspection of parts and offer advice concerning reconditioning and replacement. As a general rule, time is the primary cost of an overhaul so it does not pay to install worn or substandard parts.
9 As a final note, to ensure maximum life and minimum trouble from a rebuilt engine, everything must be assembled with care in a spotlessly clean environment.

5 Engine – removal and installation

Note: *Due to the design of the models in this range, the engine is not removed from the bike, rather the rest of the bike is removed from around the engine. The easiest way to accomplish this is to construct a support from which the guts of the bike, which include the wiring loom, the complete brake system, the fuel system (except the fuel tank), the handlebars and front frame section, can be hung (see illustration 5.35). If available, a garage roof beam can be used, or alternatively place a step ladder at each end of the bike and rest a strong plank of wood between*

them. *The various systems and components can be suspended from the support at their original level as they are disconnected or detached, and then be simultaneously raised off the engine and gearbox. When removing or detaching cables, hoses, pipes and wiring, make a careful note of the routing of each and the location of their ties, clips, guides and brackets.*

Removal

1 Position the bike on its centre stand and place a block of wood or jack under the front of the engine to provide extra support. Work can be made easier by raising the machine to a suitable working height on an hydraulic ramp or a suitable platform. Make sure the motorcycle is secure and will not topple over (see *Tools and Workshop Tips* in the Reference section).
2 If the engine is dirty, particularly around its mountings, wash it thoroughly before starting any major dismantling work. This will make work much easier and rule out the possibility of caked on lumps of dirt falling into some vital component.
3 Drain the engine oil and remove the oil filter (see Chapter 1). If work is being carried out on the gearbox, also drain the gearbox oil (see Chapter 1).
4 Remove the seat and remove all the fairing panels and body panels, according to your model (see Chapter 7).
5 Remove the fuel tank (see Chapter 3).
6 Remove the air filter (see Chapter 1). Unscrew the bolt securing the front of the air intake duct and lift the duct from the filter housing **(see illustration)**.
7 Remove the battery (see Chapter 8).
8 Remove the Motronic control unit (see Chapter 4). Note the two earth (ground) wires secured by the left-hand mounting screws.
9 Remove the spark plug cover from the valve cover **(see illustration 8.2a)**. Check that the cylinder location is marked on each plug lead, then pull the spark plug cap off each spark plug, using the tool provided in the toolkit if required, and secure them clear of the engine **(see illustrations 8.2b and c)**.
10 Remove the front wheel (see Chapter 6). Unscrew the bolt securing the front brake hose union to the underside of the lower fork bridge, then pass each brake caliper between the forks above the fork bridge and secure them to the top of the front section of the frame strut, making sure no strain is placed on the hoses **(see illustration)**. On models equipped with ABS, unscrew the two bolts securing the ABS sensor to the left-hand fork slider and withdraw the sensor, taking care not to lose any of the shims **(see illustration)**. Release the wiring from its clips on the inside of the fork leg and secure it with the brake caliper. It is advisable to secure all the shims to the sensor using wire through the bolt sockets.
11 Remove the exhaust system (see Chapter 3).

5.10a Unscrew the brake hose union bolt (arrowed)

5.10b The sensor is secured by two bolts (arrowed)

5.12a Slacken the clamp screws (arrowed) . . .

5.12b . . . and slide the air ducts into the airbox

5.12c Slacken the clamp screw (arrowed) and detach the throttle body from the cylinder head

12 Slacken the clamp screws securing the air ducts to the airbox and the throttle bodies **(see illustration)**. Carefully pull the air ducts off the throttle bodies and slide them back into the airbox **(see illustration)**. Slacken the clamp screws securing the throttle bodies to the intake manifolds on the cylinder heads and carefully pull the throttles out of the manifolds **(see illustration)**. Secure the throttles clear of the engine, making sure there is no strain on the cables. There is no need to disconnect either the cables, the fuel hoses or the injector wiring connectors, though this can be done and the throttle bodies removed from the bike if required (see Chapter 3).

13 Unscrew the two bolts securing the intake manifold to the left-hand cylinder head and remove the manifold. This provides clearance for the Telelever pivot shaft to be withdrawn **(see illustration)**.
14 On R models, remove the steering damper (see Chapter 5). On all models unscrew the lower mounting bolt nut on the front shock absorber, but do not yet remove the bolt **(see illustration)**. Remove the cap from each end of the Telelever pivot **(see illustration)**. Unscrew the threaded cap on the inside of the left-hand pivot **(see illustration)**. Remove the circlip and withdraw the inner cap from the right-hand pivot **(see**

illustration). Unscrew the pivot shaft bolt from the right-hand end of the shaft, but do not yet withdraw the shaft **(see illustration)**.
15 On RS models, slacken the fork clamp bolts in the upper fork bridge **(see illustration)**. Remove the fork caps and press in the valve on the fork to release any air pressure **(see illustration)**. Keeping the valve depressed, push the top of the fork down through the bridge and fully into the slider. Repeat for the other fork. On R and RT models, unscrew the handlebar mounting bolts and displace the bars to access the fork tops – there is no need to detach any wiring, hoses or cables. On R, RT and GS models,

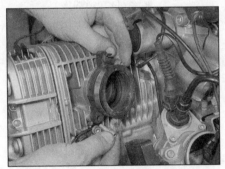

5.13 Unscrew the bolts and remove the manifold, noting the spark plug lead clip is at the top

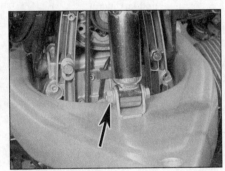

5.14a Remove the shock absorber nut (arrowed), but leave the bolt in place

5.14b Remove the blanking caps . . .

5.14c . . . then unscrew the left-hand threaded cap (arrow) . . .

5.14d . . . remove the right-hand circlip and inner cap (arrow) . . .

5.14e . . . and unscrew the right-hand bolt (arrow)

2

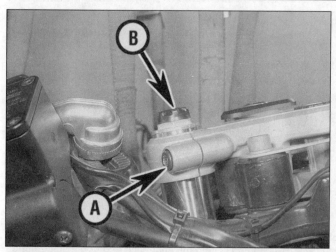

5.15a Slacken each fork clamp bolt (A) and remove each cap (B) . . .

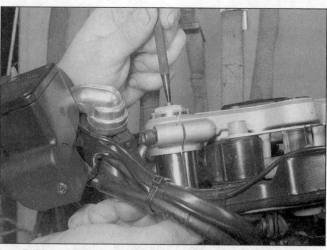

5.15b . . . then depress the valve and slide the tube down

5.16 Withdraw the pivot shaft to free the Telelever

5.17a Remove the front brake pipe clamp bolt (arrowed) . . .

5.17b . . . and the rear brake pipe clamp screw (arrowed)

remove the fork cap, then counter-hold the top bolt hex **(see illustration 6.5 in Chapter 5)** on the underside of the fork bridge and unscrew the nut. Push the top of the fork down through the bridge and fully into the slider.

16 Supporting the front suspension assembly by the lower fork bridge, withdraw the lower shock absorber bolt **(see illustration 5.14a)** and the Telelever pivot shaft, then remove the complete front suspension assembly **(see illustration)**. If the assembly proves too heavy, the forks can be removed separately and individually by removing the bolts securing them to the lower fork bridge, then the Telelever can be removed.

17 To access the front nuts securing the battery/ABS holder to the engine, it is necessary to either remove the ABS control unit, which involves detaching the hydraulic pipes from the unit, or to displace it sufficiently for the nuts to be accessible. Due to the complexity of bleeding the system, it is unwise to detach the hydraulics. Great care must be taken when displacing the unit as the pipes must be strained to an extent, and are therefore vulnerable to damage – note that there is no need to disconnect the ABS wiring. Even with the unit displaced as far as possible, access to the nuts is limited and a socket extension with a universal drive is needed. Remove the bolt securing the front

brake pipe bracket to the right-hand side of the strut, and the screw securing the rear brake pipe bracket to the gearbox housing, to provide as much flexibility as possible **(see illustrations)**. Unscrew the two bolts and the screw securing the ABS unit, then raise the unit as far as possible without putting too much strain on the pipes and unscrew the two nuts securing the front of the holder to the engine **(see illustrations)**. Lower the unit, then also remove the two nuts securing the rear of the holder to the gearbox **(see illustration)**.

18 Moving to the rear sub-frame, cut the cable ties securing all the wiring to the frame tubes. Remove the tail light unit and the rear

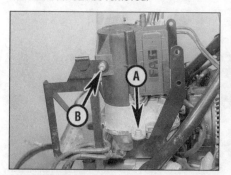

5.17c The ABS unit is secured by a bolt on each side (A) and a screw on the right (B)

5.17d Lift the ABS unit to access the front holder nuts

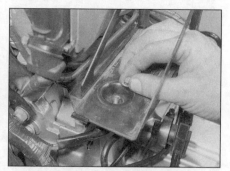

5.17e Also remove the rear holder nuts

5.19a Remove the fusebox lid . . .

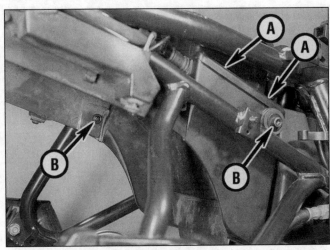

5.19b . . . followed by the holder screws (A) and the tray screws and bolts (B) on each side

turn signal units (see Chapter 8). Feed the wiring through the rear mudguard, noting its routing, and coil it on top of the fuse/relay box. All wiring must be clear of the rear sub-frame, as it has to be removed.

19 Unclip the fuse/relay box lid and remove it (see illustration). Remove the four screws securing the fuse/relay holder section of the box to the wiring tray section, then remove the four screws securing the wiring tray to the rear mudguard and the frame (see illustration). Free the wiring loom grommets from the cutouts in the tray (see illustrations). Disconnect the rear brake light

switch wiring connector and, where fitted, the rear ABS sensor wiring connector. Lift the fuse/relay holder out of the frame and position it in the air filter housing so that it is out of the way. Working around the rear frame, check that all wiring and electrical system components are clear of the rear frame. Trace the neutral switch, gear indicator, oil pressure switch and side stand wiring and disconnect them at the connectors. Feed the wiring back the components, noting its routing, and making sure it is free of any ties and clear of the frame.

20 Free the rear brake fluid reservoir from its

clips and wrap it in a plastic bag to prevent any fluid spillage (see illustration). Also free the rear brake hose-to-pipe union grommet from the bracket on the frame (see illustration).

21 On RT models, remove the bolts which secure the rear of each footrest mounting plate to the rear sub-frame.

22 Place a small block of wood between the gearbox casing and the top of the swingarm (Paralever) to prevent them folding together when the rear shock absorber upper mounting bolt is removed. Also place a jack under the gearbox to take its weight when it is separated from the engine. Check which side the rear shock upper mounting bolt is installed, then unscrew the two bolts and remove the seat support for that side (see illustration). This allows clearance for the bolt to be withdrawn. Unscrew and remove the rear shock upper mounting bolt. Where fitted, also remove the bolt securing the remote pre-load adjuster.

23 Check around the area to verify that all cables, wiring, hoses, pipes and electrical components are all free from the rear sub-frame. Remove the screws securing the airbox to the rear sub-frame. Unscrew the nut from the end of the through-bolt which secures the

2

5.19c Pull the front wiring grommet (arrowed) . . .

5.19d . . . and the rear grommet (arrowed) out of their cut-outs

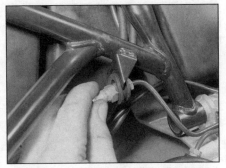

5.20a Release the reservoir from its clips . . .

5.20b . . . and the pipe union from its grommet

5.22 Remove the seat support to provide clearance for the shock absorber bolt

5.23a Front strut rails/rear sub-frame through-bolt (A), rear sub-frame mounting bolts (B) (right-hand side)

5.23b Free the three pipes from the right-hand side . . .

front of the rear sub-frame and the front strut side rails to the engine (see illustration). Tie the front strut side rails to the overhead support to prevent them from dropping (see *Note* at the beginning of the Section) and withdraw the through-bolt. Supporting the rear sub-frame, unscrew the four bolts which secure it to the engine and gearbox, then carefully manoeuvre it backwards off the engine and gearbox, making sure it is free from all cables, pipes and hoses, and remove it. Free the fuel supply pipes from the grommets at the front of the airbox (see illustrations). Remove the wiring tray and air box and support the fuse/relay holder in the battery holder.

24 Remove the oil cooler(s) with the hoses and pipes (see Section 7).
25 Trace the ignition timing sensor wiring from the front of the engine and disconnect it at the connector (see illustration).
26 Trace the oil temperature sensor wiring and disconnect it at the connector. The sensor is located on the top of the engine on the right-hand side, just ahead of the oil cooler pipe (see illustration 7.3c). Also disconnect the oil pressure switch wiring connector – the switch is located below the left-hand cylinder.
27 Remove the cap on the alternator terminal nut, then unscrew the nut and detach the

lead, and pull the other wire off its terminal (see illustration).
28 Cut the cable ties securing the wiring in the trough secured by the front battery/ABS holder mounts.
29 Release the clamp securing the engine breather hose to the top of the engine and detach the hose (see illustration). On RS models to 5/94, which use a different type breather system, also detach the hose from the bottom of the separator canister (see illustration).
30 Check that the front frame strut is securely supported, then unscrew the nut from the through-bolt securing the strut to the

5.23c . . . and the single pipe from the left

5.25 Disconnect the ignition timing sensor wiring connector

5.27 Detach the alternator wiring connectors (arrowed)

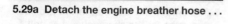

5.29a Detach the engine breather hose . . .

5.29b . . . and the canister hose (arrowed) – RS models to 5/94

5.30a Unscrew the nut . . .

5.30b . . . and withdraw the strut-to-engine bolt

5.32 Lift the battery holder and remove the earth cable bolt (arrowed)

5.34a Right-hand footrest bracket bolts (arrowed)

engine **(see illustration)**. Withdraw the bolt, using a drift if required **(see illustration)**.
31 Remove the starter motor (see Chapter 8).
32 Lift the rear of the battery/ABS unit holder and unscrew the bolt securing the earth (ground) cable to the top of the engine **(see illustration)**.
33 Slacken the clutch cable adjuster locknut at the handlebar, then screw the adjuster fully into the bracket to provide the maximum freeplay **(see illustration 29.1a)**. Detach the lower end of the inner cable from the operating lever on the gearbox, using a screwdriver to move the lever forward if there is not enough freeplay available **(see illustration 29.1b)**. Withdraw the lower end of the outer cable from the lug on the gearbox and secure the cable clear of the engine **(see illustration 29.1c)**.
34 Remove the bolts securing the right-hand footrest bracket to the gearbox **(see illustration)** on R, RS and GS models; on RT models, remove the right-hand footrest plate. Support the bracket, with the rear brake master cylinder still attached, to the overhead support so that no strain is placed on the pipes. A good way to do this is to simply tie it with string. Unscrew the bolts securing the rear brake caliper and lift the caliper off the disc **(see illustration)**. There is no need to disconnect the hose. Support the caliper in the same way that the master cylinder is supported.

35 At this point there should be nothing left attached to either the engine, gearbox or rear wheel. Check around to verify that all components, cables, wiring, hoses, pipes and electrical components are all free, and either supported by the overhead beam or resting on the engine and gearbox. Tie or hook some straps, rope or heavy string around the front frame strut, the handlebars, the ABS holder, the throttle bodies and any brake system components not already secured or tied, and pass the other ends over the beam or support. Raise the whole lot together off the engine and gearbox, in particular making sure that the brake system pipes are not bent or strained, until it is all clear and high enough to not interfere with further work **(see illustration)**. Tie off the straps, rope or string, making sure they are secure. If string is used on the front frame assembly and ABS unit, make sure it is strong enough or there is enough of it to support the weight.
36 If work is to be carried out on the gearbox as well as the engine, remove the swingarm (Paralever) pivot bolts from the gearbox and separate the rear wheel, final drive and swingarm as an assembly from the gearbox, then separate the gearbox from the engine. Refer to Chapter 5 for the procedure for removing the swingarm pivots, ignoring the

Steps which do not apply. If work is to be carried out on the engine alone, remove the bolts securing the gearbox to the engine (see Section 30) and separate the rear wheel, final drive, swingarm and gearbox as an assembly from the engine.
Caution: When separating the gearbox from the engine, make sure that the gearbox is drawn back entirely level until the clutch pushrod is clear of the clutch components. If the gearbox is raised, lowered or skewed before the pushrod is clear, the pushrod will bend.
37 The engine is now ready for disassembly and can be moved onto a suitable workbench.

⚠️ *Warning: The engine is very heavy. It is strongly recommended that you have at least one assistant to help lift the engine if it is being moved. Personal injury or damage could occur if the engine falls or is dropped.*

Installation

38 Installation is the reverse of removal, noting the following points:
a) *When installing the front frame strut, temporarily fit the through-bolt through the frame strut side rails and engine to position the frame strut correctly; tighten*

5.34b Rear brake caliper bolts (arrowed) – RS model

5.35 Raise all the components and systems off the engine, gearbox and rear drive

6.4 An engine support made from wood blocks

the nut finger-tight before tightening the frame strut-to-engine through-bolt to the torque setting specified at the beginning of the Chapter. Leave the bolt in position until the rear sub-frame is installed.

b) When installing the rear sub-frame, install all four bolts and the through-bolt that also secures the front frame strut side rails finger-tight before tightening any of them to the specified torque setting. When torquing the bolts, tighten the right-hand rear bolt first, followed by the right-hand front bolt, then the left-hand front bolt and finally the left-hand rear bolt. Finally tighten the through-bolt to the specified torque.

c) Refer to Chapter 5 when installing the front suspension (Telelever) components.

d) Refer to Section 30 when fitting the gearbox to the engine.

e) If the swingarm (Paralever) was separated from the gearbox, refer to Chapter 5 for the swingarm installation procedure.

f) Tighten all bolts to the torque settings specified either at the beginning of this Chapter, or at the beginning of the relevant Chapter elsewhere in the book.

g) Check the condition of all breather tubes, grommets and O-rings, and of all hose and pipe clamps before re-using them, and replace them if they are worn, damaged or deteriorated.

h) Make sure all wires, cables, pipes and hoses are correctly routed and connected, and secured by any clips or ties.

i) Check the operation of all electrics and lights before taking the bike on the road.

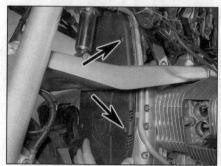

7.3a The two left-hand cover bolts (arrowed) clamp the oil pipe

j) Make a careful check of the brake system and the suspension before taking the bike on the road (see Chapter 1).

k) Refill the engine and gearbox with oil (see Chapter 1).

6 Engine disassembly and reassembly – general information

Disassembly

1 Before disassembling the engine, the external surfaces of the unit should be thoroughly cleaned and degreased. This will prevent contamination of the engine internals, and will also make working a lot easier and cleaner. A high flash-point solvent, such as paraffin (kerosene) can be used, or better still, a proprietary engine degreaser. Use old paintbrushes and toothbrushes to work the solvent into the various recesses of the engine casings. Take care to exclude solvent or water from the electrical components and intake and exhaust ports.

⚠️ **Warning: The use of petrol (gasoline) as a cleaning agent should be avoided because of the risk of fire.**

2 When clean and dry, arrange the unit on the workbench, leaving suitable clear area for working. Gather a selection of small containers and plastic bags so that parts can be grouped together in an easily identifiable manner. Some paper and a pen should be on hand to permit notes to be made and labels attached where necessary. A supply of clean rag is also required.

3 Before commencing work, read through the appropriate section so that some idea of the necessary procedure can be gained. When removing components it should be noted that great force is seldom required, unless specified. In many cases, a component's reluctance to be removed is indicative of an incorrect approach or removal method – if in any doubt, re-check with the text.

4 An engine support stand made from wood blocks bolted together into a rectangle will help support the lower section of the crankcase **(see illustration)**, although additional wood blocks will be needed to

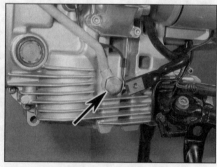

7.3b Unscrew the banjo bolt (arrowed) and detach the pipe

support the cylinders due to the engine's rather top-heavy construction. When splitting the crankcase, the engine must be on its side, and can rest on the same support so that the con-rods are clear of the bench.

5 When disassembling the engine, keep 'mated' parts together (including gears, cylinders, pistons, connecting rods, valves, etc. that have been in contact with each other during engine operation). These 'mated' parts must be reused or replaced as an assembly.

6 A complete engine/transmission disassembly should be done in the following general order with reference to the appropriate Sections.

Remove the centre stand (see Chapter 5).
Remove the valve covers
Remove the camchain tensioners
Remove the rocker/camshaft assemblies
Remove the cylinder heads
Remove the cylinders
Remove the pistons
Remove the alternator (see Chapter 8)
Remove the alternator drive and engine inner front cover
Remove the auxiliary shaft drive chain, tensioner and sprockets
Remove the oil pump
Remove the starter motor (see Chapter 8)
Remove the gearbox (if not done when removing the engine)
Remove the clutch
Separate the crankcase halves
Remove the connecting rods
Remove the crankshaft
Remove the auxiliary shaft

Reassembly

7 Reassembly is accomplished by reversing the general disassembly sequence, noting that the engine inner front cover, alternator drive and alternator should be left until after the cylinder heads and camshafts are installed to enable the timing to be set up correctly.

7 Oil cooler(s) and hoses – removal and installation

Note: The oil cooler can be removed with the engine in the frame. If the engine has been removed, ignore the steps which do not apply.

Removal

1 On R1100RS and RT models, remove the fairing panels as required (see Chapter 7).

2 Drain the engine oil and remove the oil filter (see Chapter 1).

3 On R1100RS and GS models, to remove the cooler with its hoses attached, slacken the engine front cover bolts – the two left-hand bolts clamp the left-hand oil pipe **(see illustration)**. Unscrew the banjo bolt securing the left-hand pipe to the engine below the left-hand cylinder **(see illustration)**. Slacken the right-hand fairing bracket bolt to release the

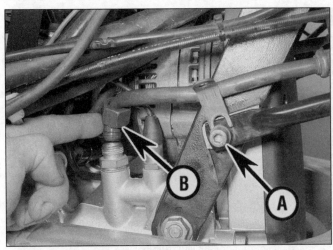

7.3c Slacken the bolt (A) and unscrew the flare nut (B)

7.3d Oil cooler mounting bolts (A), cooler hose nuts (B)

right-hand pipe clamp, then counterhold the union nut and unscrew the flare nut securing the pipe to the union on the top of the engine **(see illustration)**. Remove the bolts securing the cooler and remove the cooler, noting the routing of each hose **(see illustration)**. To remove the cooler without its hoses, unscrew the nut securing each hose to the cooler, then remove the bolts securing the cooler and remove the cooler **(see illustration 7.3d)**.

4 On R1100RT models, slacken the clamp securing each hose to the cooler and pull the hoses off their unions. Unscrew the bolts securing the cooler and remove the cooler; take note of the washer, grommet and collar arrangement at the oil cooler mounting. The oil cooler hoses connect to a union block which houses a thermostat **(see illustration)**. The thermostat opens when the engine oil reaches a certain temperature and routes oil through to the oil cooler. Remove the banjo bolt to release the hoses from the union,

noting the sealing washer on each side of the hose banjo fitting. The oil cooler inlet pipe is connected to the engine under the left-hand cylinder **(see illustration 7.3b)** and the outlet pipe is connected to the top of the engine on the right-hand side **(see illustration 7.3c)**.

5 On R850R and R1100R models, to remove the coolers with their hoses attached, slacken the engine front cover bolts – the two left-hand bolts clamp the left-hand oil pipe **(see illustration 7.3a)**. Unscrew the banjo bolt

2

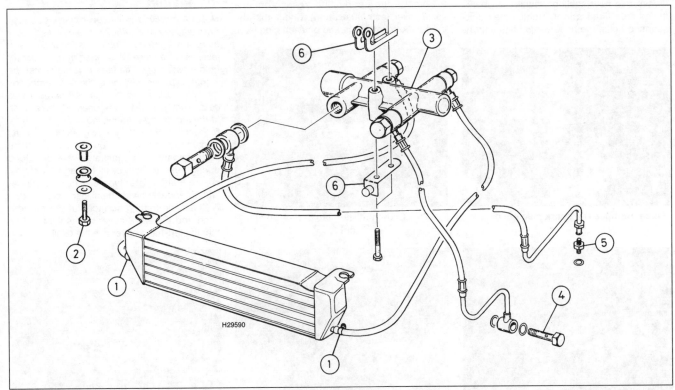

7.4 Oil cooler and thermostat – RT model

1	Hose clamps	3	Union/thermostat	5	Outlet pipe fitting
2	Cooler mounting bolts	4	Inlet pipe banjo bolt	6	Union mounting brackets

7.5a The coolers are each secured by two bolts (arrowed)

7.5b Slacken the clamps (arrowed) and detach the hoses

7.6 Fit a new sealing washer on each side of the pipe union

securing the left-hand pipe to the engine below the left-hand cylinder (see illustration 7.3b). Counterhold the union nut and unscrew the flare nut securing the pipe to the union on the top of the engine (see illustration 7.3c). Remove the bolts securing the cooler holders and remove them with the coolers attached, leaving the inter-connecting hose attached and noting the routing of each hose (see illustration). To remove the coolers without their hoses, slacken the clamp securing each hose to the coolers and pull the hoses off their unions (see illustration). Unscrew the bolts securing the coolers and remove the coolers (see illustration 7.5a).

Installation

6 Installation is the reverse of removal. Always use new sealing washers on banjo fittings (see illustration). Tighten the cooler mounting bolts, the hose banjo bolts and the

hose nuts to the torque settings specified at the beginning of the Chapter.
7 Refill the engine with oil (see Chapter 1).

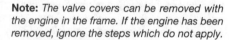

8 Valve covers – removal and installation

Note: The valve covers can be removed with the engine in the frame. If the engine has been removed, ignore the steps which do not apply.

Removal

1 On R1100RS and RT models, if required for easier access and to avoid the possibility of damage, remove the fairing side panel for the side being worked on (see Chapter 7).
2 Remove the spark plug cover from the valve cover (see illustration). Check that the cylinder location is marked on each plug lead,

then pull the spark plug cap off each spark plug, using the tool provided in the toolkit if required, and secure them clear of the engine (see illustrations).
3 Unscrew the bolts securing the cylinder protection bar and remove the bar, noting how it fits (see illustration).
4 Place a container beneath the valve cover to catch any residue oil.
5 Unscrew the four bolts securing the valve cover then pull the cover off the cylinder head (see illustration). If it is stuck, do not try to lever it off with a screwdriver. Tap it gently around the sides with a rubber hammer or block of wood to dislodge it. Also remove the outer and inner rubber gaskets from the cover if they are loose, otherwise leave them in place.

Installation

Note: A modified valve cover was introduced from engine no. 27946077. The modified cover has shorter internal ribs on its inner face and lessens valve noise transmitted from the rocker holder assembly. The later type cover can be identified by the position of the part number on the cover outer surface; the part number was cast centrally in the cover channel, rather than at the top as in illustration 8.5.
6 Examine the valve cover gaskets for signs of damage or deterioration and replace them if necessary. The gaskets were modified during 1995 to correct oil leakage problems; ensure that the later type thicker inner gasket and outer gasket with metal insert are fitted.
7 Clean the mating surfaces of the cylinder head and the valve cover with solvent. Note that the left-hand cover has the oil filler cap.

8.2a Remove the spark plug cover . . .

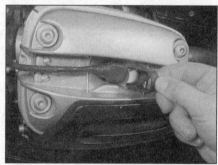

8.2b . . . then fit the plug cap removing tool . . .

8.2c . . . and remove the cap

8.3 The protector bar is secured by three bolts (arrowed)

8.5 The valve cover is secured by four bolts (arrowed)

8.8a Fit the outer gasket to the head . . .

8.8b . . . and the inner gasket to the cover

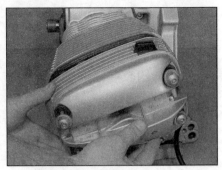

8.9a Install the cover . . .

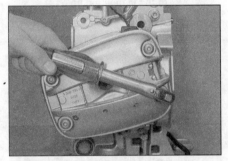

8.9b . . . and tighten the bolts to the specified torque setting

8 If new gaskets are being used, fit the outer one onto the cylinder head, making sure the half-circles locate around the pins **(see illustration)**. Fit the inner gasket onto the plug passage **(see illustration)**.
9 Position the cover on the cylinder head, making sure the gaskets stay in place **(see illustration)**. Install the cover bolts and tighten

them to the torque setting specified at the beginning of the Chapter **(see illustration)**.
10 Install the remaining components in the reverse order of removal.

9 Camchain tensioners – removal, inspection and installation

Note: *The camchain tensioners can be removed with the engine in the frame. To access the tensioner blades and guide blades the engine must be removed and the crankcases split.*

Removal

1 Drain the engine oil (see Chapter 1). On RT models and RS models with a full fairing, remove the fairing side panels (see Chapter 7).
2 Unscrew the tensioner bolt and withdraw the tensioner from the back of the cylinder block **(see illustration)**. The right-hand cylinder tensioner is on the underside of the

cylinder, the left-hand cylinder tensioner is on the top of the cylinder. Do not mix up the plungers for the right- and left-hand tensioners as they are different.
3 Remove the sealing washer and discard it as a new one must be used.

Inspection

4 Examine the tensioner components for signs of wear or damage **(see illustration)**.
5 Check that the plunger moves freely in and out of the tensioner body, and that the spring tension is good.
6 If the tensioner or any of its components are worn or damaged, they must be replaced.

Installation

7 Before installing the tensioner, turn the engine to take up the slack in the chain between the camshaft and the auxiliary shaft and transfer it to where it will be taken up by the tensioner. The engine can be turned by selecting a high gear and rotating the rear wheel by hand in its normal direction of rotation. Alternatively, remove the engine front cover and turn the engine using a 17 mm spanner or socket on the belt drive pulley bolt and turning it in a clockwise direction only **(see illustrations)**.
Caution: Be sure to turn the engine in its normal direction of rotation.
8 Place a new sealing washer on the tensioner bolt, then install it in the engine **(see illustration 9.2)**. Tighten the tensioner bolt to the torque setting specified at the beginning of the Chapter **(see illustration)**.
9 Refill the engine with oil (see Chapter 1). On RS and RT models, install the fairing panels (see Chapter 8).

9.2 Unscrew and remove the camchain tensioner – right-hand cylinder shown

9.4 Camchain tensioner components

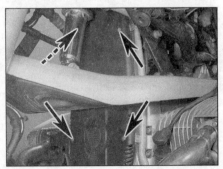

9.7a Unscrew the bolts (arrowed) and remove the front cover . . .

9.7b . . . then turn the engine using the belt pulley bolt

9.8 Tighten the tensioner to the specified torque setting

2

10.1 Remove the damper block, noting how it fits

10.2 Remove the timing hole blanking plug (arrowed)

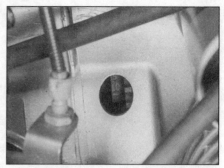

10.3 Align the "OT" mark with the middle of the inspection hole

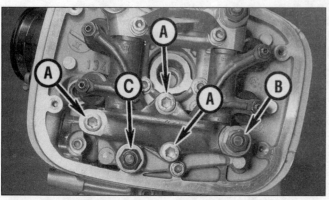

10.4a The rocker shaft cap is secured by three bolts (A) and the cylinder head nut (B). Also remove cylinder head nut (C) to provide clearance

10.4b Unscrew the shaft locating bolts and remove the contact plate

10 Rocker arms and shafts – removal, inspection and installation

Note: *The rocker arms and shafts can be removed with the engine in the frame.*

Removal

1 Remove the valve cover from the side of the engine being worked on (see Section 8). Also remove the rubber damper block which fits between the lower front cylinder head stud and the valve cover (see illustration).

2 Remove the timing inspection hole plug from behind the right-hand cylinder (see illustration). The engine can be turned by selecting a high gear and rotating the rear wheel by hand in its normal direction of rotation. Alternatively, remove the engine front cover and turn the engine using a 17 mm spanner or socket on the belt drive pulley bolt and turning it in a clockwise direction only (see illustrations 9.7a and b).

Caution: Be sure to turn the engine in its normal direction of rotation.

3 Turn the engine until the "OT" mark on the flywheel, visible via the timing inspection hole, aligns with the middle of the hole, and the valves of the cylinder being worked on are all closed (see illustration). The piston for that cylinder is now at TDC (Top Dead Centre) on the compression stroke. If any of the valves on that cylinder are open when the mark aligns, the crankshaft must be rotated one full turn until the "OT" mark again aligns with the middle of the hole. The valves will now be closed.

HAYNES HiNT *As there is no static timing mark on the timing inspection hole, mark your own using a dab of paint or Tippex to make alignment easier. Use a ruler to find the exact middle of the hole and mark the side of it accordingly.*

4 Unscrew the three bolts and the nut which secure the rocker shaft cap and remove the cap (see illustration). Also remove the lower rear cylinder head stud nut to provide clearance for the rocker shaft to be withdrawn. Unscrew the bolt securing the end of each shaft in the holder, noting how the bolt locates in the cut-out in the shaft, and remove the bolts along with the contact plate (see illustration). Withdraw the rocker shafts from the holder, using a rod inserted into the hole in the shaft end to twist them free, and remove the rocker arms as they become free (see illustration). Also remove the pushrods from the followers (see illustration). Mark each shaft, rocker arm and pushrod according to its position (ie left or right cylinder, intake or exhaust valves), and/or install each rocker arm back onto its shaft in its original position.

10.4c Withdraw the shaft and remove the arm

10.4d Lift the pushrods out of the followers

10.6 Measure the diameter of the rocker shaft

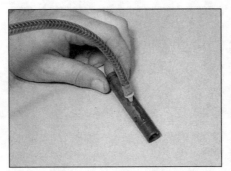

10.9a Lubricate the shafts . . .

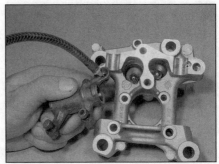

10.9b . . . and the rocker arms

Inspection

5 Inspect the rocker arm contact points for pitting, spalling, score marks, cracks and rough spots. If the rocker arms are damaged they must be replaced.

6 Measure the diameter of the rocker arm shafts in the area of contact with the rocker arms **(see illustration)**. Also measure the internal diameter of the rocker arm bores. Subtract the bore diameter from the shaft diameter to obtain the clearance and compare the result to that specified. If the clearance is excessive, compare the measurements taken to determine whether the shaft or the bore is worn, and replace either or both as required.

7 Check the pushrods for any signs of damage or wear, particularly on their ends.

8 If available, blow through any oil passages with compressed air.

Installation

9 Make sure that the cylinder being worked on is at TDC on the compression stroke (see Step 3). Place the pushrods into the followers, making sure each is returned to its original location **(see illustration 10.4d)**. Apply a smear of molybdenum disulphide grease or clean engine oil to the rocker arm shafts and to the contact faces of each rocker arm **(see illustrations)**. Position the rocker arms in the holder and slide the shafts through the arms and into the holder, making sure they are installed in their original positions and each pushrod end locates in the ball-cup in the rocker arm **(see illustration 10.4c)**. Position the shafts so that the cut-out in the end of each shaft aligns with the bolt holes in the holder, then install the bolts with the contact plate and tighten them to the torque setting specified at the beginning of the Chapter **(see illustration 10.4b)**.

10 Apply a smear of molybdenum disulphide grease or engine oil to the bearing surfaces of the rocker shaft cap. Install the cap, making sure it is the correct way round, then install the bolts and the nut and the lower rear cylinder head stud nut and tighten them to the specified torque settings **(see illustration 10.4a)**.

11 Rotate the crankshaft through 720° degrees and check that the arms move freely on the shafts. Measure the amount of end-

10.11 Measure the arm end-float using a feeler gauge

float using a feeler gauge **(see illustration)**. It should be as specified.

12 Check the valve clearances and adjust if necessary (see Chapter 1).

13 Install the valve covers (see Section 8), not forgetting the damper block **(see illustration 10.1)**.

14 Install the timing hole inspection plug **(see illustration)**.

15 Check the engine oil level and top up if necessary (see Chapter 1).

---

11 Rocker/camshaft holders – removal and installation

Note: *The rocker/camshaft holders can be removed with the engine in the frame. If no work is being carried out on the rocker arms and shafts, they can be left attached to the holder.*

Removal

1 Remove the valve cover (see Section 8) and the camchain tensioner (see Section 9) from the cylinder being worked on.

2 Remove the timing inspection hole plug from behind the right-hand cylinder **(see illustration 10.2)**. The engine can be turned by selecting a high gear and rotating the rear wheel by hand in its normal direction of rotation. Alternatively, remove the engine front cover and turn the engine using a 17 mm spanner or socket on the belt drive pulley bolt and turning it in a clockwise direction only **(see illustrations 9.7a and b)**.

11.4a The camshaft sprocket cover is secured by two bolts (arrowed)

Caution: Be sure to turn the engine in its normal direction of rotation.

3 Turn the engine until the "OT" mark on the flywheel, visible via the timing inspection hole, aligns with the middle of the hole, and the valves of the cylinder being worked on are all closed (see **Haynes Hint** – Section 10) **(see illustration 10.3)**. The piston for that cylinder is now at TDC (top dead centre) on the compression stroke. If any of the valves on that cylinder are open when the mark aligns, rotate the crankshaft one full turn until the "OT" mark again aligns with the middle of the hole. The valves will now be closed.

4 Unscrew the two bolts securing the camshaft sprocket cover and remove the cover **(see illustration)**. According to BMW, each sprocket is marked so that when at TDC on the compression stroke, the mark, either an "R" or an arrow, is in direct alignment with the tip of a sprocket tooth, which in turn is in

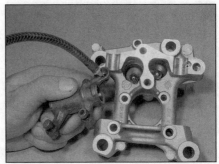

10.14 Fit the blanking plug into the timing inspection hole

11.4b Make your own alignment marks on the sprocket tooth and highlight the indent (left-hand cylinder shown)

11.5a Unscrew the sprocket bolt . . .

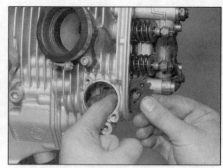

11.5b . . . and remove the sprocket

11.6 Cylinder head 6 mm bolts (A), 10 mm bolt (B), stud nuts (C), holder bolts (D)

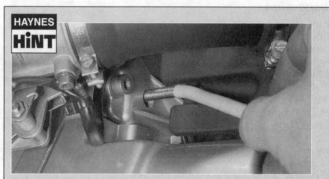

HAYNES HiNT

The engine can be locked in the TDC position by inserting a steel rod or bolt at least 100 mm long into the hole in the clutch housing on the left-hand side of the engine. There are corresponding holes in the flywheel and the rear of the crankcase. Locate the rod through all the holes, thereby locking the flywheel, and hence the engine, in the TDC position. To transfer from TDC on the compression stroke from one cylinder to the other, remove the rod and rotate the engine clockwise one full turn until the "OT" mark is in alignment and re-insert the rod.

direct alignment with the static mark on the holder. Also the locating lug indent on the sprocket should be facing down for the right-hand cylinder and up for the left-hand cylinder. However, on the bike worked on, the arrow on the sprocket was not aligned with the tip of the sprocket tooth, but the trough between two teeth, though the tip of the tooth did align with the static mark. Also, with the sprocket bolt installed, the indent is only just visible behind the bolt shoulder. It is highly advisable therefore to make your own mark on the tip of the sprocket tooth that is in direct alignment with the mark on the holder and to highlight the indent to avoid confusion on reassembly **(see illustration)**. Car touch-up paint or Tippex can be used to make the marks.
5 Unscrew the bolt securing the sprocket to the camshaft – use either a spanner or socket on the crankshaft bolt or the TDC locking pin (see **Haynes Hint**) to stop the engine rotating as the sprocket bolt is unscrewed. Work the sprocket squarely off the end of the shaft using a pair of screwdrivers, noting how the small lug on the back of the sprocket locates in the cut-out on the camshaft **(see**

illustrations). Either slip the sprocket out of the chain and remove it, in which case the chain must be secured to prevent it falling into the engine, or secure both the chain and sprocket to prevent them both falling into the engine. If both cylinders are being disassembled, remove the sprocket as it will be easier to turn the engine without it. Do not allow the chain to go slack as it could bind between the crankshaft sprocket and the crankcase.
Caution: Do not turn the camshaft or the crankshaft with the cam chain sprocket removed, as the valves could contact the pistons and will be damaged.
6 Remove the cylinder head 6 mm bolts around the cam chain tunnel, followed by the 10 mm bolt, then remove the cylinder head stud nuts evenly and in a criss-cross pattern **(see illustration)**. Remove the bolts securing the rocker/camshaft assembly holder to the cylinder head, then remove the holder, noting how the assembly fits. Mark each holder according to its cylinder (ie left or right). If required, remove the rocker arms and shafts and the camshafts from the holders (see Sections 10 and 12).

Installation

HAYNES HiNT *To prevent the pushrods falling out when installing the rocker and camshaft assembly holder, place an elastic band around the valve clearance adjusters on each rocker so that the rockers are pulled down onto the pushrods, keeping them in place.*

7 If removed, install the camshafts (see Section 12) and the rocker arms and shafts (see Section 10).
8 Make sure that the cylinder being worked on is at TDC on the compression stroke (see Steps 3 and 4). If both rocker/camshaft holders have been removed at the same time, it is necessary to remove the alternator drive belt and inner front cover (see Section 20) in order to determine which cylinder is at TDC on the compression stroke. If the engine has been disassembled, the alternator casing should not yet have been installed. The right-hand cylinder is at TDC on the compression stroke when the mark on the crankshaft

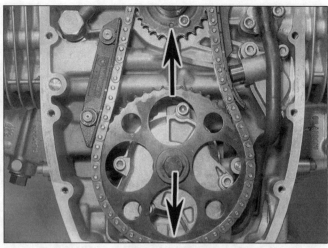

11.8a Sprocket markings – TDC right-hand cylinder

11.8b Sprocket markings – TDC left-hand cylinder

sprocket is facing down and the mark on the auxiliary shaft sprocket is facing up, and they are directly aligned **(see illustration)**. The left-hand cylinder is at TDC on the compression stroke when the mark on the crankshaft sprocket is facing down and the mark on the auxiliary shaft sprocket is facing down, and they are directly aligned **(see illustration)**. Always start with the right-hand cylinder.

9 Install the rocker/camshaft holder onto the cylinder head **(see illustration)**. Apply clean engine oil to the cylinder head stud nuts, then install them with their washers, with the collared end of the nuts facing the cylinder head **(see illustrations)**. Install the cylinder head 6 mm bolts and 10 mm bolt and the

holder bolts and tighten them all finger-tight only at this stage **(see illustration 11.6)**. Tighten the cylinder head stud nuts evenly, a little at a time and in a criss-cross sequence to the initial torque setting specified at the beginning of the Chapter **(see illustration)**. Now attach a degree disc (see *Tools and workshop tips*) to the torque wrench and tighten each nut in turn a further 90°, then repeat to a further 90° **(see illustration)**. Now tighten the 10 mm bolt to the specified torque setting, then tighten the 6 mm bolts and the holder bolts evenly, a little at a time and in a criss-cross sequence to the specified torque setting.

10 Check that the crankshaft and auxiliary shaft sprockets are positioned as described in Step 8 **(see illustrations 11.8a and b)**. Keeping the top run of the cam chain taut for the right-hand cylinder, or the bottom run for the left-hand cylinder, engage the chain on the camshaft sprocket teeth. For the right-hand cylinder (at TDC – see Step 8), the sprocket locating cut-out on the camshaft and the corresponding lug on the sprocket must be facing down and the mark on the sprocket (see Step 4) must face in and align exactly with the mark on the holder **(see illustration)**. For the left-hand cylinder (at TDC – see Step 8), the sprocket locating cut-out on the

2

11.9a Install the holder . . .

11.9b . . . then oil the studs . . .

11.9c . . . and fit the washers and nuts

11.9d Tighten the nuts to the specified torque setting . . .

11.9e . . . then to the specified angle using a degree disc

11.10 Camshaft sprocket markings – TDC right-hand cylinder

11.12 Tighten the sprocket bolt to the specified torque setting

11.15b then smear it with oil . . .

11.15a Check the cover O-ring and replace it if required . . .

11.15c . . . and fit the cover

camshaft and the corresponding lug on the sprocket must be facing up and the mark (see Step 4) on the sprocket must face in and align exactly with the mark on the holder **(see illustration 11.4b)**. With the chain engaged on the sprocket, mount the sprocket onto the camshaft, making sure the lug locates in the cut-out, and install the sprocket bolt finger-tight only **(see illustration 11.5a)**. Check that the chain is tight at the top (right-hand cylinder) or bottom (left-hand cylinder) run so that there is no slack between the auxiliary shaft sprocket and the camshaft sprocket. If any slack is evident, move the chain around the sprocket so that the slack is taken up. Any

slack in the chain must lie in the portion of the chain in the bottom (right-hand cylinder) or top (left-hand cylinder) of the run so that it is then taken up by the tensioner.
11 Before proceeding further, check that everything aligns as described in Steps 3, 4, 8 and 10. If not, the valve timing will be inaccurate and the valves could contact the pistons when the engine is turned over.
12 Tighten the camshaft sprocket bolt to the specified torque setting, using either a spanner or socket on the crankshaft bolt or the TDC locking pin to stop the engine from rotating (see **Haynes Hint**) **(see illustration)**.
13 Check that the valve timing marks still

align (see Steps 3, 4, 8 and 10). If fitted, removed the TDC locating pin. Check that each camshaft is not pinched by turning it a few degrees in each direction using a suitable spanner or socket on the crankshaft bolt.
14 Install the cam chain tensioners (see Section 9).
15 Rotate the crankshaft through 720° and re-check that the valve timing for both cylinders is correct (see Step 3, 4, 8 and 10). Install the camshaft sprocket cover using a new O-ring if necessary, and tighten the bolts to the specified torque setting **(see illustrations)**.
16 Check the valve clearances and adjust if necessary (see Chapter 1).
17 If removed, install the inner front cover and alternator drive belt (see Section 20).
18 Install the valve covers (see Section 7).
19 Install the timing inspection hole plug **(see illustration 10.14)**.
20 Check the engine oil level and top up if necessary (see Chapter 1).

12 Camshafts and followers – removal, inspection and installation

Note: *The camshafts can be removed with the engine in the frame.*

Removal

1 Remove the rocker/camshaft holders (see Section 11).
2 Unscrew the bolts securing the camshaft bearing cap and base and remove the cap, noting the alignment marks between the cap, base and the holder **(see illustrations)**. Remove the camshaft and the bearing base together **(see illustration)**. Remove the followers from their bores, either by tipping them out or, if the rocker assemblies have been removed, by pushing them up from the other side of the holder **(see illustration)**.

12.2a Note the alignment marks (A), then unscrew the cap bolts (B) . . .

12.2b . . . and remove the cap

12.2c Remove the camshaft and base together

12.2d Lift the followers out of the holder

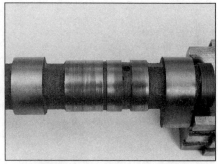

12.3 Check the journal surfaces of the camshaft for scratches or wear

Mark each shaft, bearing, follower and pushrod according to its position (ie left or right cylinder, intake or exhaust valves).

Inspection

3 Inspect the bearing surfaces of the holder and the bearing cap and the corresponding journals on the camshaft. Look for score marks, deep scratches and evidence of spalling (a pitted appearance) **(see illustration)**.
4 Check the camshaft lobes for heat discoloration (blue appearance), score marks, chipped areas, flat spots and spalling **(see illustration)**. If damage is noted or wear is excessive, the camshaft must be replaced. Also, be sure to check the condition of the followers, as described later in this Section.
5 Steps 6 to 13 detail measurement of the main camshaft journal (between the two

12.4 Check the lobes of the camshaft for wear – here's an example of damage requiring camshaft repair or renewal

12.8 Place a strip of Plastigauge on each bearing journal

lobes) and the oil clearance between journal and base/cap. No specifications are available with which to assess wear of the camshaft end journal which locates in the guide bore in the camshaft holder.

> **HAYNES HiNT** *Refer to Tools and Workshop Tips in the Reference section for details of how to read a micrometer and dial gauge.*

6 The camshaft journal oil clearance can be measured using two methods. The camshaft journal can be measured with a micrometer and the reading subtracted from the bore diameter (see Step 13), or you can use a product known as Plastigauge (see Steps 7 to 12) to determine oil clearance.
7 To measure the oil clearance using Plastigauge, first clean the camshaft, cap and base with a clean, lint-free cloth. Install the journal base in the holder, aligning the marks, then lay the camshaft in place in the holder and journal base.
8 Cut a strip of Plastigauge and lay it on the bearing journal parallel with the camshaft centreline **(see illustration)**. Install the cap, making sure the camshaft does not rotate at all, and tighten the journal bolts evenly and a little at a time, until the specified torque setting is reached.
9 Now unscrew the bolts evenly, a little at a time, and carefully lift off the cap, again making sure the camshaft does not rotate.
10 To determine the oil clearance, compare the crushed Plastigauge (at its widest point) on each journal to the scale printed on the

12.10 Compare the width of the crushed Plastigauge to the scale provided with it to obtain the clearance

Plastigauge container **(see illustration)**. Compare the results to this Chapter's Specifications.
11 On completion carefully scrape away all traces of the Plastigauge material from the camshaft and journal cap using a fingernail or other object which is unlikely to score the metal.
12 If the Plastigauge test indicates that the oil clearance is excessive, measure the diameter the camshaft journal diameter and bore diameter as described in Step 13. Comparison with the specifications at the beginning of this chapter will indicate whether the camshaft or bore is worn. Note that the camshaft can be replaced individually, but the holder, base, cap and rocker shaft cap are only available as an assembly.
13 To measure the oil clearance using a micrometer and a telescoping bore gauge (a Vernier caliper will do the job of both, but is less accurate), measure the diameter of the journal in two different planes **(see illustration)** and the internal bore of the bearing housing (holder, base and cap assembled and the cap bolts tightened to the specified torque). Calculate the difference to determine the clearance and compare the result to the clearance specified at the beginning of this chapter. If it is greater than specified, the measurements taken will tell you whether the camshaft or the holder and base are worn. Note that the camshaft can be replaced individually, but the holder, base, cap and rocker shaft cap are only available as an assembly.

2

12.13 Measure the cam bearing journals with a micrometer

12.18a Install the followers . . .

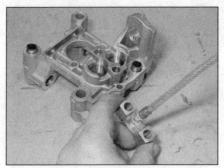

12.18b . . . then lubricate all the journals . . .

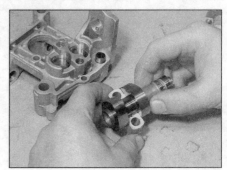

12.18c . . . and fit the camshaft into the base

14 Except in cases of oil starvation, the cam chains wear very little. If the chains have stretched excessively, which makes it difficult to maintain proper tension, they must be replaced (see Section 27).

15 Check the sprockets for cracks and other damage, replacing them if necessary. Note that if new sprockets are installed, new cam chains must also be installed. If the sprockets are worn, the cam chains are also worn, and also the sprockets on the auxiliary shaft (which can only be remedied by replacing it). If wear this severe is apparent, the entire engine should be disassembled for inspection.

16 Inspect the outer surfaces of the cam followers for evidence of scoring or other damage. If a follower is in poor condition, it is probable that the bore in which it works is also damaged. Check for clearance between the followers and their bores by measuring the external diameter of the followers with a micrometer and the internal diameter of the bores with a telescoping gauge and micrometer. Calculate the oil clearance by subtracting one measurement from the other and compare with the specifications at the beginning of this chapter to determine the extent of wear.

17 If available, blow through any oil passages with compressed air.

Installation

18 Install the followers into the holder, making sure each is returned to its original location **(see illustration)**. Apply a smear of molybdenum disulphide grease or clean engine oil to the camshaft journals. Fit the camshaft into the journal base, making sure the alignment mark on the base will align with that on the holder when it is installed **(see illustration)**. Install the camshaft and journal base in the holder, aligning the marks **(see illustration 12.2c)**. Install the journal cap **(see illustration 12.2b)**, again aligning the marks **(see illustration 12.2a)**, and tighten the journal bolts evenly and a little at a time, until the specified torque setting is reached.

19 Install the rocker/camshaft holders (see Section 11).

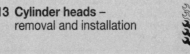

13 Cylinder heads – removal and installation

Caution: The engine must be completely cool before beginning this procedure or the cylinder head may become warped.
Note: *The cylinder heads can be removed with the engine in the frame. If no work is being carried out on the rockers or camshafts, the holders can be left attached to the cylinder heads. Otherwise, the holders can be removed either with the head in place or after it has been removed (see Sections 10, 11 and 12).*

Removal

1 Remove the throttle bodies (see Chapter 3).
2 Remove the exhaust system (see Chapter 3).
3 Remove the valve covers (see Section 8) and the camchain tensioners (see Section 9).
4 Remove the timing inspection hole plug from behind the right-hand cylinder **(see illustration 10.2)**. The engine can be turned by selecting a high gear and rotating the rear wheel by hand in its normal direction of rotation. Alternatively, remove the engine front cover and turn the engine using a 17 mm spanner or socket on the belt drive pulley bolt and turning it in a clockwise direction only **(see illustrations 9.7a and b)**.
Caution: Be sure to turn the engine in its normal direction of rotation.
5 Turn the engine until the "OT" mark on the flywheel, visible via the timing inspection hole, aligns with the middle of the hole, and the valves of the cylinder being worked on are all closed (see *Haynes Hint* – Section 10) **(see illustration 10.3)**. The piston for that cylinder is now at TDC (top dead centre) on the compression stroke. If any of the valves on that cylinder are open when the mark aligns, rotate the crankshaft clockwise one full turn until the "OT" mark again aligns with the middle of the hole. The valves will now be closed.
6 Unscrew the two bolts securing the camshaft sprocket cover and remove the cover **(see illustration 11.4a)**. According to

BMW, each sprocket is marked so that when at TDC on the compression stroke, the mark, either an "R" or an arrow, is in direct alignment with the tip of a sprocket tooth, which in turn is in direct alignment with the static mark on the holder. Also the locating lug indent on the sprocket should be facing down for the right-hand cylinder and up for the left-hand cylinder. However, on the bike worked on, the arrow on the sprocket was not aligned with the tip of the sprocket tooth, but the trough between two teeth, though the tip of the tooth did align with the static mark. Also, with the sprocket bolt installed, the indent is only just visible behind the bolt shoulder. It is highly advisable therefore to make your own mark on the tip of the sprocket tooth that is in direct alignment with the mark on the holder and to highlight the indent to avoid confusion on reassembly **(see illustration 11.4b)**. Car touch-up paint or Tippex can be used to make the marks.
7 Unscrew the bolt securing the sprocket to the camshaft – use either a spanner or socket on the crankshaft bolt or the TDC locking pin (see *Haynes Hint*) to stop the engine rotating as the sprocket bolt is unscrewed. Work the sprocket squarely off the end of the shaft using a pair of screwdrivers, noting how the small lug on the back of the sprocket locates in the cut-out on the camshaft **(see illustrations 11.5a and b)**. Either slip the sprocket out of the chain and remove it, in which case the chain must be secured to prevent it falling into the engine, or secure both the chain and sprocket to prevent them both falling into the engine. If both cylinders are being disassembled, remove the sprocket as it will be easier to turn the engine without it. Do not allow the chain to go slack as it could bind between the crankshaft sprocket and the crankcase. The best method of securing the chains is to cable-tie them to the tensioner and guide blades.
Caution: Do not turn the camshaft or the crankshaft with the cam chain sprocket removed, as the valves could contact the pistons and will be damaged.
8 Each cylinder head is secured by four nuts and four bolts **(see illustration)**. Slacken the nuts and bolts evenly and a little at a time in a

13.8 Cylinder head 6 mm bolts (A), 10 mm bolt (B) and nuts (C)

13.9 Draw the head off the studs

criss-cross sequence until they are all slack. Remove the nuts and bolts and their washers, noting the position of the longer bolt. Note that the cylinder head nuts and bolts may already have been slackened or removed if the rocker/camshaft holders have been removed.

9 Pull the cylinder head up off the studs (see illustration). If it is stuck, tap around the joint faces of the cylinder head with a soft-faced mallet to free the head. Do not attempt to free the head by inserting a screwdriver between the head and cylinder – you'll damage the sealing surfaces. Remove the old cylinder head gasket and discard it as a new one must be used.

10 If they are loose, remove the dowels from the cylinder or the underside of the cylinder head. Check the length of the four cylinder studs from the cylinder surface. The shorter stud (lower rear stud) should project 132 mm and the other three 152.5 mm. If necessary, screw the studs further into the crankcase using two nuts locked together on their threaded ends or a knurled wheel type tool. **Note:** *Loose cylinder studs can cause oil leakage at the cylinder head gasket.*

11 Check the cylinder head gasket and the mating surfaces on the cylinder head and cylinder for signs of leakage, which could indicate warpage.

12 Clean all traces of old gasket material from the cylinder head and cylinder. If a scraper is used, take care not to scratch or gouge the soft aluminium. Be careful not to let any of the gasket material fall into the crankcase, the cylinder bores or the oil passages.

Installation

Note: *If the rocker/camshaft holder assemblies have been removed from the cylinder head, install them onto the head before installing the head (see Sections 10, 11 and 12).*

13 If removed, install the two dowels into the cylinder. Lubricate the cylinder bores with engine oil.

14 Ensure both cylinder head and cylinder mating surfaces are clean, then lay the new head gasket in place on the cylinder, making sure all the holes are correctly aligned (see illustration). Never re-use the old gasket.

15 Carefully fit the cylinder head onto the cylinder, making sure the cam chain and sprocket fit into the cavity (see illustration 13.9).

16 Apply clean engine oil to the cylinder head stud nuts, then install them with their washers, with their collared end facing the cylinder head (see illustration 11.9b and c). Install the bolts and tighten them all finger-tight only at this stage (see illustration 13.8). Tighten the nuts evenly, a little at a time and in a criss-cross sequence to the initial torque setting specified at the beginning of the Chapter. Now attach a degree disc (see *Tools and workshop tips*) to the torque wrench and tighten each nut in turn a further 90°, then repeat to a further 90° (see illustration 11.9d and e). Now tighten the 10 mm bolt to the specified torque setting, then tighten the 6 mm bolts to the specified torque setting. If previously removed, install the

13.14 Fit the new gasket onto the studs and dowels

rocker/camshaft holder bolts and tighten them to the specified torque setting.

17 The remainder of the installation procedure is as described in Section 11, Steps 10 to 20. Install the exhaust system and throttle bodies (see Chapter 3).

18 Note that it is essential to check the valve clearances following cylinder head dismantling, especially if any of the valve or rocker components have been renewed (see Chapter 1 for details).

14 Valves/valve seats/valve guides – servicing

1 Because of the complex nature of this job and the special tools and equipment required, most owners leave servicing of the valves, valve seats and valve guides to a professional.

2 The home mechanic can, however, remove the valves from the cylinder head, clean and check the components for wear and assess the extent of the work needed, and, unless a valve service is required, grind in the valves (see Section 15).

3 The BMW dealer or engine specialist will remove the valves and springs, replace the valves and guides, recut or replace the valve seats, check and replace the valve springs, spring retainers and collets (as necessary), replace the valve seals with new ones and reassemble the valve components.

4 After the valve service has been performed, the head will be in like-new condition. When the head is returned, be sure to clean it again very thoroughly before installation on the engine to remove any metal particles or abrasive grit that may still be present from the valve service operations. Use compressed air, if available, to blow out all the holes and passages.

15.5a Make sure the compressor locates squarely onto the valve head and spring retainer

15 Cylinder head and valves –
disassembly, inspection and reassembly

1 As mentioned in the previous section valve seat re-cutting and valve seat or guide replacement should be left to a BMW dealer or engine specialist. However, disassembly, cleaning and inspection of the valves and related components can be done (if the necessary special tools are available) by the home mechanic. This way no expense is incurred if the inspection reveals that overhaul is not required at this time.
2 To disassemble the valve components without the risk of damaging them, a valve spring compressor is absolutely necessary.

Disassembly

3 Before proceeding, arrange to label and store the valves along with their related components in such a way that they can be returned to their original locations without getting mixed up. A good way to do this is to obtain a container which is divided into eight compartments, and to label each compartment with the identity of the valve which will be stored in it (ie R or L cylinder, intake top or bottom and exhaust top or bottom valve). Alternatively, labelled plastic bags will do just as well.

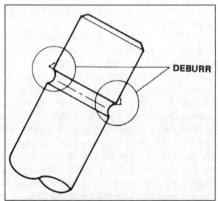

15.5d If the valve stem won't pull through the guide, deburr the area above the collet groove

15.5b Remove the collets with needle-nose pliers, tweezers, a magnet or a screwdriver with a dab of grease on it

4 If not already done, clean all traces of old gasket material from the cylinder head. If a scraper is used, take care not to scratch or gouge the soft aluminium.

> **HAYNES HiNT** *Refer to Tools and Workshop Tips for details of gasket removal methods.*

5 Compress the valve spring on the first valve with a spring compressor, making sure it is correctly located onto each end of the valve assembly **(see illustration)**. Do not compress the spring any more than is absolutely necessary. Remove the collets, using either needle-nose pliers, tweezers, a magnet or a screwdriver with a dab of grease on it **(see illustration)**. Carefully release the valve spring compressor and remove the spring retainer, noting which way up it fits, the spring, spring seat, noting which way up it fits, and the valve from the head **(see illustration)**. Push the valve down into the head and withdraw it from the underside. If the valve binds in the guide (won't pull through), push it back into the head and deburr the area around the collet groove with a very fine file or whetstone **(see illustration)**.
6 Repeat the procedure for the remaining valves. Remember to keep the parts for each valve together and in order so they can be reinstalled in the same location.
7 Once the valves have been removed and labelled, pull the valve stem seals off the top

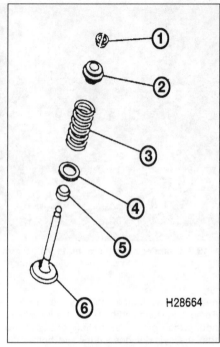

15.5c Valve components

1 Collets	4 Spring seat
2 Spring retainer	5 Valve stem seal
3 Spring	6 Valve

of the valve guides using either a special removing tool, pliers or screwdrivers **(see illustrations)**. Discard the seals – they should never be reused.
8 Next, clean the cylinder head with solvent and dry it thoroughly. Compressed air will speed the drying process and ensure that all holes and recessed areas are clean.
9 Clean all of the valve springs, collets, retainers and spring seats with solvent and dry them thoroughly. Do the parts from one valve at a time so that no mixing of parts between valves occurs.
10 Scrape off any deposits that may have formed on the valve, then use a motorised wire brush to remove deposits from the valve heads and stems. Again, make sure the valves do not get mixed up.

15.7a Removing the valve stem seal using a special tool

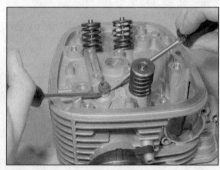

15.7b Removing the valve stem seal using screwdrivers

15.12 Measure the valve seat width with a ruler (or for greater precision use a Vernier caliper)

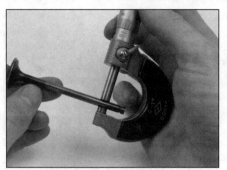

15.13a Measure the valve stem diameter with a micrometer

15.13b Insert a small hole gauge into the valve guide and expand it so there's a slight drag when it's pulled out

Inspection

11 Inspect the head very carefully for cracks and other damage. If cracks are found, a new head will be required. Note that due to modifications having been made to the combustion chamber area and the piston crown profile, it is essential that engine and frame number details are provided when ordering a new head. As further confirmation, state whether a 9-fin or 10-fin cylinder head is fitted and provide the piston size code (A, B or AB) stamped in the piston crown.

12 Examine the valve seats in the combustion chamber. If they are pitted, cracked or burned, the head will require work beyond the scope of the home mechanic. Measure the valve seat width and compare it to this Chapter's Specifications (see illustration). If it exceeds the service limit, or if it varies around its circumference, valve overhaul is required.

13 Measure the valve stem diameter (see illustration). Clean the valve guide bores to remove any carbon build-up, then measure the inside diameters of the guides (at both ends and the centre of the guide) with a small hole gauge and micrometer (see illustrations).The guides are measured at the ends and at the centre to determine if they are worn in a bell-mouth pattern (more wear at the ends).

14 Subtract the stem diameter from the valve guide diameter to obtain the valve stem-to-guide clearance. If the stem-to-guide clearance is greater than listed in this

Chapter's Specifications, the guides and valves will have to be replaced with new ones. If the valve stem or guide is worn beyond its limit, or if the guide is worn unevenly, it must be replaced. Note that valve guide replacement is a task for a BMW dealer or engine specialist.

15 Carefully inspect each valve face for cracks, pits and burned spots. Check the valve stem and the collet groove area for cracks (see illustration). Rotate the valve and check for any obvious indication that it is bent. Check the end of the stem for pitting and excessive wear. The presence of any of the above conditions indicates the need for valve servicing.

16 Check the end of each valve spring for wear and pitting. Measure the spring free length and compare it to that listed in the specifications (see illustration). If any spring is shorter than specified it has sagged and must be replaced. Also place the spring upright on a flat surface and check it for bend by placing a ruler against it (see illustration). If the bend in any spring is excessive, it must be replaced.

17 Check the spring retainers and collets for obvious wear and cracks. Any questionable parts should not be reused, as extensive damage will occur in the event of failure during engine operation.

18 If the inspection indicates that no overhaul work is required, the valve components can be reinstalled in the head.

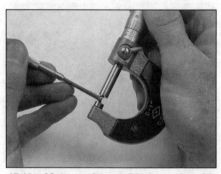

15.13c Measure the small hole gauge with a micrometer

Reassembly

19 Unless a valve service has been performed, before installing the valves in the head they should be ground in (lapped) to ensure a positive seal between the valves and seats. This procedure requires coarse and fine valve grinding compound and a valve grinding tool. If a grinding tool is not available, a piece of rubber or plastic hose can be slipped over the valve stem (after the valve has been installed in the guide) and used to turn the valve.

20 Apply a small amount of coarse grinding compound to the valve face, then oil the valve stem and insert it into the guide (see illustration). Note: Make sure each valve is installed in its correct guide and be careful not to get any grinding compound on the valve stem.

2

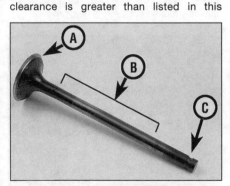

15.15 Check the valve face (A), stem (B) and collet groove (C) for signs of wear and damage

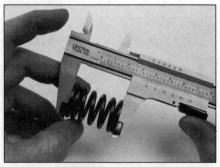

15.16a Measure the free length of the valve springs

15.16b Check the valve springs for squareness

15.20 Apply the lapping compound very sparingly, in small dabs, to the valve face only

15.21a Rotate the valve grinding tool back and forth between the palms of your hands

15.21b The valve face and seat should show a uniform unbroken ring . . .

15.21c . . . and the seat (arrowed) should be the specified width all the way round

15.24a Fit the new seal onto the guide . . .

15.24b . . . and press it on using a socket

21 Attach the grinding tool (or hose) to the valve and rotate the tool between the palms of your hands. Use a back-and-forth motion (as though rubbing your hands together) rather than a circular motion (ie so that the valve rotates alternately clockwise and anti-clockwise rather than in one direction only) **(see illustration)**. Lift the valve off the seat and turn it at regular intervals to distribute the grinding compound properly. Continue the grinding procedure until the valve face and seat contact area is of uniform width and unbroken around the entire circumference of the valve face and seat **(see illustrations)**.

22 Carefully remove the valve from the guide and wipe off all traces of grinding compound. Use solvent to clean the valve and wipe the seat area thoroughly with a solvent soaked cloth.

23 Repeat the procedure with fine valve grinding compound, then repeat the entire procedure for the remaining valves.

24 Install new valve stem seals onto the top of each of the guides **(see illustration)**. Use an appropriate size deep socket to push the seals over the end of the valve guide until they are felt to clip into place **(see illustration)**. Don't twist or cock them, or they will not seal properly against the valve stems. Also, don't remove them again or they will be damaged.

25 Coat the valve stems with molybdenum disulphide grease or oil, then install one of them into its guide, rotating it slowly to avoid damaging the seal as the end pushes through **(see illustrations)**. Check that the valve moves up and down freely in the guide.

26 Lay the spring seats for all the valves in place in the cylinder head with their shouldered

side facing up so that they fit into the base of the springs (the spring seat can be identified from the spring retainer by its larger internal diameter – be sure not to mix up the two) **(see illustration)**. Next, install the springs, with their closer-wound coils facing down into the cylinder head, followed by the spring retainer, with its shouldered side facing down so that it fits into the top of the springs **(see illustrations)**.

27 Apply a small amount of grease to the collets to help hold them in place as the pressure is released from the springs **(see illustration)**. Compress the springs with the valve spring compressor and install the collets **(see illustration)**. When compressing the springs, depress them only as far as is absolutely necessary to slip the collets into place. Make certain that the collets are securely locked in their retaining grooves.

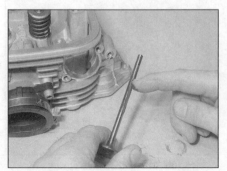

15.25a Lubricate the valve stem . . .

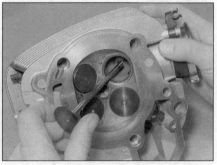

15.25b . . . and slide it into the guide

15.26a Fit the spring seat . . .

15.26b . . . the spring . . .

15.26c . . . and the spring retainer

15.27a A small dab of grease will help to keep the collets in place on the valve while the spring is released

28 Support the cylinder head on blocks so the valves can't contact the workbench top, then tap each of the valve stems with a soft-faced hammer **(see illustration)**. This will help seat the collets in their grooves.

 Check for proper sealing of the valves by pouring a small amount of solvent into each of the valve ports. If the solvent leaks past any valve into the combustion chamber area the valve grinding operation on that valve should be repeated.

15.27b Make sure each collet locates in the groove in the stem

15.28 Tap the stem to ensure the collets are locked

16 Cylinders –
removal, inspection and installation

Note: *The cylinders can be removed with the engine in the frame.*

Removal

1 Remove the cylinder head (see Section 13).
2 Unscrew the cam chain guide blade pivot bolt and remove it with its sealing washer **(see illustration)**.
3 Unscrew the four bolts which secure the cylinder to the crankcase **(see illustration)**.
4 If not already done, it is advisable to cable-tie each cam chain to its tensioner and guide blades to prevent it slipping into the crankcase. Draw the cylinder off the studs,

supporting the connecting rod as you do to prevent it or the piston hitting the crankcase **(see illustration)**. If the cylinder is stuck, tap around the joint faces with a soft-faced mallet to free it from the crankcase. Don't attempt to free it by inserting a screwdriver between it and the crankcase – you'll damage the sealing surfaces. When the cylinder is removed, stuff clean rags around the piston to prevent anything falling into the crankcase.
5 Where fitted, remove the two oil passage O-rings from either the cylinder or the crankcase and discard them as new ones must be used. **Note:** *O-rings were fitted to certain early engines. Exact details of which engines should have O-rings fitted are not available, but if they are found on removal and the crankcase surface is recessed to accept them, new O-rings should be installed during the rebuild.*

6 Remove the two dowels from either the cylinder or crankcase if they are loose Clean all traces of sealant from the cylinder block and crankcase mating surfaces. If a scraper is used, take care not to scratch or gouge the soft aluminium. Be careful not to let any of the material fall into the crankcase or the oil passages.

Inspection

7 The bores are coated with a special nickel silicon substance, called Gilnisel, which has a high resistance to wear. The bore coating should last the life of the engine unless catastrophic engine damage, such as seizure, has occurred. Note that the cylinder liners should not be separated from the cylinder blocks.
8 Check the cylinder walls carefully for scratches and score marks. If damage is

16.2 Unscrew the camchain guide blade pivot bolt (arrowed)

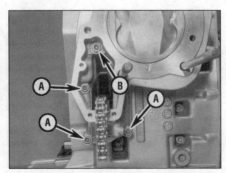

16.3 Remove the 6 mm bolts (A) and the 8 mm bolt (B) . . .

16.4 . . . and draw the cylinder off the studs

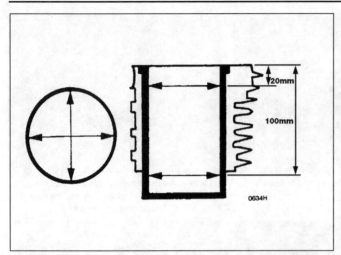

16.9 Measure the cylinder bore in the directions shown with a telescoping gauge, then measure the gauge with a micrometer

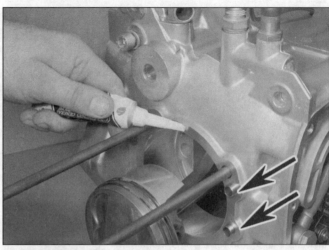

16.13 Apply the correct sealant to the crankcase mating surface. Make sure the dowels (arrowed) are fitted

noted, yet the bore diameter is still within the service limit, seek the advice of a BMW dealer or engine specialist as to its suitability for continued use.

9 Using telescoping gauges and a micrometer (see *Tools and working facilities* in the Reference section), check the dimensions of each cylinder to assess the amount of wear and ovality. Take four measurements, at 20 mm and 100 mm from the top of the bore, parallel with the piston pin and at 90° to it **(see illustration)**. Calculate any differences between the measurements taken to determine ovality in the bore. Compare the results to the specifications at the beginning of the Chapter, noting that the bore size coding must be established – the code letter is stamped in the liner's machined top surface. If the cylinders are worn beyond the service limit, or bore ovality exceeds the service limits, that cylinder must be renewed. **Note:** *Always renew the piston if you are renewing the cylinder.*

10 If the cylinder and piston are to be renewed, make sure you supply the dealer with the size code letter of each component when purchasing new parts. The cylinder will be coded A or B and the piston will be coded A, B or AB. Cylinders marked A must be fitted with pistons marked A, and cylinders marked

B must be fitted with pistons marked B. Pistons marked AB can be installed in either cylinder. Note that it is essential that the pistons are of the same weight to ensure smooth running of the engine. The weight code will be either a + (indicating 10g over standard weight), no mark at all (standard weight), or a – (indicating 10g under standard weight). Inspect the piston surface for a weight marking, and have a BMW determine the piston's weight category if you are in any doubt.

Installation

11 Check that the mating surfaces of the cylinder and crankcase are clean and free from oil or old sealant. Check the length of the four cylinder studs from the crankcase surface. The shorter stud (lower rear stud) should project 212 mm and the other three 232.5 mm. If necessary, screw the studs further into the crankcase using two nuts locked together on their threaded ends or a knurled wheel type tool. **Note:** *Loose cylinder studs can cause oil leakage at the cylinder head gasket.*

12 Where fitted (see Step 5), install a new oil passage O-ring into each of the recesses in the cylinder/crankcase mating surface.

13 Remove the rags from around the piston. Apply a small amount of 3-bond 1209 or

equivalent sealant to the mating surface of the upper crankcase half **(see illustration)**. Check that the dowels are installed.

Caution: Do not apply an excessive amount of sealant as it will ooze out and may obstruct oil passages.

14 If required, install a piston ring clamp onto the piston to ease its entry into the bore as the block is installed. This is not essential as the cylinder has a good lead-in enabling the piston rings to be hand-fed into the bore.

15 Lubricate the cylinder bore, piston and piston rings, and the connecting rod big- and small-ends, with clean engine oil **(see illustrations)**. Install the cylinder over the studs until the piston crown fits into the bore **(see illustration 16.4)**. Make sure the cam chain is positioned so that it enters the tunnel without causing an obstruction.

16 Gently push on the cylinder, making sure the piston enters the bore squarely and does not get cocked sideways. Unless a piston ring clamp is being used, carefully compress and feed each ring into the bore as the cylinder is lowered **(see illustration)**. If necessary, use a soft mallet to gently tap the cylinder down, but do not use force if it appears to be stuck as the piston and/or rings will be damaged. If a clamp was used, remove it once the piston is in the bore.

16.15a Lubricate the cylinder bore . . .

16.15b . . . piston and rings with clean engine oil

16.16 Feed the rings carefully into the bore as the cylinder is installed

17 When the piston is correctly installed in the bore, press the cylinder down onto the crankcase, making sure the dowels and where fitted, O-rings, locate correctly. Install the bolts which secure the cylinder to the crankcase and tighten them to the torque settings specified at the beginning of the Chapter **(see illustration 16.3)**.

18 Install the cam chain guide blade pivot bolt using a new sealing washer and tighten it to the specified torque setting **(see illustrations)**.

19 Install the cylinder head (see Section 11).

17 Pistons –
removal, inspection and installation

Note: *The pistons can be removed with the engine in the frame.*

Removal

1 Remove the cylinder from the side being worked on (see Section 16).

2 Before removing the piston from the connecting rod, use a sharp scriber or felt marker pen to write the cylinder identity (ie R or L) on the piston crown (or on the inside of the skirt if the piston is dirty and going to be cleaned). Each piston also has a slightly raised square cast on its underside which should face the exhaust side of the bore **(see illustration)** and also an arrow on its crown which should also face the exhaust side. If these marks are not clear, mark the piston with an arrow pointing to the exhaust (front),

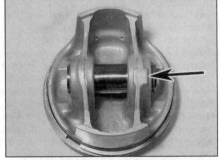

17.2 Note the raised square (arrowed) which marks the exhaust side of the piston

17.3a Remove the circlips . . .

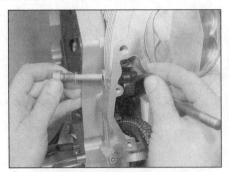

16.18a Install the guide blade bolt . . .

so that it can be installed the correct way round. The piston crown is marked with a size identification letter, either A, B, or AB. The letter is very faint and the piston will have to be cleaned to see it.

3 Carefully remove the circlip from each end of the piston pin using external circlip pliers **(see illustration)**. Discard them as new ones must be used. Push the piston pin out to free the piston from the connecting rod **(see illustration)**. When the piston has been removed, install its pin back into the piston.

 HAYNES HiNT *If a piston pin is a tight fit in the piston bosses, soak a rag in boiling water then wring it out and wrap it around the piston – this will expand the alloy piston sufficiently to release its grip on the pin. If the piston pin is particularly stubborn, extract it using a drawbolt tool, but be careful not to damage the rings or piston body.*

Inspection

4 Before the inspection process can be carried out, the piston must be cleaned and the old piston rings removed.

5 Using your thumbs or a piston ring removal and installation tool, carefully remove the rings from the piston **(see illustration)**. Do not nick or gouge the piston in the process. Carefully note which way up each ring fits and in which groove as they must be installed in their original positions if being re-used. The upper surface of each ring is marked "TOP".

17.3b . . . and withdraw the piston pin

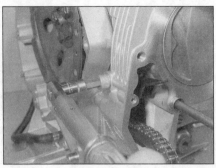

16.18b . . . and tighten it to the specified torque setting

6 Scrape all traces of carbon from the piston crown. A hand-held wire brush or a piece of fine emery cloth can be used once most of the deposits have been scraped away, although be careful not to remove the piston size code letter. Do not, under any circumstances, use a wire brush mounted in a drill motor to remove deposits from the piston; the piston material is soft and will be eroded away by the wire brush.

7 Use a piston ring groove cleaning tool to remove any carbon deposits from the ring grooves. If a tool is not available, a piece broken off an old ring will do the job. Be very careful to remove only the carbon deposits. Do not remove any metal and do not nick or gouge the sides of the ring grooves.

8 Once the deposits have been removed, clean the piston with solvent and dry it thoroughly. If the identification previously marked on the piston is cleaned off, be sure to re-mark it with the correct identity. Make sure the oil return holes below the oil ring groove are clear.

9 Carefully inspect each piston for cracks around the skirt, at the pin bosses and at the ring lands. Normal piston wear appears as even, vertical wear on the thrust surfaces of the piston and slight looseness of the top ring in its groove. If the skirt is scored or scuffed, the engine may have been suffering from overheating and/or abnormal combustion, which caused excessively high operating temperatures. The oil pump should be checked thoroughly.

10 A hole in the piston crown, an extreme to be sure, is an indication that abnormal

2

17.5 Removing the piston rings using a ring removal and installation tool

17.11 Measure the piston ring-to-groove clearance with a feeler gauge

17.12 Measure the piston diameter with a micrometer 6 mm from the bottom of the skirt

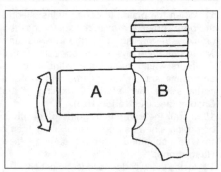

17.13a Slip the pin (A) into the piston (B) and try to rock it back and forth. If it's loose, replace the piston and pin

combustion (pre-ignition) was occurring. Burned areas at the edge of the piston crown are usually evidence of spark knock (detonation). If any of the above problems exist, the causes must be corrected or the damage will occur again.

11 Measure the piston ring-to-groove clearance by laying each piston ring in its groove and slipping a feeler gauge in beside it **(see illustration)**. Make sure you have the correct ring for the groove (see Step 5). Check the clearance at three or four locations around the groove. If the clearance is greater than specified, replace both the piston and rings as a set. If new rings are being used, measure the clearance using the new rings. If the clearance is greater than that specified, the piston is worn and must be replaced.

12 Check the piston-to-bore clearance by measuring the bore (see Section 16) and the piston diameter. Measure the piston 6.0 mm

up from the bottom of the skirt and at 90° to the piston pin axis **(see illustration)**. Subtract the piston diameter from the bore diameter to obtain the clearance. If it is greater than the specified figure, the piston must be replaced (assuming the bore itself is within limits, otherwise a new cylinder is also necessary).

13 Apply clean engine oil to the piston pin, insert it into the piston and check for any freeplay between the two **(see illustration)**. Measure the pin external diameter, the pin bore in the piston and the connecting rod small-end bore **(see illustrations)**. Calculate the difference between the measurements to obtain the piston pin-to-bore clearance and the pin-to-small-end clearance. Compare the measurements to the specifications at the beginning of the Chapter and replace components that are worn beyond the specified limits.

14 If the pistons are to be replaced, ensure the correct size of piston is ordered. Each piston crown is marked A, B or AB. Each cylinder is correspondingly marked A or B. Cylinders marked A must be fitted with pistons marked A, and cylinders marked B must be fitted with pistons marked B. Pistons marked AB can be installed in either an A or B cylinder. Note that it is essential that the pistons are of the same weight to ensure smooth running of the engine. The weight code will be either a + (indicating 10g over standard weight), no mark at all (standard weight), or a – (indicating 10g under standard weight). Inspect the piston surface for a weight marking, and have a BMW determine the piston's weight category if you are in any doubt.

Installation

15 Inspect and install the piston rings (see Section 18).
16 Lubricate the piston pin, the piston pin bore and the connecting rod small-end bore with clean engine oil **(see illustration)**.
17 When installing the piston onto the connecting rod, make sure that the raised square on the underside of the piston and the arrow on its crown face the exhaust side (front) **(see illustration 17.2)**. Line up the piston on the connecting rod, and insert the piston pin **(see illustration 17.3b)**. Secure the pin with new circlips. When installing the circlips, expand them only just enough to fit them on the pin, and make sure they are properly seated in their grooves **(see illustration)**.
18 Install the cylinder (see Section 16).

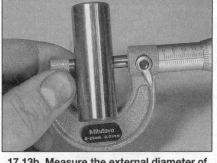

17.13b Measure the external diameter of the pin . . .

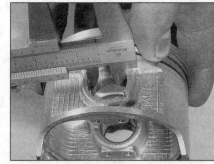

17.13c . . . the internal diameter of the bore in the piston . . .

17.13d . . . and the internal diameter of the connecting rod small-end

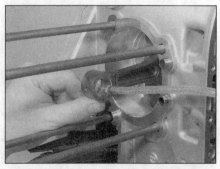

17.16 Lubricate the small-end bore and other components with clean engine oil

17.17 Do not over-expand the circlips when fitting them onto the piston pin and make sure they are correctly seated

18.3 Measuring piston ring installed end gap

18.5 Ring end gap can be enlarged by clamping a file in a vice and filing the ring ends

18.9 Make sure the side rails locate over the expander on the oil ring

18 Piston rings – inspection and installation

1 It is good practice to replace the piston rings when an engine is being overhauled. Before installing the new piston rings, the ring end gaps must be checked with the rings installed in the cylinder.
2 If you are working on both pistons, lay out the pistons and the new ring sets so the rings will be matched with the same piston and cylinder during the end gap measurement procedure and engine assembly.
3 To measure the installed ring end gap, insert the top ring into the top of the cylinder and square it up with the cylinder walls by pushing it in with the top of the piston. The ring should be about 20 mm below the top edge of the cylinder. To measure the end gap, slip a feeler gauge between the ends of the ring and compare the measurement to the specifications at the beginning of the Chapter **(see illustration)**.
4 If the gap is larger or smaller than specified, double check to make sure that you have the correct rings before proceeding.
5 If the gap is too small, it must be enlarged or the ring ends may come in contact with each other during engine operation, which can cause serious damage. The end gap can be increased by filing the ring ends very carefully with a fine file. When performing this operation, file only from the outside in **(see illustration)**.
6 Excess end gap is not critical unless it exceeds the service limit. Again, double-check to make sure you have the correct rings for your engine and check that the bore is not worn.
7 Repeat the procedure for each ring that will be installed in the cylinders. Remember to keep the rings, pistons and cylinders matched up.
8 Once the ring end gaps have been checked/corrected, the rings can be installed on the pistons.
9 The oil control ring (lowest on the piston) is installed first. It is composed of two separate

components, namely the expander and the side rails. Slip the expander into the groove, then install the side rail. Do not use a piston ring installation tool on the oil ring side rails as they may be damaged. Instead, place one end of the side rail into the groove over the expander. Hold it firmly in place and slide a finger around the piston while pushing the rail into the groove, making sure it fits over the expander **(see illustration)**.
10 After the oil ring components have been installed, check to make sure that the side rail can be turned smoothly in the ring groove.
11 The upper surface of each compression ring is marked with "TOP". Both rings are identical. Install the second (middle) ring next. Make sure that the "TOP" mark near the end gap is facing up. Fit the ring into the middle groove in the piston. Do not expand the ring any more than is necessary to slide it into place. To avoid breaking the ring, use a piston ring installation tool.
12 Finally, install the top ring in the same manner into the top groove in the piston. Make sure the "TOP" mark near the end gap is facing up.
13 Once the rings are correctly installed, check they move freely without snagging. Viewed as installed on the connecting rods, stagger the ring end gaps 120° apart, with the oil control ring side rail gaps facing up (the split in the expander ring should face down), the second (middle) ring gap facing forward and down, and the top ring gap facing back and down.

19 Connecting rods – removal, inspection and installation

Note: *The connecting rods can be removed with the engine in the frame, but note that access to the bolts for one cylinder is via the cylinder aperture in the crankcase for the other cylinder. Therefore both cylinders and pistons must be removed, even if only one connecting rod is to be removed. Alternatively, the crankcase halves can be separated (following engine removal) and the crankshaft and connecting rods removed.*

Removal

1 Remove the pistons (see Section 17). Turn the crankshaft to provide the best access to the connecting rod being removed.
2 Before removing the rod from the crankshaft, measure its side clearance with a feeler gauge **(see illustration)**. If the clearance is greater than the service limit listed in this Chapter's Specifications, the rod must be renewed.
3 Using paint or a felt marker pen, mark the cylinder identity (R or L) on the connecting rod and cap. There should already be some lettering on the upwards facing side of the connecting rod – if not, mark across the cap-to-connecting rod join and note which side of the rod faces up to ensure that the cap and rod are fitted the correct way around on reassembly **(see illustration)**.

19.2 Measure the connecting rod side clearance using a feeler gauge

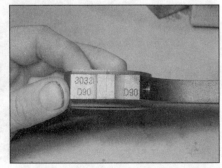

19.3 Note the markings on the upper face of the rod and cap

2

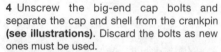

19.4a Unscrew the big-end cap bolts (arrowed) . . .

19.4b . . . and remove the cap and shell

19.5 Withdraw the con-rod and shell from the other side

4 Unscrew the big-end cap bolts and separate the cap and shell from the crankpin **(see illustrations)**. Discard the bolts as new ones must be used.
Caution: Make sure the bearing shells do not fall out of either the cap or the rod and into the crankcase.
5 Moving to the opposite side of the engine, remove the connecting rod and shell from the crankshaft **(see illustration)**. Keep the rod, cap and (if they are to be reused) the bearing shells together in their correct positions to ensure correct installation.
6 Note that the big-end cap bolts are of the stretch type, and must be renewed every time they are disturbed.

Inspection

7 Check the connecting rods for cracks and other obvious damage.
8 If not already done (see Section 17), apply clean engine oil to the piston pin, insert it into the connecting rod small-end and check for any freeplay between the two **(see illustration)**. Measure the pin external diameter and the small-end bore diameter **(see illustrations 17.13b and 17.13d)**. Calculate the difference between the measurements to obtain the pin-to-small-end clearance. Compare the measurements to the specifications at the beginning of the Chapter and replace components that are worn beyond the specified limits.
9 Refer to Section 25 and examine the connecting rod bearing shells. If they are scored, badly scuffed or appear to have seized, new shells must be installed. Always replace the shells in the connecting rods as a set. If they are badly damaged, check the corresponding crankpin. Evidence of extreme heat, such as discoloration, indicates that lubrication failure has occurred. Be sure to thoroughly check the oil pump and pressure regulator as well as all oil holes and passages before reassembling the engine.
10 Have the rods checked for twist and bend by a BMW dealer if you are in doubt about their straightness. Note that the sintered-metal type connecting rods used in this engine cannot be straightened due to risk of fracture.

Oil clearance check

11 Whether new bearing shells are being fitted or the original ones are being re-used, the connecting rod oil clearance should be checked prior to reassembly. Connecting rod oil clearance can be measured with either a product known as Plastigauge (see Steps 12 to 16), or by using a micrometer and telescoping bore gauge (see Step 19). Due to the cost of accurate measuring equipment, Plastigauge is more readily available to many people. It is, however, not as accurate or reliable and therefore it is worth having any doubtful results verified by a BMW dealer or specialist with the required equipment. Also, the check using Plastigauge can only be properly carried out with the crankshaft removed, which means removing the engine and separating the crankcase halves.
12 To measure the oil clearance using Plastigauge, first clean the backs of the bearing shells and the bearing locations in both the connecting rod and cap.
13 Press the bearing shells into their locations, ensuring that the tab on each shell engages the notch in the connecting rod/cap **(see illustration)**. Make sure the bearings are fitted in the correct locations and take care not to touch any shell's bearing surface with your fingers.
14 Cut a length of the appropriate size Plastigauge (it should be slightly shorter than the width of the crankpin). Place a strand of Plastigauge on the (cleaned) crankpin journal and fit the (clean) connecting rod and cap. Make sure the cap is fitted the correct way around so the previously made markings align, and that the rod is facing the right way

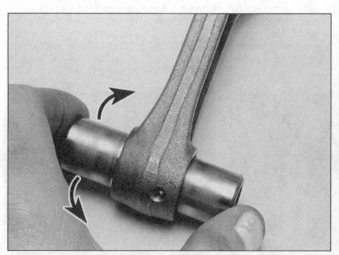

19.8 Slip the piston pin into the rod's small-end and rock it back and forth to check for looseness

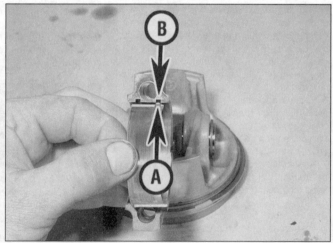

19.13 Make sure the tab (A) locates in the notch (B)

with the lettering facing up (see illustration 19.3). Oil the bearing cap bolts and tighten them evenly to the initial torque setting specified at the beginning of the Chapter, whilst ensuring that the connecting rod does not turn on the crankshaft. Now attach a degree disc (see *Tools and workshop tips*) to the torque wrench and tighten each nut in turn a further 80°. Slacken the bolts and remove the connecting rod, again taking great care not to rotate the crankshaft.

15 Compare the width of the crushed Plastigauge on the crankpin to the scale printed on the Plastigauge envelope to obtain the connecting rod bearing oil clearance (see illustration 26.15). Compare the reading to the specifications at the beginning of the Chapter.

16 On completion carefully scrape away all traces of the Plastigauge material from the crankpin and bearing shells using a fingernail or other object which is unlikely to score the shells.

17 If the clearance is within the range listed in this Chapter's Specifications and the bearings are in perfect condition, they can be reused. If the clearance is beyond the service limit, replace the bearing shells with new ones. Check the oil clearance once again (the new shells may be thick enough to bring bearing clearance within the specified range). Always replace the bearing shells in both connecting rods at the same time.

18 If the clearance is still greater than the service limit listed in this Chapter's Specifications, the crankpin is worn; this can be confirmed by measuring the diameter of the journal using a micrometer and compare it to the specifications (see illustration). If it is worn, it can be reground one stage only and fitted with oversize (+0.25 mm) bearing shells available from a BMW dealer. If it has already been ground once, it will have to be rehardened and finished off to its original specification. Due to the precision required in these processes, and the fact that it is often not achieved, crankshaft replacement is often regarded as the best long-term course of action.

19 To measure the oil clearance with the crankshaft in the engine, you will need a micrometer and a telescoping bore gauge. A Vernier caliper will do the job of both, but is less accurate. Refer to *Tools and Workshop Tips* in the Reference section for more information. Measure the diameter of the crankpin in two different planes and the internal bore of the connecting rod big-end perpendicular to the join, with the bearing shells in place and the cap tightened down to the specified torque (see Step 14). Calculate the difference to obtain the clearance and compare the result to the clearance specified. If it is greater than specified, the measurements taken will determine whether the shells or the crankpin are worn, and the appropriate action can be taken (see Steps 17 and 18).

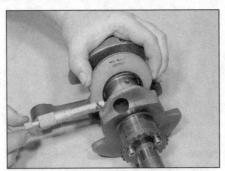

19.18 Measuring the crankpin journal diameter using a micrometer

Installation

20 Install the bearing shells in the connecting rods and caps, aligning the notch in the bearing with the groove in the rod or cap (see illustration 19.13). Lubricate the shells and the crankpin with clean engine oil (see illustration).

21 Fit the rod, making sure the shell remains in place, onto the crankshaft, making sure it is the right way round according to the marks made earlier and those already on the rod (see Step 3) (see illustrations 19.3 and 19.5).

22 Fit the bearing cap, making sure the shell remains in place, onto the connecting rod (see illustration 19.4b). Make sure the cap is fitted the correct way around so the connecting rod and bearing cap markings are correctly aligned.

23 Using new bolts, apply a smear of clean engine oil to the threads and underside of the heads (see illustration). Install the bolts and tighten them evenly to the initial torque setting specified. Now attach a degree disc (see *Tools and workshop tips*) to the torque wrench and tighten each nut in turn a further 80° (see illustration).

24 Check that the crankshaft is free to rotate easily, then install the other rod in the same way. Check to make sure that all components have been returned to their original locations using the marks made on disassembly.

25 Check that the rods rotate smoothly and freely on the crankpin. If there are any signs of roughness or tightness, slacken the bolts and re-tighten them, making sure the correct torque setting and angle is achieved.

19.23a Lubricate the new connecting rod bolts . . .

19.20 Lubricate the shells and crankpin with clean engine oil

Otherwise, remove the rods and re-check the bearing clearance. Sometimes tapping the bottom of the connecting rod cap will relieve tightness, but if in doubt, recheck the clearances. Note that new connecting rod bolts must be used each time they are disturbed.

20 Alternator drive and engine inner front cover – removal and installation

Note: *The alternator drive components and the inner front engine cover can be removed with the engine in the frame. If required, the crankshaft oil seal in the cover can be replaced without removing the cover (see Tools and Workshop Tips).*

Removal

1 Remove the alternator (see Chapter 8).

2 On all except RS models to 5/94, unscrew the two bolts securing the breather pipe and remove the pipe. Discard the O-ring from the lower fitting and the sealing washers from the banjo bolt as new ones must be used.

3 Unscrew the bolt securing the alternator belt drive pulley and remove the pulley, noting how it fits (see illustration). It will be necessary to lock the engine to prevent the crankshaft turning whilst slackening the bolt. This can be done in a number of ways: BMW provide a tool, part no. 11 5 640 (see illustration 4.4b in Chapter 4), which locks the clutch housing, and can be fitted after removing the starter motor (see Chapter 8).

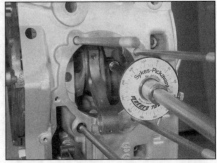

19.23b . . . and tighten them as specified

20.3 Unscrew the pulley bolt (arrowed) and remove the pulley

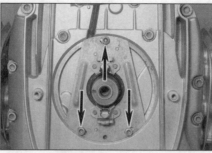

20.4a The timing plate is secured by three bolts (arrowed). Only remove the plate if absolutely necessary

20.4b Slacken the bolt (arrowed) to free the sensor wiring grommet

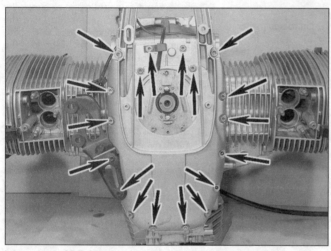

20.5 Inner cover mounting bolts (arrowed)

20.7a Apply the sealant to the mating surfaces . . .

20.7b . . . then install the cover

Alternatively, if the engine is in the frame, select a gear and have an assistant press on the rear brake pedal or use the TDC locking pin to stop the engine from rotating (see *Haynes Hint* – Section 11).

4 If required, unscrew the bolts securing the timing sensor plate and remove the plate, but note that the cover can be removed with it in place **(see illustration)**. It is better to leave the plate undisturbed if possible, as the timing will be upset if the plate is not returned to its exact original position. If the plate is removed, make some accurate alignment marks between it and the cover to ensure it is

correctly installed. Note how the sensor wiring grommet locates in the cut-out and is secured by the tab – slacken the bolt to free it and disconnect the wiring at the connector inside the rubber boot **(see illustration)**.

5 Unscrew the bolts securing the cover to the crankcase and remove the cover **(see illustration)**. Clean off all traces of old sealant from the mating surfaces of the cover and the crankcase. Remove the dowels from either the cover or the crankcase if they are loose.

Installation

Note: *If the engine has been disassembled, do not install the inner front cover until the cylinder heads and camshafts have been installed. This is so that the timing marks on the crankshaft and auxiliary shaft sprockets can be seen.*

6 Check the condition of the crankshaft and rotary breather oil seals (except RS models to 5/94) in the cover and replace them if there are any signs of leakage or damage. The crankshaft seal should be levered out from the front of the cover, having first removed the timing sensor plate. The new oil seal needs to have its inner lip pre-formed before it is installed. Do this by pressing the lip in with your fingers, making sure it is even all round, and

pressed in far enough to allow the shaft through it, but not so far that it becomes wider than the shaft itself. Install the oil seal into the front of the cover using either a large socket which bears on the outer edge of the seal or the BMW tool no. 11 5 680. The rotary breather seal (except RS models to 5/94) is levered out from the rear of the cover. Pre-shape the new seal as described above for the crankshaft seal and drive it into the rear of the cover using a socket which bears on its outer edge or the BMW tools nos. 00 5 550 and 11 5 650.

7 If removed, fit the dowels into the crankcase. Make sure the mating surfaces of the cover and crankcase are clean and free from oil, grease and old sealant. Apply a 3-Bond 1209 or equivalent sealant to the mating surfaces, then install the cover and tighten its bolts to the torque setting specified at the beginning of the Chapter, making sure you distinguish correctly between the 6 mm and 8 mm bolts **(see illustrations)**.

8 If removed, install the timing sensor plate, making sure the marks made on removal are exactly in alignment, and tighten its bolts securely **(see illustration 20.4a)**. Fit the grommet into its cut-out and secure it with the tab, then connect the wiring at the connector **(see illustration 20.4b)**.

20.9a Locate the tab (A) in the notch (B) . . .

20.9b . . . then install the bolt with washer . . .

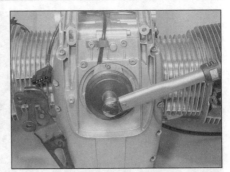

20.9c . . . and tighten it to the specified torque setting

21.2 Align the sprocket marks as shown

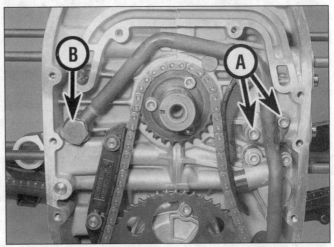

21.3 Unscrew the bracket bolts (A) and the banjo bolt (B), then swing the pipe down

9 Install the pulley onto the end of the crankshaft, making sure the tab on the inside of the pulley locates in the notch in the end of the crankshaft (see illustration). Install the bolt with washer and tighten it to the specified torque setting, using the method employed on removal to prevent the crankshaft turning (see illustrations).
10 On all except RS models to 5/94, install the breather pipe using a new O-ring and sealing washers and tighten the bolts to the specified torque settings.

11 Install the alternator (see Chapter 8). Check the drive belt along its entire length for splits, cracks, worn or damaged teeth, frays and any other damage or deterioration. Be careful not to bend the belt excessively or to get oil or grease on it. Replace the belt if it is in any way worn, damaged or deteriorated, or if renewal is due under the maintenance schedule (see Chapter 1).

21 Auxiliary shaft drive chain, tensioner and sprockets – removal, inspection and installation

Removal

1 Remove the alternator drive and engine inner front cover (see Section 20).
2 Turn the engine in a clockwise direction, using a spanner or socket on the lower (auxiliary shaft) sprocket bolt, until the mark on the upper (crankshaft) sprocket is facing down, and the mark on the lower sprocket is facing up, and they are directly in line (see illustration).
3 Unscrew the oil pipe bracket bolts and banjo bolt, then swing the pipe forwards and

down so that it is clear of the chain tensioner assembly (see illustration). Be prepared to catch any residue oil. Discard the banjo bolt sealing washers as new ones must be used.
4 Unscrew the three bolts securing the chain tensioner to the crankcase and remove the tensioner, taking care not to let the tensioner piston fly out under pressure of the spring (see illustration). If required, remove the E-clips securing the tensioner blade and the guide blade and slide the blades off their lugs (see illustration).

21.4a Unscrew the three bolts (arrowed) and remove the tensioner

21.4b If required, prise off the E-clips (arrowed) and remove the tensioner and guide blade

21.5 Lock the engine as described and unscrew the sprocket bolt (arrowed)

21.11a Fit the sprocket, making sure the pin (A) locates in the hole (B) . . .

5 Unscrew the bolt securing the auxiliary shaft sprocket and slide the sprocket and chain off the end of the shaft, noting how they fit (see illustration). It will be necessary to lock the engine to prevent the shaft turning whilst slackening the bolt. This can be done in a number of ways: BMW provide a tool, part no. 11 5 640 (see illustration 4.4b in Chapter 4), which locks the clutch housing, and can be fitted after removing the starter motor (see Chapter 8). Alternatively, if the engine is in the frame, select a gear and have an assistant press on the rear brake pedal or use the TDC locking pin to stop the engine from rotating

21.11b . . . then install the bolt . . .

(see *Haynes Hint* – Section 11). Note the rotary breather mounted on the front of the sprocket on all models after 5/94.
6 If required, unscrew the three bolts securing the upper sprocket to the crankshaft and remove the sprocket, noting how the alignment mark on the sprocket aligns with the mark on the crankshaft. There is no need to remove the sprocket unless it is being replaced or if the crankshaft is being replaced.

Inspection

7 Check that the tensioner piston moves freely in its bore and that the spring has not sagged.
8 Check the tensioner blade and the guide blade for wear or damage and replace them if necessary.
9 If the chain appears slack and the tensioner components are good, the chain may have stretched beyond the range of the tensioner. If this is the case, check the sprockets for worn or damaged teeth. BMW provide no specifications for the chain, but if it is obviously slack on the sprockets, it should be replaced.

Installation

10 If removed, install the upper sprocket onto the end of the crankshaft, aligning the marks, and tighten the bolts to the torque setting

specified at the beginning of the Chapter. If the crankshaft and/or auxiliary shaft have been removed or turned since the sprockets were removed, turn them so that when the sprockets are installed the marks will align as described in Step 2.
11 Fit the lower sprocket into the chain, then fit the chain around the upper sprocket. Mount the lower sprocket with the breather rotor onto the end of the auxiliary shaft, locating the pin on the back of the sprocket in the hole in the end of the crankshaft, and install the bolt (see illustration). Check that the marks on the sprockets align as described in Step 2. Using the method employed on removal to prevent the engine turning, tighten the lower sprocket bolt to the specified torque setting (see illustration).
12 If removed, slide the guide blade and tensioner blade onto their lugs, then fit the washers and secure them with the E-clips (see illustrations). Fit the guide blade with the marked side facing out. Check that the tensioner blade pivots freely on its lug.
13 Fit the spring and piston into the tensioner and compress the piston, then install the tensioner, making sure the piston sits squarely against the blade, and tighten the bolts to the specified torque setting (see illustrations).

21.11c . . . and tighten it to the specified torque setting

21.12a Fit the guide blade with the marked side facing out . . .

21.12b . . . and the tensioner blade – secure them with the E-clips

21.13a Fit the spring and piston into the tensioner body . . .

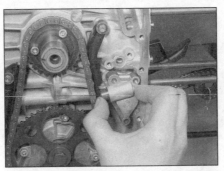

21.13b . . . then compress the piston and fit the tensioner

21.13c Tighten the bolts to the specified torque setting

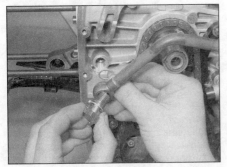

21.14a Fit new sealing washers with the oil pipe banjo bolt . . .

21.14b . . . then swing the pipe up and fit the bolts . . .

21.14c . . . and tighten them to the specified torque setting

14 Install the oil pipe banjo bolt using new sealing washers on each side of the union, then swing the oil pipe back into position and locate it with the bracket bolts (see

22.2a The pump is secured by six bolts (arrowed). Note the position of the longer bolts (A)

illustrations). Tighten the banjo bolt and bracket bolts to the specified torque settings (see illustration).
15 Install the alternator drive and engine inner front cover (see Section 20).

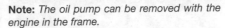

22 Oil pump –
removal, inspection and installation

Note: *The oil pump can be removed with the engine in the frame.*

Removal

1 Remove the auxiliary shaft drive chain, tensioner and the lower sprocket (see Section 21).
2 Unscrew the six bolts securing the oil pump, noting which length fits where (see illustration). Remove the pump cover, then

either draw the complete pump housing, with the oil pipe attached, off the end of the auxiliary shaft, or remove the outer pump rotors, housing and inner pump outer rotor separately (see illustration). Take care not to mix up the outer rotors as they are marginally different in thickness. Remove the Woodruff key from the end of the shaft if it is loose, noting how it locates in the cut-out in the inner rotor of the outer pump (see illustration 22.12a). The outer pump drives the cooling oil circuit, while the inner pump drives the lubrication circuit.
3 If required, withdraw the oil pipe from the housing then remove the O-ring and discard it as a new one must be used (see illustrations).

Inspection

4 If not already done, remove the inner and outer rotors from the outer pump, and the outer rotor from the inner pump, noting which

2

22.2b Remove the pump as an assembly or as separate parts

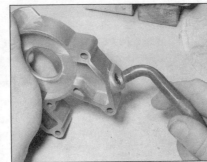

22.3a Withdraw the pipe . . .

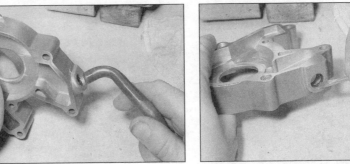

22.3b . . . and remove the O-ring

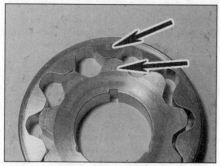

22.4 The square marked on each rotor (arrows) must face out

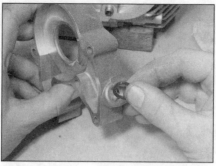

22.10 Fit a new O-ring into the pipe orifice

22.11a Fit the inner pump outer rotor . . .

way round they fit as they should be installed in their original positions. Each rotor has a square marked on its outer face to ensure correct installation (see illustration). Take care not to mix up the outer rotors as they are marginally different in thickness.

Caution: Do not remove the oil temperature regulator from the pipe union in the housing.

5 Clean all the components in solvent. If your marks come off the rotors, re-mark them.

6 Inspect the pump body and rotors for scoring and wear. If any damage, scoring or uneven or excessive wear is evident, replace the components as required.

7 Measure the depth of each pump housing and the corresponding thickness of the rotors for that pump. Calculate the difference to determine the amount of clearance (end-float) between the rotors and the housing. Compare the results to the specifications. If the clearance is greater than specified, use the measurements taken to determine whether the rotors or the housing, or both, are worn, and replace the components as required.

8 If the measuring equipment is not available to determine the clearance as described above, lay a straight-edge across the rotors and the pump housing and, using a feeler gauge, measure the clearance (the gap between the rotors and the straight-edge.

9 If the clearance measured is greater than the maximum listed, have the components measured as described above to determine which are worn. Alternatively, replace the whole pump.

Installation

10 If the pump is good, make sure all the components are clean, then lubricate them with new engine oil. Fit a new O-ring into the housing, making sure it seats properly in its groove (see illustration). Fit the oil pipe into the housing (see illustration 22.3a).

11 Fit the inner pump outer rotor into the housing, making sure the square mark faces out, then fit the housing over the end of the auxiliary shaft, making sure it locates squarely onto the dowels (see illustrations).

12 If removed, fit the Woodruff key into its slot in the end of the auxiliary shaft (see illustration). Install the outer pump rotors, making sure the square marks face out, and the square cut-out in the inner rotor locates over the Woodruff key (see illustrations). Install the pump cover, making sure the different length bolts are in their correct positions (see illustration 22.2a), and tighten the bolts to the torque setting specified at the beginning of the Chapter (see illustrations).

13 Install the auxiliary shaft drive chain, tensioner and the lower sprocket (see Section 21).

22.11b . . . followed by the housing, making sure it locates onto the dowels

22.12a Fit the Woodruff key . . .

22.12b . . . then locate the square cut-out in the inner rotor over the key

22.12c Fit the outer rotor . .

22.12d . . . and the cover . . .

22.12e . . . and tighten the bolts to the specified torque setting

22.14 Oil pressure relief valve location (arrowed)

Oil pressure relief valve

14 The pressure relief valve cap is located on the engine right-hand side, in the lower part of the crankcase (see illustration). Unscrew the cap and washer, then withdraw the spring and relief valve. On installation, tighten the cap to the specified torque setting.

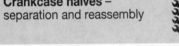

23 Crankcase halves –
separation and reassembly

Note 1: *To separate the crankcase halves, the engine must be removed from the frame.*
Note 2: *The crankshaft rear oil seal can be renewed without separating the crankcases (see Section 28).*

Separation

1 To access the crankshaft, bearings and auxiliary shaft, the crankcase must be split into two parts.
2 To enable the crankcases to be separated, the engine must be removed from the frame (see Section 5). Before the crankcases can be separated the following components must be removed:

a) Oil cooler(s) (Section 7).
b) Valve covers (Section 8).
c) Camchain tensioners (Section 9).
d) Rocker/camshaft assemblies (Sections 10 and 11).
e) Cylinder heads (Section 13).
f) Cylinders (Section 16).
g) Pistons (Section 17).
h) Alternator (see Chapter 8).
i) Alternator drive and engine inner front cover (Section 20).
j) Auxiliary shaft drive chain, tensioner and sprockets (Section 21).
k) Oil pump (Section 22).
l) Starter motor (see Chapter 8).
m) Gearbox (if not done when removing the engine) (Section 30).
n) Clutch and flywheel (Section 28).

3 Before separating the crankcase halves, measure the amount of crankshaft end-float using a dial gauge, mounted so that its pointer sits against the front end of the crankshaft (see illustration). Pull the crankshaft in and out of the casing and record the end-float. Compare the reading to the specifications If the amount of end-float exceeds the service limit, refer to the specifications and take the appropriate crankshaft measurements once it has been removed to determine what has worn. Also note the depth of the crankshaft rear oil seal in its housing as the new one must be fitted to the same depth.
4 Remove the two rubber battery/ABS holder mounts which secure the cable trough (see illustration). This can be done by threading two nuts onto each stud and tightening them together. Counter-holding the upper nut, unscrew the nut using a spanner on the lower nut. If the mount is very tight in the crankcase, this will not work and the mount will just twist due to the rubber. If this is the case, carefully knock the base of the mount round using a drift and a hammer.

23.3 Mount a dial gauge as shown to measure crankshaft end-float

5 Carefully lay the engine onto its left-hand side, making sure it is properly supported using blocks of wood. Unscrew the four 8 mm bolts followed by the two 10 mm bolts (see illustration). Unscrew the bolts evenly, a little at a time and in a criss-cross sequence until they are loose, then remove them. **Note:** *As each bolt is removed, store it in its relative position in a cardboard template of the right-hand crankcase half. This will ensure all bolts are installed in the correct location on reassembly.*
6 Turn the engine over so that it rests on its right-hand side, making sure it is properly supported using blocks of wood.
7 Unscrew the nineteen 6 mm bolts, followed by the two 10 mm bolts (see illustrations). Unscrew the bolts evenly, a little at a time and in a criss-cross sequence until they are loose, then remove them. **Note:** *As each bolt is removed, store it in its relative position in a cardboard template of the left-hand crankcase half. This will ensure all bolts are installed in the correct location on reassembly*
8 Carefully lift the left-hand crankcase half off the right-hand half; use a soft-faced hammer to tap around the joint to initially separate the halves if necessary (see illustration). **Note:** *If*

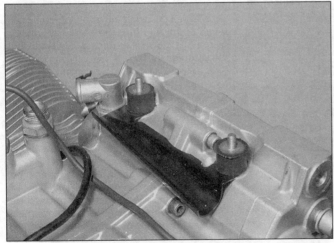

23.4 Remove the rubber mounts and cable trough as described

23.5 Right-hand crankcase half 8 mm bolts (A) and 10 mm bolts (B)

2

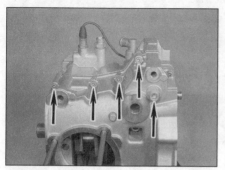

23.7a Left-hand upper crankcase half 6 mm bolts (arrowed)

23.7b Left-hand lower crankcase half 6 mm bolts (arrowed)

23.7c Left-hand rear crankcase half 6 mm bolts (arrowed)

23.7d Left-hand crankcase half 10 mm bolts (arrowed)

23.8 Carefully separate the crankcase halves

23.14 Crankcase half dowel locations (arrows)

the halves do not separate easily, make sure all fasteners have been removed. Do not try and separate the halves by levering against the crankcase mating surfaces as they are easily scored and will leak oil. Tap around the joint faces with a soft-faced mallet.

9 Remove the four locating dowels from the crankcase if they are loose (they could be in either crankcase half), noting their locations.

10 Refer to Sections 24 to 27 for the removal and installation of the components housed within the crankcases.

Reassembly

11 Remove all traces of sealant from the crankcase mating surfaces.

12 Ensure that all components are in place in the right- and left-hand crankcase halves (Step 10).

13 Generously lubricate the crankshaft, particularly around the bearings, with clean

engine oil, then use a rag soaked in high flash-point solvent to wipe over the mating surfaces of both crankcase halves to remove all traces of oil.

14 If removed, install the four locating dowels in the upper crankcase half (see illustration).

15 Apply a small amount of 3-bond 1209 or equivalent sealant to the mating surface of one crankcase half (see illustration) and to the rear end of the upper tensioner and guide blades' pivot pin to prevent oil leaking into the clutch housing (see illustration 27.8b).

Caution: Do not apply an excessive amount of sealant as it will ooze out when the case halves are assembled and may obstruct oil passages. Do not apply the sealant on or too close to any of the bearing inserts or surfaces.

16 Check again that all components are in position, particularly that the bearing shells are still correctly located in the left-hand

crankcase half. Make very sure that the camchains are fitted correctly onto the auxiliary shaft sprockets without any kinked links, and that they do not slip off the sprocket creating a kinked link as the crankcase halves are joined. If this happens, it is not possible to manually feed any kink around the sprocket by turning the shaft due to the limited clearance between the chain and the tensioner and guide blades. The only remedy is to separate the crankcase halves again. It is well worth having an assistant to feed both the con-rod and the camchain into the left-hand crankcase half as it is fitted, and especially to make sure the chain does not become kinked.

17 Carefully install the left-hand crankcase half down onto the right-hand half, making sure the dowels all locate correctly (see illustration 23.8).

18 Check that the left-hand crankcase half is correctly seated. **Note:** *The crankcase halves should fit together without being forced. If the casings are not correctly seated, remove the left-hand crankcase half and investigate the problem. Do not attempt to pull them together using the crankcase bolts as the casing will crack and be ruined.*

19 Clean the threads of all the crankcase bolts and apply clean engine oil to the threads and under the heads of the 10 mm and 8 mm bolts (see illustration). Insert them in their original locations, not forgetting the washers (using new ones if necessary) (see illustrations 23.5 and 23.7a, b, c and d). Secure all bolts finger-tight at first, then

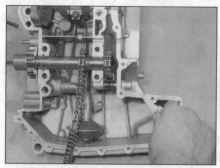

23.15 Apply the sealant to the mating surfaces

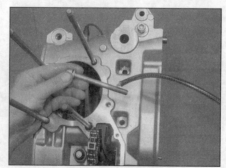

23.19 Oil the threads and heads of the 10 mm and 8 mm bolts

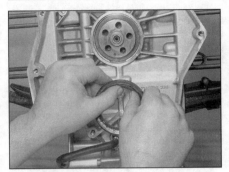

23.21a Pre-form the oil seal as described . . .

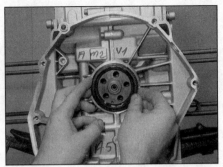

23.21b . . . then press it into place to the precise depth

23.22a Install the battery/ABS holder bracket . . .

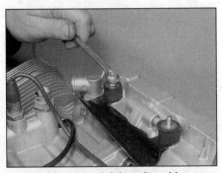

23.22b . . . and tighten its rubber mountings

tighten the 10 mm bolts first, followed by the 8 mm bolts, and finally the 6 mm bolts. Tighten each set of bolts evenly a little at a time in a criss-cross sequence to the torque settings specified at the beginning of the Chapter.

20 With all crankcase fasteners tightened, check that the crankshaft and auxiliary shaft rotate smoothly and easily. If there are any signs of undue stiffness, tight or rough spots, or of any other problem, the fault must be rectified before proceeding further.

21 The new crankshaft rear oil seal needs to have its inner lip pre-formed before it is installed. Do this by pressing the lip in with your fingers, making sure it is even all round, and pressed in far enough to allow its shaft through it, but not so far that it becomes wider than the shaft itself (see illustration). Install the new oil seal, using a suitable sized socket as a drift which bears on the seal's outer edge, or the BMW tool no. 11 5 660. Ensure that the seal is set to the depth noted on removal (see *Tools and Workshop Tips*) (see illustration).

22 Install the two rubber battery/ABS holder mounts with the cable trough and tighten them down using the nuts threaded onto them for removal (see illustrations).

23 Install all other removed assemblies in the reverse of the sequence given in Step 2, noting that the engine inner front cover, alternator drive and alternator should be left off until after the cylinder heads and camshafts are installed to enable the timing to be set up correctly.

24 Crankcase halves – inspection and servicing

1 After the crankcases have been separated, remove the crankshaft (with the connecting rods if not already removed), auxiliary shaft and camchains, camchain tensioner blades and camchain guide blades, referring to the relevant Sections of this Chapter. Remove the oil pressure and temperature switches as described in Chapter 8. Unscrew the bolt securing the oil pickup strainer and remove the strainer and its O-ring (see illustrations). Discard the O-ring as a new one should be used. If required, for instance if it is difficult to see the oil level through it, the oil level window in the left-hand crankcase half can be drifted or levered out (by piercing the plastic window with a screwdriver).

2 The crankcases should be cleaned thoroughly with new solvent and dried with compressed air. All oil passages should be blown out with compressed air.

3 All traces of old gasket sealant should be removed from the mating surfaces. Minor damage to the surfaces can be cleaned up with a fine sharpening stone or grindstone. *Caution: Be very careful not to nick or gouge the crankcase mating surfaces or oil leaks will result. Check both crankcase halves very carefully for cracks and other damage.*

4 Small cracks or holes in aluminium castings may be repaired with an epoxy resin adhesive as a temporary measure. Permanent repairs can only be effected by argon-arc welding, and only a specialist in this process is in a position to advise on the economy or practical aspect of such a repair. If any damage is found that can't be repaired, replace the crankcase halves as a set.

5 Damaged threads can be economically reclaimed by using a diamond section wire insert, of the Heli-Coil type, which is easily fitted after drilling and re-tapping the affected thread.

6 Sheared studs or screws can usually be removed with screw extractors, which consist of a tapered, left-hand thread screw of very hard steel. These are inserted into a pre-drilled hole in the stud, and usually succeed in dislodging the most stubborn stud or screw.

24.1a Unscrew the bolt (arrowed) . . .

24.1b and remove the oil pickup strainer . . .

24.1c . . . and its O-ring

24.7 Tighten the oil pickup bolt to the specified torque setting

Check the length of the four cylinder studs from the crankcase surface. The shorter stud (lower rear stud) should project 212 mm and the other three 232.5 mm. If necessary, screw the studs further into the crankcase using two nuts locked together on their threaded ends or a knurled wheel type tool. **Note:** *Loose cylinder studs can cause oil leakage at the cylinder head gasket.*

 Refer to Tools and Workshop Tips for details of installing a thread insert and using screw extractors.

7 Install the oil pickup strainer using a new O-ring, making sure it seats properly in the groove, and tighten the bolt to the torque setting specified at the beginning of the Chapter **(see illustration)**. If the oil level window was removed, install a new one (do not re-use the old one), and coat the outer sealing rim with clean engine oil before drifting it squarely in. Install all other components and assemblies (see Step 1), referring to the relevant Sections of this Chapter and to Chapter 8, before reassembling the crankcase halves.

25 Main and connecting rod bearings – general information

1 Even though main and connecting rod bearings are generally replaced with new ones during the engine overhaul, the old bearings

26.2 Lift the crankshaft out of the casing

should be retained for close examination as they may reveal valuable information about the condition of the engine.
2 Bearing failure occurs mainly because of lack of lubrication, the presence of dirt or other foreign particles, overloading the engine and/or corrosion. Regardless of the cause of bearing failure, it must be corrected before the engine is reassembled to prevent it from happening again.
3 When examining the connecting rod bearings, remove them from the connecting rods and caps and lay them out on a clean surface in the same general position as their location on the crankshaft journals. This will enable you to match any noted bearing problems with the corresponding crankshaft journal.
4 Dirt and other foreign particles get into the engine in a variety of ways. It may be left in the engine during assembly or it may pass through filters or breathers. It may get into the oil and from there into the bearings. Metal chips from machining operations and normal engine wear are often present. Abrasives are sometimes left in engine components after reconditioning operations, especially when parts are not thoroughly cleaned using the proper cleaning methods. Whatever the source, these foreign objects often end up imbedded in the soft bearing material and are easily recognised. Large particles will not imbed in the bearing and will score or gouge the bearing and journal. The best prevention for this cause of bearing failure is to clean all parts thoroughly and keep everything spotlessly clean during engine reassembly. Frequent and regular oil and filter changes are also recommended.
5 Lack of lubrication or lubrication breakdown has a number of interrelated causes. Excessive heat (which thins the oil), overloading (which squeezes the oil from the bearing face) and oil leakage or throw off (from excessive bearing clearances, a worn oil pump or high engine speeds) all contribute to lubrication breakdown. Blocked oil passages will also starve a bearing and destroy it. When lack of lubrication is the cause of bearing failure, the bearing material is wiped or extruded from the steel backing of the bearing. Temperatures may increase to the point where the steel backing and the journal turn blue from overheating.

26.3a Remove the front . . .

 Refer to Tools and Workshop Tips in the Reference section for bearing fault finding.

6 Riding habits can have a definite effect on bearing life. Full throttle low speed operation, or labouring the engine, puts very high loads on bearings, which tend to squeeze out the oil film. These loads cause the bearings to flex, which produces fine cracks in the bearing face (fatigue failure). Eventually the bearing material will loosen in pieces and tear away from the steel backing. Short trip riding leads to corrosion of bearings, as insufficient engine heat is produced to drive off the condensed water and corrosive gases produced. These products collect in the engine oil, forming acid and sludge. As the oil is carried to the engine bearings, the acid attacks and corrodes the bearing material.
7 Incorrect bearing installation during engine assembly will lead to bearing failure as well. Tight fitting bearings which leave insufficient bearing oil clearances result in oil starvation. Dirt or foreign particles trapped behind a bearing insert result in high spots on the bearing which lead to failure.
8 To avoid bearing problems, clean all parts thoroughly before reassembly, double check all bearing clearance measurements and lubricate the new bearings with clean engine oil during installation.

26 Crankshaft and main bearings – removal, inspection and installation

Note: *To remove the crankshaft the engine must be removed from the frame and the crankcase halves separated. The crankshaft rear oil seal can be replaced without disturbing the crankshaft. Remove the clutch for access to it (see Section 28).*

Removal

1 Remove the engine from the frame (see Section 5) and separate the crankcase halves (see Section 23).
2 Lift the crankshaft out of the right-hand crankcase half, taking care not to dislodge the main bearing shells **(see illustration)**. If not already done during crankcase separation, measure the depth of the rear oil seal in the casing and record this for installation. Remove the seal from the end of the shaft and discard it.
3 Remove the main bearing (front) shells from each crankcase half by pushing their centres to the side, then lifting them out **(see illustration)**. Remove the guide bearing (rear) shells by pressing down on the side thrust sections on one end of the shell so that it slips round in the housing, then lift it out by the protruding end **(see illustration)**. Keep the shells in order to allow accurate oil clearance

26.3b . . . and rear shells from each crankcase half

measurement and wear inspection, and to enable them to be returned to their original locations if renewal is not necessary.

4 If not already done, and if required, separate the connecting rods from the crankshaft (see Section 19).

Inspection

5 Clean the crankshaft with solvent, using a rifle-cleaning brush to scrub out the oil passages. If available, blow the crank dry with compressed air, and also blow through the oil passages. Check the drive gear for the auxiliary shaft chain for wear or damage. If any of the gear teeth are excessively worn, chipped or broken, the sprocket must be unbolted and renewed – see Section 21 for details.

6 Refer to Section 25 and examine the main bearing shells. If they are scored, badly scuffed or appear to have been seized, new bearings must be installed. Always replace the main bearings as a set of four. If they are badly damaged, check the corresponding crankshaft journals. Evidence of extreme heat, such as discoloration, indicates that lubrication failure has occurred. Be sure to thoroughly check the oil pump as well as all oil holes and passages before reassembling the engine.

7 The crankshaft journals should be given a close visual examination, paying particular attention where damaged bearings have been discovered. If the journals are scored or pitted in any way, the crankshaft can be reground one stage only and fitted with oversize (+0.25 mm) bearing shells available from BMW. If it has already been ground once, it will have to be rehardened and finished off to its original specification. Due to the precision required in these processes, and the fact that it is often not achieved, crankshaft replacement is often regarded as the best long-term course of action.

Oil clearance check

8 Whether new bearing shells are being fitted or the original ones are being re-used, the main bearing oil clearance should be checked before the engine is reassembled. Main bearing oil clearance can be measured with either a product known as Plastigauge (see Steps 9 to 16), or by using a micrometer and telescoping bore gauge (see Step 19). Due to the cost of accurate measuring equipment, Plastigauge is more readily available to many people. It is, however, not as accurate or reliable and therefore it is worth having any doubtful results verified by a BMW dealer or specialist with the required equipment.

9 To measure the oil clearance using Plastigauge, first clean the backs of the bearing shells and the bearing housings in both crankcase halves.

10 Press the bearing shells into their cut-outs, ensuring that the tab on each shell engages in the notch in the crankcase **(see illustration)**. Make sure the bearings are fitted in the correct locations and take care not to touch any shell's bearing surface with your fingers.

11 Ensure the shells and crankshaft are clean and dry. Lay the crankshaft in position in the right-hand crankcase.

12 Cut two lengths of the appropriate size Plastigauge (they should be slightly shorter than the width of the crankshaft journals). Place a strand of Plastigauge on each (cleaned) journal **(see illustration)**. Make sure the crankshaft is not rotated.

13 Carefully install the left-hand crankcase half on to the right half. Check that the left-hand crankcase half is correctly seated. **Note:** *Do not tighten the crankcase bolts if the casing is not correctly seated.* Install the crankcase 10 mm and 8 mm bolts in their original locations and tighten them evenly a little at a time in a criss-cross sequence to the torque setting specified at the beginning of the Chapter **(see illustrations 23.5 and 23.7d)**. Make sure that the crankshaft is not rotated as the bolts are tightened.

14 Slacken each bolt evenly a little at a time in a criss-cross sequence until they are all finger-tight, then remove the bolts. Carefully separate the crankcase halves, making sure the Plastigauge is not disturbed.

15 Compare the width of the crushed Plastigauge on each crankshaft journal to the scale printed on the Plastigauge envelope to obtain the main bearing oil clearance **(see illustration)**. Compare the reading to the specifications at the beginning of the Chapter.

16 On completion carefully scrape away all traces of the Plastigauge material from the crankshaft journal and bearing shells; use a fingernail or other object which is unlikely to score them.

17 If the oil clearance falls into the specified range, no bearing shell replacement is required (provided they are in good condition). If the clearance is beyond the service limit, install new shells and check the oil clearance once again (the new shells may bring bearing clearance within the specified range). Check the colour coding on the crankshaft to determine which replacement shells are required; the main bearing colour codes are green or yellow and

2

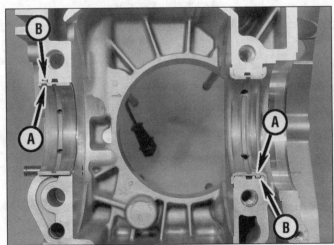

26.10 Make sure the tab (A) on each shell locates in the notch (B) in each housing

26.12 Lay a strip of Plastigauge (arrowed) on each journal parallel to the crankshaft centreline

26.15 Measure the width of the crushed Plastigauge (be sure to use the correct scale – metric and imperial are included)

the guide bearing colour codes are green, yellow or white **(see illustrations)**. Always replace all four shells at the same time.

18 If the clearance is still greater than the service limit listed in this Chapter's Specifications (even with replacement shells), the crankshaft journal is worn. If possible, confirm this by measuring the diameter of the journal using a micrometer and comparing it to the specifications **(see illustration 26.19)**. If it

is worn, it can be reground one stage only and fitted with oversize (+0.25 mm) bearing shells available from BMW. If it has already been ground once, it will have to be rehardened and finished off to its original specification. Due to the precision required in these processes, and the fact that it is often not achieved, crankshaft replacement is often regarded as the best long-term course of action.

19 To measure the oil clearance using direct measurement, two readings are taken. Use a micrometer to measure the diameter of the crankshaft journal, taking measurements in two different planes **(see illustration)**. Using a telescoping bore gauge and micrometer, measure the internal bore of the bearing housing, with the bearing shells in place and the crankcase 10 mm and 8 mm bolts tightened to the specified torque (see Section 23); take the reading in line with the cylinder bore openings, where the greatest loading is taken. Calculate the difference to determine the clearance and compare the result to the clearance specified. If it is greater than specified, the measurements taken will tell you whether the shells or the crankpin are worn, and the appropriate action can be taken (see Steps 17 and 18).

20 If crankshaft end-float, measured in Section 23, was excessive, measure the guide bearing shell diameter and crankshaft guide bearing journal width to determine which component is worn. Subtract one reading from the other to arrive at the end-float.

Installation

21 Clean the backs of the bearing shells and the bearing cut-outs in both crankcase halves. If new shells are being fitted, ensure that all traces of the protective grease are cleaned off using paraffin (kerosene). Wipe the shells and crankcase halves dry with a lint-free cloth. Make sure all the oil passages and holes are clear, and blow them through with compressed air if it is available. Press the bearing shells into their locations **(see illustrations 26.3a and b)**. Make sure the tab on each shell engages in the notch in the casing **(see illustration 26.10)**. Make sure the bearings are fitted in the correct locations and take care not to touch any shell's bearing surface with your fingers. Lubricate each shell, including the side thrust sections of the rear guide shells, with clean engine oil **(see illustration)**.

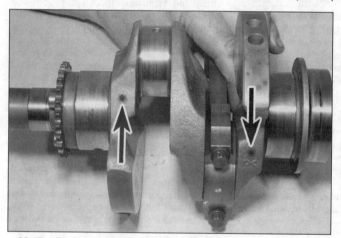

26.17a Each crankshaft main bearing journal is marked with a colour coding (arrows) . . .

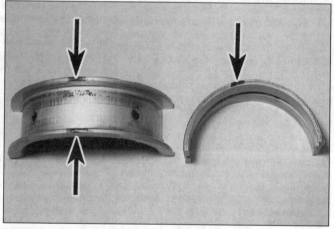

26.17b . . . which must correspond with the colour coding marked on each shell (arrows)

26.19 Measure the main bearing journal diameter using a micrometer

26.21 Lubricate the shells generously with clean engine oil

27.2a Remove the upper tensioner and guide blades with the pivot pin

22 Lower the crankshaft into position in the right-hand crankcase half, making sure all bearings remain in place **(see illustration 26.2)**. The rear oil seal is fitted after the crankcase halves have been joined.
23 Fit the connecting rods onto the crankshaft (see Section 19).
24 Reassemble the crankcase halves (see Section 24).

27 Auxiliary shaft, camchains and tensioner/guide blades – removal, inspection and installation

Note: *To remove the auxiliary shaft the engine must be removed from the frame.*

Removal

1 Remove the engine (see Section 5) and

27.3 Lift the auxiliary shaft and camchains out of the crankcase

27.7b Make sure the camchains are engaged properly on the sprockets

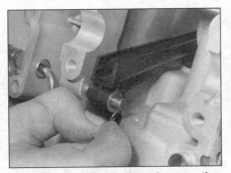

27.2b Prise off the E-clip and remove the lower guide blade

separate the crankcase halves (Section 23).
2 Remove the two upper tensioner and guide blades with their pivot pin, noting which fits on which side **(see illustration)**. Remove the E-clip securing the lower guide blade in the left-hand crankcase half and slide the blade off the pivot pin **(see illustration)**. Unscrew the pivot bolt securing the lower tensioner blade – the bolt is accessed from the rear of the right-hand crankcase half **(see illustration)**. Do not get the various blades mixed up as they must be installed in their original positions.
3 Remove the auxiliary shaft from the engine and slip the camchains off their sprockets **(see illustration)**.

Inspection

4 Except in cases of oil starvation, the cam chains wear very little. If the chains have

27.7a Lubricate the auxiliary shaft bearing surfaces with clean engine oil

27.8a Tighten the pivot bolt to the specified torque setting

27.2c Unscrew the bolt and remove the lower tensioner blade

stretched excessively, which makes it difficult to maintain proper tension, they must be replaced.
5 Check the sprockets for cracks and other damage, replacing the shaft if necessary. Note that if a new shaft is installed, new cam chains must also be installed. If the sprockets are worn, the cam chains are also worn, and also the sprockets on the camshafts. If wear this severe is apparent, the entire engine should be inspected for wear.
6 Check the tensioner and guide blades and their pivots for signs of wear and damage and replace them if necessary.

Installation

7 Apply clean engine oil to the auxiliary shaft bearing surfaces in the crankcase **(see illustration)**. Slip the cam chains around their sprockets, making sure the chains are properly and fully engaged with the sprocket teeth **(see illustration)**. Lay the auxiliary shaft in the right-hand crankcase half **(see illustration 27.3)**.
8 Install the right-hand cylinder tensioner blade and its pivot bolt **(see illustration 27.2c)**; tighten the pivot bolt to the torque setting specified at the beginning of the Chapter **(see illustration)**. Install the left-hand cylinder tensioner blade and the right-hand cylinder guide blade with the upper pivot pin **(see illustration)**. Moving to the left-hand crankcase half, slide the left-hand cylinder guide blade onto its pivot pin and secure it with the E-clip **(see illustration 27.2b)**.
9 Join the crankcase halves (see Section 23).

2

27.8b Install the left cylinder tensioner blade and right cylinder guide blade with the upper pin

28.2 Make your own alignment marks between all the clutch components

28.3 Insert a bolt through the hole in the flywheel and into the hole in the crankcase to lock the clutch

28 Clutch –
removal, inspection and installation

Note: *The clutch can be removed with the engine in the frame. If the engine has been removed, ignore the steps which don't apply.*

Removal

1 Remove the gearbox (see Section 30).
2 Before removing the clutch components, check for the manufacturer's painted balance marks on the cover plate, pressure plate and flywheel. The marks should be spaced at 120° intervals. In practice the marks were difficult to identify, and it is suggested that you make your own mark across each component to ensure that they are installed in the same relative position to each other **(see illustration)**. This is very important for the smoothness of the engine, as any imbalance

between these heavy components could cause severe vibration.
3 It will be necessary to lock the flywheel to prevent the clutch from turning while slackening the bolts. This can be done using the BMW tool no. 11 5 640 **(see illustration 4.4b in Chapter 4)**. Alternatively, turn the engine until the hole in the flywheel aligns with the corresponding hole in the crankcase, and locate a suitable rod or bolt through them to lock the engine **(see illustration)**.
4 Slacken the six cover plate bolts evenly and a little at a time in a criss-cross sequence until spring pressure has been released, then remove the bolts and their washers **(see illustration)**. Discard the bolts as new ones must be used. The cover plate, friction plate and pressure plate will probably come away together joined by the locating pins between the cover and pressure plate **(see illustration)**. These pins can be a very tight fit, making the components difficult to separate.

Carefully lever them apart using a large screwdriver and separate the components, noting which way round they fit. Withdraw the diaphragm spring.
5 If required, unscrew the five bolts securing the flywheel to the end of the crankshaft and remove the flywheel and its bolt washer, noting the alignment of the pin on the crankshaft with the hole in the flywheel **(see illustration)**. Discard the flywheel bolts as new ones must be used.

Inspection

Note: *Modified clutch components were fitted to early RS models, from frame no. 0 297 746, due to high wear of the friction plate and the need for adjustment. Ensure that the later type clutch components are installed if problems are experienced.*
6 After an extended period of service the friction plate will wear and promote clutch slip. Measure the thickness of the plate using a

28.4a Unscrew the six bolts (arrowed) . . .

28.4b . . . and remove the clutch cover, friction plate and pressure plate as an assembly

28.5 Unscrew the five bolts (arrowed) and remove the flywheel

28.6 Measuring friction plate thickness

28.11 Press in on one side of the crankshaft seal and lever it out on the other side

vernier caliper (see illustration). If the thickness is less than the service limit given in the Specifications at the beginning of the Chapter, the plate must be replaced. Also, if the plate smells burnt, is contaminated with oil or grease or is glazed, it must be replaced. Note: If there is oil contamination on the clutch plate or in the clutch housing, check the crankshaft rear oil seal and gearbox oil seal for leakage.

7 Check the friction plate splines and the corresponding splines on the gearbox input shaft for signs of wear and damage and replace them if required. Fit the friction plate on the input shaft splines and check that it is able to move freely on the splines, but without undue freeplay – note that wear between the two components may result in a rattle when the machine is in neutral.

8 Check the pressure plate surface for distortion using a straight-edge and feeler gauges.

9 If the clutch has been slipping and the friction plate thickness is satisfactory, then either the spring has lost its pressure or the pressure plate has become excessively worn. Replace the components as required.

10 Withdraw the clutch pushrod from the gearbox, noting which way round it fits (see illustration 30.24). Check that it is straight by rolling it on a flat surface such as a piece of

glass or setting it up in vee-blocks and measuring runout with a dial gauge. If bent, the pushrod must be replaced. Note that the pushrod can be damaged if care wasn't taken to withdraw the gearbox squarely from the engine.

11 Check around the clutch housing for signs of any oil leakage from either the crankcase or the gearbox, which indicates either a blown crankshaft rear oil seal or gearbox input shaft seal. The crankshaft oil seal can be replaced without removing the crankshaft. Make a note of its depth in the housing before removing it. Press down on one side of it using a large flat-bladed screwdriver so that the opposite side is raised up, then lever the seal out (see illustration). Take great care not to slip and damage the sealing surfaces on either the crankshaft or the casing. The crankshaft rear oil seal needs to have its inner lip pre-formed before it is installed. Do this by pressing the lip in with your fingers, making sure it is even all round, and pressed in far enough to allow the shaft through it, but not so far that it becomes wider than the shaft itself (see illustration 23.21a). Install the new oil seal, using a suitable sized socket or drift to knock it in, and making sure it is set to the depth noted on removal (see illustration 23.21b). Alternatively, the crankcase halves can be separated to access

the crankshaft rear oil seal (see Section 23). Refer to Section 33 for replacement of the gearbox input shaft seal. Thoroughly degrease all components before assembling the clutch.

> **HAYNES HINT** If the crankshaft oil seal is difficult to remove, drill a very small hole into it and thread in a suitable self-tapping screw. Use the screw to pull the seal out. If required, drill two holes and use two screws.

Installation

12 Make sure all the components are clean, dry and free of all oil and grease. Always use new bolts when installing the flywheel and clutch cover plate as they are of a type which deform on tightening and can therefore only be used once. If the engine has been removed from the frame, note clutch installation is much easier if the engine is turned onto its front end and supported by wooden blocks. If the engine is in the frame, it is very useful to have an assistant at hand.

13 Install the flywheel onto the end of the crankshaft, making sure the small pin on the flywheel locates in the hole in the crankshaft (see illustration). Lightly oil the threads of the

 2

28.13a Locate the pin (A) into the hole (B)

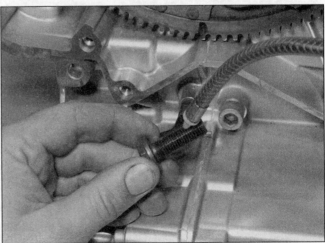

28.13b Oil the flywheel bolts . . .

28.13c . . . then install them with the washer . . .

28.13d . . . and tighten them evenly to the specified torque . . .

28.13e . . . and then to the specified angle using a degree disc

28.14a Apply grease to the contact points on the spring . . .

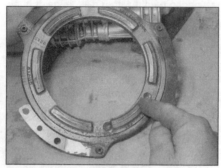

28.14b . . . and the pressure plate as described

28.15a Install the spring . . .

28.15b . . . the pressure plate . . .

28.15c . . . the friction plate . . .

28.15d . . . and the cover, aligning all the marks, bolt holes and pins

28.15e Assemble the substitute nuts, bolts and washers as shown. Counter-hold the bolts and tighten the nuts to press the pins (arrowed) into the flywheel

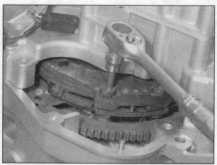

28.16a Tighten the bolts via the starter motor aperture . . .

28.16b . . . then remove the gearbox and tighten them to the specified torque setting

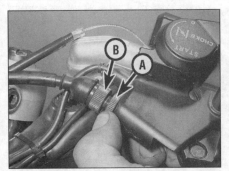

29.1a Slacken the locknut (A) and screw the adjuster (B) fully in

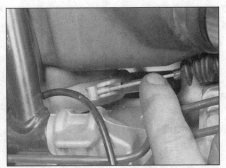

29.1b Push the operating lever forward and slip the cable out

new flywheel bolts, then fit the large washer and the bolts (see illustrations). Using the method employed on removal to prevent the flywheel turning, tighten the bolts evenly and in a criss-cross pattern to the torque setting specified at the beginning of the Chapter, then fit a degree disc to the torque wrench and tighten them further to the angle specified (see illustrations).

14 Apply a smear of Optimoly MP3 or equivalent paste to all points of contact between the diaphragm spring and the flywheel, and between the spring and the pressure plate (see illustrations). Use only a small amount of the paste, and make sure none gets onto the friction material.

15 Install the diaphragm spring, the pressure plate, the friction plate and the cover plate in that order, making sure they are all the correct way round (see illustrations). If the engine is in the frame, it is easier to assemble the components on the bench and then install them as an assembly. Make sure that the manufacturer's painted marks on the flywheel, pressure plate and cover plate are spaced 120° apart, or that your own marks align correctly (see Step 2), and make sure the locating pins in the cover locate through the holes in the pressure plate. The locating pins are a very tight fit in the flywheel, probably too tight to be fitted by hand, meaning they will have to be tightened down onto the flywheel. As the bolts are not long enough to reach between the cover and into the flywheel until the pins are located, longer bolts will have to be

substituted to start the tightening procedure. Assemble six suitable bolts, washers and nuts. Thread the nuts onto the bolts, then fit the washers and thread the bolts through the cover and pressure plate and into the flywheel (see illustration). Make sure the bolts are not threaded in too far as they will contact the solid inner part of the flywheel, causing possible damage. Once the bolts are fully into, but not beyond, the threaded holes in the flywheel, counter-hold them and tighten the nuts down onto the cover. Working in a criss-cross pattern, tighten the nuts evenly and a little at a time making sure the pins enter the holes in the flywheel simultaneously. Continue to tighten the nuts until the clutch components are secure on the flywheel, but do not overtighten them as the friction plate must be free enough to move independently of the cover and pressure plate to allow it to be centred. Remove each of the long bolts and nuts in turn, replacing them with the proper *new* clutch bolts.

16 At this point the friction plate must be centred. This is best achieved using BMW tool no. 21 3 680. Alternatively, it is possible to centralise the plate by temporarily installing the gearbox (having first removed the flywheel locking bolt or rod), so that the input shaft boss splines align with those in the clutch friction plate, then aligning the gearbox with the engine and installing some of the bolts. To avoid having to remove the gearbox before tightening the clutch bolts, thereby risking decentralising the friction plate, tighten the

bolts via the starter motor aperture (see illustration). Tighten them evenly and a little at a time in a criss-cross sequence, accessing the bolts in turn by moving the flywheel round. When the bolts are tight enough for the friction plate to be stable, remove the gearbox, then install the locking bolt or rod (see illustration 28.3) and tighten the bolts to the specified torque setting (see illustration).

17 Install the gearbox (see Section 30).

29 Clutch cable and operating mechanism – removal, inspection and installation

Note: *To remove the clutch pushrod, the swingarm must be removed. The operating mechanism components can be accessed with a little dexterity; on RS (with full fairing) and all RT models, remove the fairing right-hand side panel for access.*

Removal

1 Slacken the clutch cable adjuster locknut at the handlebar, then screw the adjuster fully into the bracket to provide the maximum freeplay (see illustration). Detach the lower end of the inner cable from the operating lever on the gearbox, using a screwdriver to move the lever forward if there is not enough freeplay available (see illustration). If the cable is being removed, free the lower end of the outer cable from the lug on the gearbox and unscrew the adjuster from the bracket, then slip the nipple on the upper end of the inner cable out of the lever and off the inner cable (see illustrations). Withdraw the cable from the bike.

> **HAYNES HINT** *Before removing the cable from the bike, tape the lower end of the new cable to the upper end of the old cable. Slowly pull the lower end of the old cable out, guiding the new cable down into position. Using this method will ensure the cable is routed correctly.*

2 Unscrew the pivot bolt securing the release lever and remove the lever (see illustration).

29.1c Withdraw the cable from its lug . . .

29.1d . . . and fully unscrew the adjuster . . .

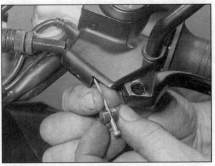

29.1e . . . then detach the inner cable from the lever and remove the nipple

2

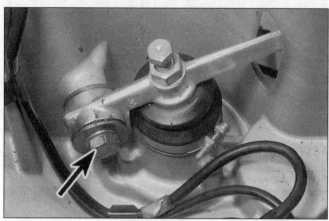

29.2 Unscrew the pivot bolt (arrowed) and remove the operating lever

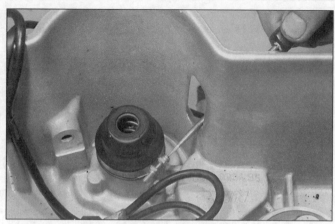

29.3a Slacken the clamp screw . . .

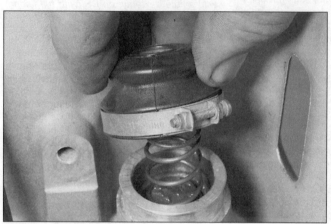

29.3b . . . then remove the boot and spring . . .

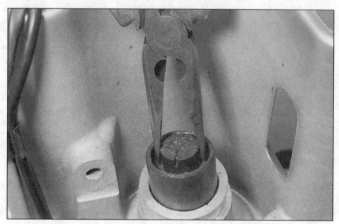

29.3c . . . and withdraw the piston

3 Slacken the clamp securing the rubber boot, then remove the boot, the spring and the operating piston from the housing (see illustrations). The spring will probably come away with the boot.

Inspection

4 Check all components for signs of wear and damage, and replace them if necessary.
5 If removed, check that the pushrod is straight by rolling it on a flat surface such as a piece of glass or setting it up in vee-blocks

and measuring runout with a dial gauge. If bent, the pushrod must be replaced.
6 Check the needle roller bearing in the operating lever and replace it if there is excessive play in the lever, or if the lever action is not smooth and free.
7 Check along the entire cable for kinks, splits or damage to the outer cable and for frays or kinks in the visible portions of the inner cable. Also check that the inner cable moves smoothly and freely and lubricate it if required (see Chapter 1).

Installation

8 Make sure all components are clean, then apply molybdenum grease to the contact points between the pushrod, the piston, the spring, the operating lever and its pivot bolt (see illustration).
9 If removed, install the pushrod with its longer pointed end towards the engine.
10 Install the piston with its open end facing in followed by the spring (see illustrations). Fit the rubber boot and secure it with the clamp (see illustrations 29.3b and 29.3a).

29.8 Grease the components as specified

29.10a Install the piston . . .

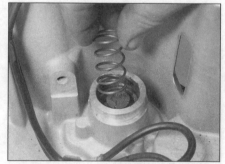

29.10b . . . and the spring

29.11a Install the operating lever with its pivot bolt and washers . . .

29.11b . . . and tighten it to the specified torque setting

11 Install the operating lever and tighten the pivot bolt to the torque setting specified at the beginning of the Chapter (see illustrations). Check that the lever moves freely and smoothly.

12 If removed, install the cable, making sure it is correctly routed. Apply grease to the upper nipple and fit it onto the end of the inner cable (see illustration 29.1e). Fit the cable and nipple into the clutch lever (see illustration). Thread the adjuster fully into the clutch lever bracket (see illustration 29.1d). Grease the nipple on the lower end of the cable, then fit the outer cable into its lug and fit the inner cable nipple into the operating lever, using a screwdriver to angle it forward if necessary (see illustration 29.1c and b).

13 Refer to Chapter 1 and adjust the amount of freeplay.

30 Gearbox –
removed and installation

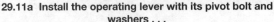

Note 1: *If the gearbox is being separated from the engine in order to carry out work on either the engine or clutch, the rear wheel and swingarm (Paralever) can be left attached to the gearbox. However, due to the size and*

29.12 Fit the cable into the lever and screw the adjuster fully in

weight of the assembly, this is not recommended unless at least one assistant is available.
Note 2: *Refer to Section 5 of this Chapter where directed for the relevant photographs.*

Removal

1 Position the bike on its centre stand and place a block of wood or jack under the front of the engine to provide extra support. Work can be made easier by raising the machine to a suitable working height on an hydraulic ramp or a suitable platform. Make sure the motorcycle is secure and will not topple over (see *Tools and Workshop Tips* in the Reference section).

2 If the gearbox is dirty, particularly around its mountings, wash it thoroughly before starting any major dismantling work. This will make work much easier and rule out the possibility of caked on lumps of dirt falling into some vital component.

3 Drain the gearbox oil (see Chapter 1).

4 Remove the seat and remove all the fairing side panels and body panels, according to your model (see Chapter 7).

5 Remove the fuel tank (see Chapter 3).

6 Remove the air filter (see Chapter 1). Unscrew the bolt securing the front of the air intake duct and lift the duct from the filter housing (see illustration 5.6).

7 Remove the battery (see Chapter 8). Remove the two nuts securing the rear of the holder to the gearbox (see illustration 5.17e).

8 Remove the exhaust silencer (see Chapter 3).

9 Slacken the clamp screws securing the air ducts to the airbox and the throttle bodies (see illustration 5.12a). Carefully pull the air ducts off the throttle bodies and slide them back into the airbox (see illustration 5.12b). Slacken the clamp screws securing the throttle bodies to the intake manifolds on the cylinder heads and carefully pull the throttles out of the manifolds (see illustration 5.12c).

Secure the throttles clear of the engine, making sure there is no strain on the cables. There is no need to disconnect either the cables, the fuel hoses or the injector wiring connectors, though this can be done and the throttle bodies removed from the bike if required (see Chapter 3).

10 On ABS models, remove the screw securing the rear brake pipe bracket to the swingarm (see illustration 5.17b).

11 Cut the cable ties securing all the wiring to the rear sub-frame tubes. Remove the tail light unit and the rear turn signal units (see Chapter 8). Feed the wiring through the rear mudguard, noting its routing, and coil it on top of the fuse/relay box. All wiring must be clear of the rear sub-frame, as it is to be removed.

12 Unclip the fuse/relay box lid and remove it (see illustration 5.19a). Remove the four screws securing the fuse/relay holder section of the box to the wiring tray section, then remove the four screws securing the wiring tray to the rear mudguard and the frame (see illustration 5.19b). Free the wiring loom grommets from the cutouts in the tray (see illustrations 5.19c and d). Disconnect the rear brake light switch wiring connector and, where fitted, the rear ABS sensor wiring connector. Lift the fuse/relay holder out of the frame and position it in the air filter housing so that it is out of the way. Working around the rear frame, check that all wiring and electrical system components are clear of the rear frame. Trace the neutral switch, gear indicator, oil pressure switch and side stand wiring and disconnect them at the connectors. Feed the wiring back the components, noting its routing, and making sure it is free of any ties and clear of the frame.

13 Free the rear brake fluid reservoir from its clips and wrap it in a plastic bag to prevent any fluid spillage (see illustration 5.20a). Also free the rear brake hose-to-pipe union grommet from the bracket on the frame (see illustration 5.20b).

2

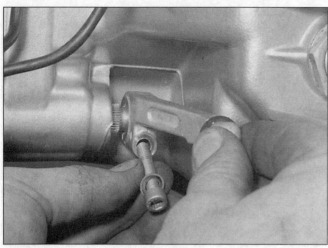

30.19a Remove the pinch bolt and slide the gearchange arm off the shaft

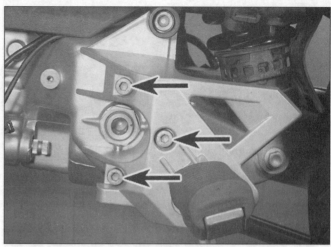

30.19b Unscrew the three bolts (arrowed) and remove the footrest bracket

30.23a Six bolts secure the gearbox, three on the left (arrowed) . . .

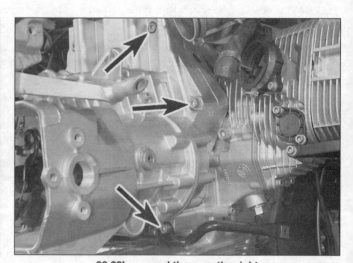

30.23b . . . and three on the right

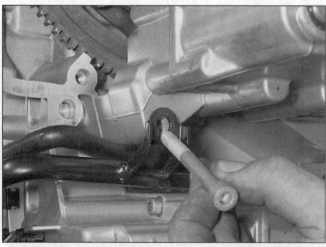

30.23c Note the fairing bracket (where fitted) secured by the lower bolts

30.23d Draw the gearbox away from the engine, making sure it is kept straight and level

14 On RT models, remove the bolts which secure the rear of each footrest mounting bracket to the rear sub-frame.

15 Place a small block of wood between the top of the gearbox casing and the swingarm (Paralever) to prevent them folding together when the rear shock absorber upper mounting bolt is removed. Also place a jack under the gearbox to take the weight when it is separated from the engine. Check which side the rear shock upper mounting bolt is installed, then unscrew the two bolts and remove the seat support for that side **(see illustration 5.22)**. This allows clearance for the bolt to be withdrawn. Unscrew and remove the rear shock upper mounting bolt. Where fitted, also remove the bolt securing the remote pre-load adjuster.

16 Check around to verify that all cables, wiring, hoses, pipes, electrical components and anything else are all free from the rear sub-frame. Remove the screws securing the airbox to the rear sub-frame. Unscrew the nut from the end of the through-bolt which secures the front of the rear sub-frame and the front strut side rails to the engine **(see illustration 5.23a)**. Supporting the rear sub-frame, unscrew the four bolts which secure it to the engine and gearbox, then carefully manoeuvre it backwards off the engine and gearbox, making sure it is free from all cables, pipes and hoses, and remove it. Free the fuel supply pipes from the grommets at the front of the airbox **(see illustration 5.23b and c)**. Remove the wiring tray and airbox and support the fuse/relay holder in the battery holder.

17 Remove the starter motor (see Chapter 8).

18 Slacken the clutch cable adjuster locknut at the handlebar, then screw the adjuster fully into the bracket to provide the maximum freeplay **(see illustration 29.1a)**. Detach the lower end of the inner cable from the operating lever on the gearbox, using a screwdriver to move the lever forward if there is not enough freeplay available **(see illustration 29.1b)**. Withdraw the lower end of the outer cable from the lug on the gearbox and secure the cable clear of the engine **(see illustration 29.1c)**.

19 Remove the bolts securing the right-hand footrest bracket to the gearbox on R, RS and GS models **(see illustration 5.34a)**; on RT

models, remove the right-hand footrest plate. Support or tie the bracket, with the rear brake master cylinder still attached, so that no strain is placed on the pipes. On RT models, remove the left-hand footrest bracket plate. Note the alignment of the gearchange pedal linkage arm on the shaft, then remove the bolt and slide the arm off the shaft **(see illustration)**. On R, RS and GS models remove the bolts securing the left-hand footrest bracket to the gearbox and remove the bracket **(see illustration)**.

20 Unscrew the bolts securing the rear brake caliper and lift the caliper off the disc **(see illustration 5.34b)**. There is no need to disconnect the hose. Support the caliper in the same way that the master cylinder is supported.

21 At this point there should be nothing left attached to the gearbox, swingarm or rear wheel. Check around to verify that all components, cables, wiring, hoses, pipes, electrical components and anything else are all free.

22 Remove the swingarm pivot bolts from the gearbox and separate the rear wheel, final drive and swingarm as an assembly from the gearbox, then remove the driveshaft. Refer to Chapter 5 for the removal procedure, ignoring the Steps which do not apply.

23 Remove the bolts securing the gearbox to the engine and separate it from the engine **(see illustrations)**. Note the fairing bracket (where fitted) secured by the lower bolts.
Caution: When separating the gearbox from the engine, make sure that the gearbox is drawn back entirely straight and

30.24 Withdraw the pushrod from the gearbox

level until the clutch pushrod is clear of the clutch components. If the gearbox is raised, lowered or skewed before the pushrod is clear, the pushrod will bend and have to be replaced with a new one.

24 The gearbox is now ready for disassembly and can be moved onto a suitable workbench. Withdraw the clutch pushrod, noting which way round it fits **(see illustration)**.

Installation

25 Installation is the reverse of removal, noting the following points:

a) *Check that the clutch pushrod is straight by rolling it on a flat surface such as a piece of glass. Lubricate the pushrod ends with molybdenum disulphide grease **(see illustration)**. Apply a smear of Optimoly MP3 or equivalent paste to the clutch plate and gearbox input shaft splines **(see illustrations)**. Refer to the Caution above before installing the gearbox.*

b) *Tighten the gearbox mounting bolts to the torque setting specified at the beginning of the Chapter **(see illustration)**.*

c) *When installing the rear sub-frame, install all four bolts and the through-bolt that also secures the front frame strut side rails finger-tight before tightening any of them to the specified torque setting. When torquing the bolts, tighten the right-hand rear bolt first, followed by the right-hand front bolt, then the left-hand front bolt and finally the left-hand rear bolt. Then tighten the through-bolt to the specified torque.*

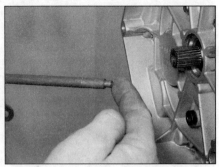

30.25a Lubricate the clutch pushrod ends . . .

30.25b . . . the clutch splines . . .

30.25c . . . and the input shaft splines

30.25d Tighten the gearbox bolts to the specified torque setting

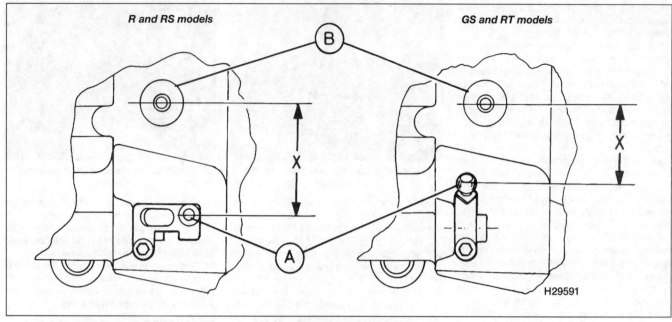

30.25e Gearchange lever height

| A | Ball end of lever | B | Boss on gearbox housing | X | Lever height |

d) Refer to Chapter 5 for the swingarm (Paralever) installation procedure.
e) When fitting the gearchange lever to the splined end of the gearchange shaft, match up the alignment marks made on removal and tighten the pinch bolt to the specified torque setting. Note that the lever should be positioned horizontally on the shaft, so that the height from the ball end of the lever to the threaded boss on the gearbox housing measures 51.9 ± 1.8 mm on GS and RT models and 61.2 ± 1.8 mm on R and RS models **(see illustration)**.
f) Tighten all bolts to the torque settings specified either at the beginning of this Chapter, or at the beginning of the relevant Chapter elsewhere in the book.
g) Check the condition of all breather tubes, grommets and O-rings, and of all hose and pipe clamps before re-using them, and replace them if they are worn, damaged or deteriorated.

31.2 Remove the circlip (arrowed) securing the selector drum in the bearing

h) Make sure all wires, cables, pipes and hoses are correctly routed and connected, and secured by any clips or ties.
i) Check the operation of all electrics and lights before taking the bike on the road.
j) Make a careful check of the brake system and the suspension before taking the bike on the road (see Chapter 1).
k) Refill the gearbox with oil (see Chapter 1).

31 Selector drum and forks – removal, inspection and installation

Note: To remove the selector drum and forks the gearbox must first be removed.

Removal

1 Remove the gearbox (see Section 30).
2 Remove the neutral/gear indicator switch (see Chapter 8). Remove the small external circlip from the rear end of the selector drum **(see illustration)**.
3 Support the gearbox on its rear end, making sure it is secure. Unscrew the neutral detent ball bolt, then remove the spring and ball bearing by turning the gearbox upside down and tipping them out **(see illustration)**.
4 Wrap some insulating tape around the input shaft splines to avoid damaging the oil seal as the gearbox cover is removed. Unscrew the bolts securing the front cover to the gearbox then carefully lift the cover off the gearbox **(see illustration 31.3)**. Use a soft-faced hammer to tap around the joint to initially separate the cover if necessary. **Note:** If the halves do not separate easily, make sure all

fasteners have been removed. Do not try and separate the halves by levering against the casing mating surfaces. If the cover still proves difficult to remove, heat it around the transmission shaft bearing housings to approximately 100°C and try again. Remove the dowels if they are loose. Note how the gearchange shaft stopper arm spring locates against the cover.
5 Remove the shim(s) from the end of the intermediate transmission shaft and the shim(s) and oil guide plate from the end of the transmission output shaft **(see illustrations)**. Do not get the shims mixed up.
6 Remove the large oil deflector plate, noting carefully how it fits **(see illustration)**.
7 Before removing the selector forks, mark each one using a felt pen according to its location and which way up it fits as an aid to installation **(see illustration 31.9b)**.
8 Move the stopper arm off the selector drum stopper plate and position it clear. Supporting the selector forks, withdraw the fork shafts from the casing, then move the forks aside so that the guide pin rollers are clear of the selector drum tracks **(see illustrations)**.
9 Note how the teeth of the gearchange shaft selector arm engage with the pins on the selector drum. Turn the selector drum until the cut-out in the pin plate aligns with the selector arm pawl, then remove the drum **(see illustration)**. There is a good chance that the pins will fall out of the end of the drum – take care not to lose them. With the drum removed, remove the selector forks, making sure that the guide pin rollers stay in place. Once removed, slide the forks back onto their shafts in the correct order and way round **(see illustration)**.

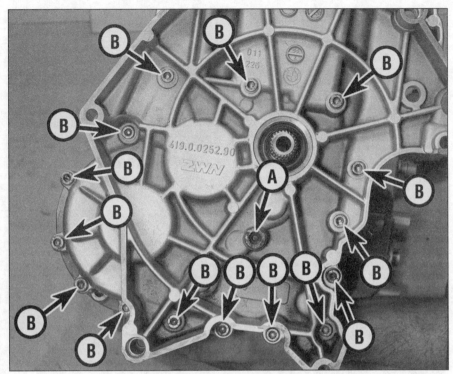

31.3 Neutral detent ball bolt (A), gearbox cover bolts (B)

31.5a Remove the shims from the intermediate shaft . . .

31.5b . . . and the shims and oil plate from the output shaft

31.6 Remove the large oil deflector plate

31.8a Withdraw the shafts . . .

31.8b . . . and move the forks aside

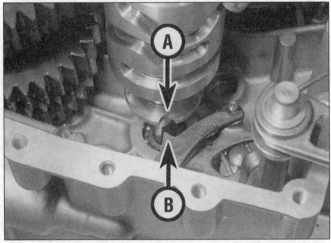

31.9a Align the cut-out (A) with the pawl (B) and remove the drum

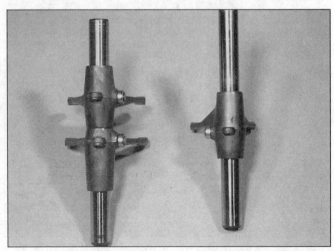

31.9b Correct positioning of selector forks on the shafts

2

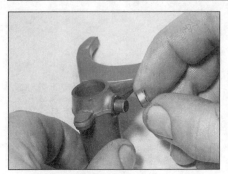

31.12 The guide pin rollers can be replaced if required

31.13 Selector drum bearing location in gearbox housing

31.15 Slide the forks into their pinions

Inspection

10 Inspect the selector forks for any signs of wear or damage, especially around the fork ends where they engage with the groove in the pinion. Check that each fork fits correctly in its pinion groove. Check closely to see if the forks are bent. If the forks are in any way damaged they must be replaced.

11 Check that the forks fit correctly on their shaft. They should move freely with a light fit but no appreciable freeplay. Replace the forks and/or shaft if excessive wear is noted. Check that the fork shaft holes in the casing and cover are not worn or damaged. The selector fork shaft can be checked for trueness by rolling it along a flat surface. A bent shaft will cause difficulty in selecting gears and make the gearshift action heavy. Replace either shaft if it is bent.

12 Inspect the selector drum tracks and selector fork guide pin rollers for signs of wear or damage **(see illustration)**. The guide pin rollers can be replaced, but if the drum tracks are worn or damaged, the drum must be replaced.

13 Check that the selector drum bearing in the rear of the gearbox housing rotates freely **(see illustration)**. The bearing is secured by a circlip accessed from outside the housing (see the larger circlip in **illustration 31.2**). To

renew the bearing, remove the circlip and washer and drive the bearing out from the inside of the gearbox housing outwards; note that BMW advise that the housing is heated to 100°C to aid bearing removal. When installing the new bearing, heat the housing to 100°C and use a socket or bearing driver which contacts the outer race of the bearing to drive it into the housing from the outside inwards; provide support on the inside of the housing and make sure the bearing rests on its seat in the base of the housing. Install the washer and circlip, make sure that the circlip fully engages its groove in the housing.

14 The front end of the selector drum locates in a plain bearing set in the gearbox cover; if the bearing is worn seek the advice of a BMW dealer concerning replacement bearings.

Installation

15 Fit each selector fork into its groove in the relevant pinion, making sure it is in its correct location and the right way round (see Step 7) **(see illustration)**.

16 Fit the pins into the selector drum, using a dab of grease to keep them in place as the drum is installed **(see illustration)**. Make sure the pins locate correctly in the holes. Slide the selector drum into position in the gearbox housing bearing, turning it as necessary to align the cut-out in the pin plate with the

31.16 Use grease to retain the pins in the drum

selector arm pawl **(see illustration 31.9a)**.

17 Locate the guide pin roller on the end of each fork into its track in the drum **(see illustration 31.8b)**. Lubricate the selector fork shafts with clean engine oil and slide them through the forks and into the bores in the crankcase **(see illustration 31.8a)**. Place the stopper arm against the neutral detent on the stopper plate, turning the selector drum as required to locate it **(see illustration)**.

18 Install the large oil deflector plate **(see illustration 31.6)**. Make sure it locates against the chamfered edge of each of the lugs on the gearbox housing and fits correctly between the various components **(see illustration)**.

31.17 Locate the stopper arm against the neutral detent on the stopper plate (arrowed)

31.18 Locate the plate against the chamfered face of each lug (arrowed)

31.20a Apply the sealant to one mating surface . . .

31.20b . . . then fit the cover, making sure the spring end locates against the inside of the cover (arrowed) . . .

31.20c . . . and tighten the bolts to the specified torque setting

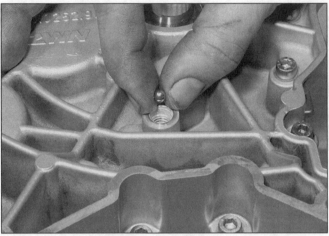

31.21a Insert the ball . . .

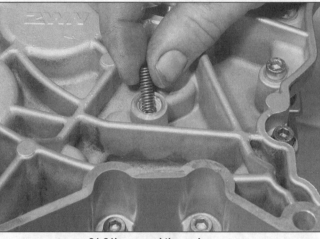

31.21b . . . and the spring . . .

19 Fit the shim(s) onto the end of the intermediate transmission shaft and the shim(s) and oil guide plate onto the end of the transmission output shaft **(see illustrations 31.5a and b)**. Smear them with grease to help locate them centrally on each shaft, otherwise they could slip out of position as the cover is installed. Do not get the shims mixed up.

20 Make sure the mating surfaces of the gearbox and cover are clean and free from oil and grease, then apply a thin coat of Loctite 573 or equivalent to one surface **(see illustration)**. If removed, fit the dowels into the casing, then install the cover. Make sure the stopper arm spring end locates against the inside of the cover by pressing it in, and

that the stopper arm remains located against the stopper plate **(see illustration)**. If the cover is a tight fit onto the bearings, heat the cover around the transmission shaft bearing housings to approximately 100°C and then try again. Tighten the bolts to the torque setting specified at the beginning of the Chapter **(see illustration)**. Fit the gearchange lever onto the shaft and check that all gears can be selected and that the shafts rotate smoothly in every gear. Note that there is a certain amount of built-in resistance between the shafts due to a pre-load on the input shaft bearing. Remove the tape from around the input shaft splines.

21 Fit the neutral detent ball and spring into the end of the selector drum **(see**

illustrations). Clean the bolt threads and apply Loctite 243 or equivalent, then tighten the bolt to the specified torque setting **(see illustrations)**.

22 Fit the circlip onto the rear end of the selector drum **(see illustration)**. Install the neutral/gear indicator switch (see Chapter 8). Before fitting the gearbox to the engine, check again the operation of the transmission in each gear. If there are any signs of tight or rough spots (taking into account the pre-load), or of any other problem, the fault must be rectified before proceeding further.

23 Refit the gearbox (see Section 30) and refill with oil (see Chapter 1).

2

31.21c . . . then apply the threadlock . . .

31.21d . . . and tighten the bolt to the specified torque setting

31.22 Fit the circlip, making sure it fits properly in the groove

32.3a Withdraw the gearchange shaft and remove the washer

32.3b Remove the bush (arrowed) if required

32.4a Check the selector arm return springs . . .

32.4b . . . the shaft centralising spring . . .

32.4c . . . and the stopper arm return spring, making sure they are all correctly located

32 Gearchange mechanism – removal, inspection and installation

Note: *The gearchange mechanism can only be removed with the gearbox removed.*

Removal

1 Remove the gearbox (see Section 30).
2 Remove the selector drum and forks (see Section 31).
3 Wrap some insulating tape around the gearchange shaft splines to avoid damaging the oil seal as the shaft is removed, then remove the shaft from the gearbox, noting how the centralising spring ends locate **(see illustration)**. Check that the washer is on the end of the shaft – if not, remove it from the casing. Also remove the shaft bush, if required, noting how it fits **(see illustration)**.

Inspection

4 Inspect the selector arm return springs, the shaft centralising spring and the stopper arm return spring **(see illustrations)**. If they are fatigued, worn or damaged they must be replaced.

5 Check the gearchange shaft for straightness and damage to the splines. If the shaft is bent you can attempt to straighten it, but if the splines are damaged the shaft must be replaced. Also check the condition of the shaft oil seal in the casing and replace it if damaged or deteriorated. Lever the old seal out using a screwdriver and drive the new seal in squarely.
6 Inspect the selector arm claw and the pins in the selector drum, and the stopper arm roller and the stopper plate on the drum. If they are worn or damaged they must be replaced.

Installation

7 If removed, fit the washer onto the end of the shaft and check that the shaft assembly components are correctly positioned **(see illustrations 32.3a and 32.4a, b, and c)**. If removed, install the bush into the casing, then install the shaft, making sure the centralising spring ends locate on either side of the pin **(see illustrations)**. Remove the tape from around the splines.
8 Install the selector drum and forks (see Section 31).
9 Fit the gearchange linkage arm onto the end of the shaft and check that the mechanism works correctly before installing the gearbox.

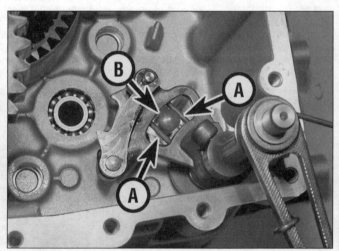

32.7a If removed, fit the bush into the casing

32.7b Locate the spring ends (A) on each side of the pin (B)

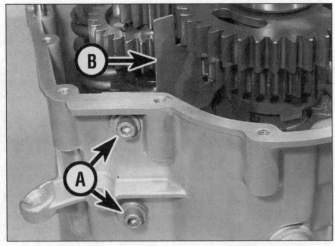

33.3 Unscrew the two bolts (A) and remove the plate (B)

33.5 Input shaft (A), intermediate shaft (B), output shaft (C)

33 Transmission shafts – removal and installation

Note: *To remove the transmission shafts the gearbox must be removed from the engine. If the engine has already been removed, ignore the steps which do not apply.*

Removal

1 Remove the gearbox (see Section 30).
2 Remove the selector drum and forks (see Section 31).
3 Unscrew the two bolts securing the small oil deflector plate and remove the plate, noting how it fits **(see illustration)**.
4 Wrap some insulating tape around the output shaft splines to avoid damaging the oil seal as the shaft is withdrawn.
5 Note the relative positions of the three shafts and how they fit together **(see illustration)**. On 1993 models, lift the input shaft out, then lift the intermediate and output shafts together. On 1994-on models, lift the input shaft, intermediate shaft and output shaft out of the casing together. If they are stuck, heat the gearbox casing around the output shaft bearing to approximately 100°C and/or use a soft-faced hammer and gently tap on the ends of the shafts to free them. If necessary, the input shaft and output shaft can be disassembled and inspected for wear or damage (see Section 34).

Bearings and seals

6 Referring to *Tools and Workshop Tips* (Sections 5 and 6) in the Reference Section, examine the transmission shaft bearings and oil seals. Refer to the method of fitting given in Steps 7 to 9 below, and remove the bearings and seals using the techniques in *Tools and Workshop Tips*. Note that BMW advise that the cover or housing (as applicable) be heated to 100°C to aid bearing removal, and it many well be found that the bearings drop out under

their own weight once heated. When installing new oil seals, note that the type supplied will not be pre-formed and their sealing lips will require shaping. Do this by pressing the lip in with your fingers, making sure it is even all round, and pressed in far enough to allow its shaft through it, but not so far that it becomes wider than the shaft itself.
7 On 1993 to early 1997 models, the tapered roller bearings at each end of the input shaft are a press-fit on the shaft – refer to Section 34 for details. The tapered roller bearing outer races are a press fit in the gearbox cover and housing. On later models, the sealed ball bearings are a press-fit on the shaft – refer to Section 34 for removal details. The input shaft oil seal set in the gearbox cover can be levered out of position using a large-flat bladed screwdriver or driven out from the inside of the cover outwards. The clutch pushrod oil seal which runs through the input shaft has an oil seal in the gearbox housing; either lever the seal out or drive it from the outside of the housing inwards.
8 On 1993 to early 1997 models, the intermediate shaft runs in a ball bearing at its front end, which is a press fit on the shaft (see Section 34), and a caged roller bearing at its rear end; the caged roller bearing is a press fit in the gearbox housing. On later models, a sealed ball bearing is fitted to each end of the

intermediate shaft. If heat is used to aid removal of the sealed bearings, check that the bearing sealing surfaces are not damaged by excessive heat.
9 The output shaft runs in ball bearings at each end, which are a press fit on the shaft (see Section 34). The output shaft oil seal can be levered out of the gearbox housing from the outside using a large flat-bladed screwdriver or driven out from the inside outwards.

Installation

Note: *If any of the gearbox shafts or bearings have been renewed or the input shaft disassembled, the shaft end-float must be checked and if necessary adjusted using shims. Furthermore on 1993 to early 1997 models (with tapered-roller bearings) the input shaft pre-load will require checking after re-shimming. These tasks are best left to a BMW dealer equipped with the necessary equipment and experience.*

10 On 1993 models, install the input shaft, then install the intermediate and output shafts together **(see illustrations)**. On 1994-on models, install the input shaft, intermediate shaft and output shaft together **(see illustration 33.5)**. If the shafts are difficult to fit, heat the gearbox casing around the output shaft bearing to approximately 100°C.

33.10a On early models, install the input shaft first . . .

33.10b . . . followed by the intermediate and output shafts

2

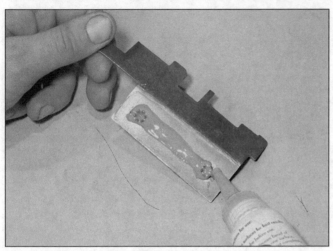

33.12a Apply the locking compound as specified . . .

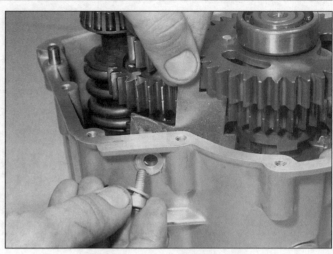

33.12b . . . then install the plate

Make sure the output shaft end does not turn the inner lip of the oil seal inside out as it is installed – lubricate the seal lip with grease and rotate the shaft as you install it, if possible. Make sure the shafts are correctly seated in their bearings and their related pinions are still correctly engaged. Tap the output shaft and intermediate shaft ends with a soft-faced hammer to ensure the bearings are seated – do not strike the input shaft end otherwise there is a danger of the circlip and bearing position on the shaft lower end being disturbed.

11 Referring to the **Note** above, check shaft end-float where necessary.

12 Apply a thin smear of Loctite 573 or equivalent to the small oil deflector plate bolt threads and to the contact points between the plate and the casing, then install the plate and tighten the bolts to the specified torque setting **(see illustrations)**.

13 Thoroughly lubricate all the transmission components with the correct grade of clean gear oil.

14 Position the gears in the neutral position and check the shafts are free to rotate easily and independently (ie the input shaft can turn whilst the output shaft is held stationary) before proceeding further.

15 Install the selector drum and forks (see Section 31).

34.3 Remove the circlip securing the bearing . . .

34 Transmission shafts – disassembly, inspection and reassembly

Note: *When ordering parts for the gearbox it is important to provide full engine number details to ensure that the correct parts are supplied. The gearbox was modified at the start of the 1994 model year to improve gearchanging by altering the gear ratios, primary ratio and final drive ratio; noise was lessened by reducing the tolerance between the clutch friction plate and input shaft splines, and installing O-rings on the output shaft. A further change was made to the gear pinions and shift forks for the 1997 model year to improve gear selection. Following this, the input shaft tapered-roller bearings and intermediate shaft roller bearing were replaced by sealed ball bearings.*

1 Remove the transmission shafts from the casing (see Section 33). Always disassemble the transmission shafts separately to avoid mixing up the components. There should be no real need to disassemble the input shaft unless any of the splines are worn or if the bearings need replacing. The intermediate shaft cannot be disassembled, except for the bearings.

34.4 . . . then remove the bearing and rear cam using a puller as shown

Input shaft

Note: *If new input shaft components are fitted, shaft end-float must be checked following reassembly.*

2 Check the shaft splines and the corresponding splines of the cam for signs of wear or damage and replace them if necessary. If the cam splines cannot be seen well enough with the shaft assembled, disassemble it as described in Steps 5 and 6. Check the circlip and its groove in the shaft for signs of wear or damage, in particular rounded edges, and replace it if necessary. On 1993 to early 1997 models, examine the tapered-roller bearings on each end of the shaft. If the rollers or the cage are damaged the bearing must be replaced together with its outer race in the housing or cover; use the later model ball bearing if replacement is necessary. On later models check the ball bearings on each end of the shaft for wear or damage, replacing them with new ones if necessary.

3 To dismantle the input shaft, remove the circlip and washer (shim) from the rear end of the shaft **(see illustration)**. If the circlip is very tight in its groove, gently tap on the inner race of the bearing to relieve the pressure.

4 Using a bearing puller attached to the rear cam gear, pull the rear cam and bearing off the shaft, noting that spring tension will be released on the damper spring **(see illustration)**. Remove the front cam, spring and spring seat from the shaft. The bearing can be removed from the front end of the input shaft using the puller.

5 To reassemble the input shaft, install the spring seat over the rear end of the shaft so that its chamfered edge faces the front of the shaft. Lubricate the shaft splines, then fit the spring, front cam and rear cam **(see illustration)**.

6 On 1993 to early 1997 models, assemble a spring compressor on the front end of the shaft so that its edges pull against the rear cam gear. Compress the damper spring until

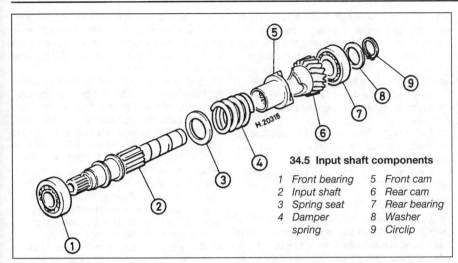

34.5 **Input shaft components**

1	Front bearing	5	Front cam
2	Input shaft	6	Rear cam
3	Spring seat	7	Rear bearing
4	Damper spring	8	Washer
		9	Circlip

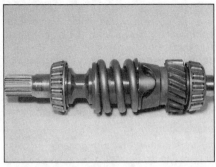

34.7 **The assembled input shaft assembly**

the shaft step is exposed. Install the bearing on the rear end of the input shaft using a press or tubular drift which bears only on the bearing's inner race, not the cage; note that the bearing can be heated to 80°C to aid installation. Drive the bearing down until it contacts the shoulder, at which point the circlip groove should be exposed. Fit the washer (shim) and new circlip and use a bearing puller (see illustration 34.4) to draw the bearing back into contact with it. Note, there must be no clearance between the bearing, shim and the circlip or the pre-load setting will be incorrect. On later models, with a ball bearings, clamp the shaft front end in a vice with soft jaws so that the shaft is held securely upright but cannot be marked or damaged. Install the bearing on the rear end of the input shaft using a press or tubular drift which bears only on the bearing's inner race, not the balls or outer race; note that the bearing can be heated to 80°C to aid installation, although take care that the bearing's sealed surfaces are not damaged by the heat. Drive the bearing down until it

contacts the shoulder, at which point the circlip groove should be exposed. Fit the washer (shim) and new circlip and use a bearing puller (see illustration 34.4) to draw the bearing back into contact with it. Note, there must be no clearance between the bearing, shim and the circlip or the end-float will be incorrect.

7 The front bearing can be installed as described for the rear bearing (see illustration). On 1993 to early 1997 models, with tapered-roller bearings, if new input shaft components have been fitted, the front bearing should be left off until input shaft end-float has been checked.

Intermediate shaft

8 The intermediate shaft cannot be disassembled and with the exception of the bearings, no replacement parts are available (see illustration). If the shaft gears are damaged or worn, the complete intermediate shaft must be renewed.

9 The rear bearing is retained in the gearbox housing and can be removed as described in Section 33. The front bearing can be drawn

off the end of the intermediate shaft using a knife-edged bearing puller which locates on the shaft end and behind the bearing. **Note:** *Only remove the bearing if renewal is required.* Use a drift, such as a socket which bears only on the bearing's inner race to tap the new bearing fully onto the shaft. The bearing can be pressed on cold, or heated to 80°C to aid installation.

Output shaft

Disassembly

 HAYNES HINT *When disassembling the output shaft, place the parts on a long rod or thread a wire through them to keep them in order and facing the proper direction.*

10 Remove the bearings from each end of the shaft using a knife-edged bearing puller which locates on the shaft end and behind the bearing (see illustration). Remove the washer from behind each bearing.

11 Slide the helical 5th gear pinion and its bush off the rear of the shaft, followed by the thrust washer and the 3rd gear pinion (see illustration). Remove the circlip securing the

34.8 **Intermediate shaft assembly**

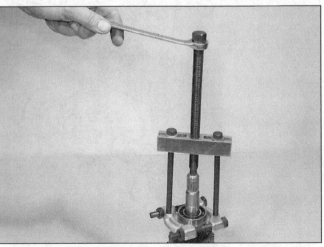

34.10 **Use a bearing puller to draw off the output shaft bearings**

2nd gear pinion, then slide off the splined washer followed by the pinion. Remove the split bearing halves from the output shaft, followed on 1994-on models, by the O-ring.

12 Slide the 1st gear pinion and its bush off the front of the shaft. Remove the O-ring (1994-on models only). Remove the spacer and the 4th gear pinion, then remove the circlip and the splined washer.

Inspection

13 Wash all of the components in clean solvent and dry them off. Check the gear teeth for cracking, chipping, pitting and other obvious wear or damage. Any pinion that is damaged as such must be replaced. If damage is noted, check for wear on the corresponding intermediate shaft gears.

14 Inspect the dogs and the dog holes in the gears for cracks, chips, and excessive wear especially in the form of rounded edges. Make sure mating gears engage properly. Check for signs of wear on the selector fork grooves.

15 The shaft is unlikely to sustain damage unless the engine has seized, placing an unusually high loading on the transmission, or the machine has covered a very high mileage. Check the surface of the shaft, especially where a pinion turns on it, and replace the shaft if it has scored or picked up, or if there

are any cracks. Damage of any kind can only be cured by renewal. Similarly, check the 1st and 5th gear pinion bushes and the 2nd gear pinion split needle roller bearing for damage.

17 Check the washers and circlips and replace any that are bent or appear weakened or worn. Note that it is good policy to renew circlips as a matter of course when rebuilding a gearbox.

18 Referring to *Tools and Workshop Tips* in the Reference section, check the shaft ball bearings for wear. If the bearings were a tight fit on the shaft they may have been damaged during removal – renewal is the safest option.

Reassembly

19 During reassembly, apply gearbox oil to the mating surfaces of the shaft, pinions and bushes. When installing the circlips, do not expand the ends any further than is necessary, and install them so that the chamfered side faces the pinion it secures (see *correct fitting of a stamped circlip* illustration in Tools and Workshop Tips of the Reference section).

20 Fit the front circlip for the 2nd gear pinion, making sure it fits properly into its groove. Slide the front splined washer onto the rear of the shaft so that it rests against the circlip. On 1994-on models, install a new O-ring on the

shaft seat. Install the split bearing halves, using a small dab of grease to hold them in place. Slide the 2nd gear pinion onto the bearing halves and secure it with the rear splined washer and circlip.

21 Slide the 3rd gear pinion, with the selector fork groove facing the 2nd gear pinion onto the shaft, followed by the thrust washer and the 5th gear pinion bush. Note that BMW advise that the bush is first heated to 80°C to aid installation. Fit the 5th gear pinion over the bush so that its dogs face the 3rd gear pinion. Install the shim and rear bearing on the rear end of the shaft. Use a tubular drift which bears only on the bearings' inner race to drive it on the shaft; if necessary heat the bearing to 80°C to aid installation. Ensure that the bearing seats fully on the shaft

22 Slide the 4th gear pinion with its dogs facing the 2nd gear pinion onto the front of the shaft, followed by the spacer.

23 On 1993 models, use a vernier gauge to measure the distance from the spacer to the rear face of the rear bearing – this distance should be 125.7 to 125.8 mm **(see illustration)**; if necessary obtain a different thickness spacer washer from a BMW dealer to obtain the correct specification. Slide the 1st gear pinion bush onto the shaft so that it abuts the spacer, noting that BMW advise

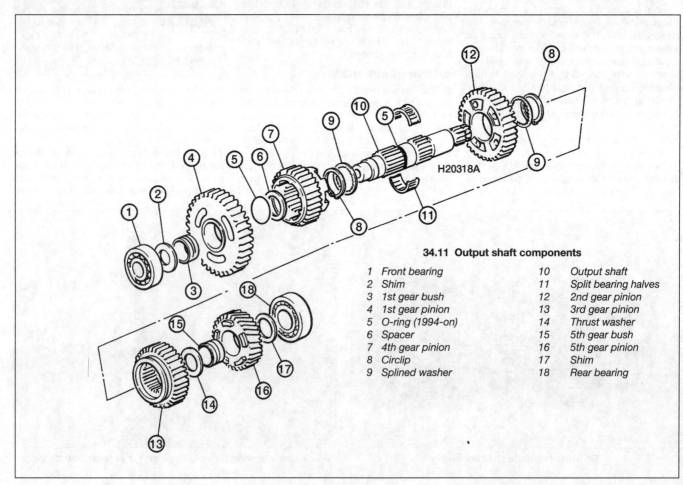

34.11 Output shaft components

1	Front bearing	10	Output shaft
2	Shim	11	Split bearing halves
3	1st gear bush	12	2nd gear pinion
4	1st gear pinion	13	3rd gear pinion
5	O-ring (1994-on)	14	Thrust washer
6	Spacer	15	5th gear bush
7	4th gear pinion	16	5th gear pinion
8	Circlip	17	Shim
9	Splined washer	18	Rear bearing

H20318A

34.23 Output shaft length – 1993 models

*Length is measured from face of spacer to end of rear bearing, with
front bearing, shim, 1st gear pinion and bush removed*

that the bush is first heated to 80°C to aid installation. Fit the 1st gear pinion over the bush.

24 On 1994-on models, fit a new O-ring onto its seat on the shaft. Slide the 1st gear pinion bush onto the shaft so that it is flush with the step in the shaft, noting that BMW advise that the bush is first heated to 80°C to aid installation. Fit the 1st gear pinion over the bush.

25 Install the bearing washers and the bearings in their original locations on the output shaft. Use a tubular drift which bears only on the bearings' inner race to drive them on the shaft; if necessary heat the bearings to 80°C to aid installation. Ensure that each bearing seats fully on the shaft.

35 Initial start-up after overhaul

1 Make sure the engine and gearbox oil levels are correct (see *Daily (pre-ride) checks* and Chapter 1).

2 Remove the spark plug covers and disconnect the HT leads from the spark plugs (see Chapter 1). Connect each lead to a spare spark plug and lay each plug on the engine with the threads contacting the engine. Make sure the plug threads are in contact with the engine otherwise the Motronic system will be damaged. Check that the kill switch is in the RUN position, the gearbox is in neutral and the side stand is

up, then turn the ignition switch ON and crank the engine over with the starter until the oil pressure indicator light goes off (which indicates that oil pressure exists). Turn the ignition off, remove the test spark plugs and install the plug caps.

 Warning: Do not remove either of the spark plugs from the engine to perform this check – atomised fuel being pumped out of the open spark plug hole could ignite, causing severe injury! Also ensure that there is no fuel vapour present in the working area.

3 Make sure there is fuel in the tank and set the choke as required.

4 Start the engine and allow it to run at a moderately fast idle until it reaches operating temperature.

 Warning: If the oil pressure indicator light doesn't go off, or it comes on while the engine is running, stop the engine immediately.

5 Check carefully for oil leaks and make sure the transmission and controls, especially the brakes, function properly before road testing the machine. Refer to Section 36 for the recommended running-in procedure.

6 Upon completion of the road test, and after the engine has cooled down completely, recheck the valve clearances (see Chapter 1) and check the engine and gearbox oil levels (see *Daily (pre-ride) checks* and Chapter 1).

36 Recommended running-in procedure

1 Treat the machine gently for the first few miles to make sure oil has circulated throughout the engine and any new parts installed have started to seat.

2 Even greater care is necessary if new pistons, rings or bearing shells have been fitted. In the case of a new pistons or cylinders, the bike will have to be run in as when new. This means greater use of the transmission and a restraining hand on the throttle until at least 600 miles (1000 km) have been covered. There's no point in keeping to any set speed limit – the main idea is to keep from labouring the engine and to gradually increase performance up to the 600 mile (1000 km) mark. Experience is the best guide, since it's easy to tell when an engine is running freely. The following maximum engine speed limitations, which BMW provide for new motorcycles, can be used as a guide.

Up to 600 miles (1000 km)	**Do not exceed 4000 rpm or two thirds throttle**
600 to 1200 miles (1000 to 2000 km)	**Vary throttle position/speed. Avoid full throttle use**

3 If a lubrication failure is suspected, stop the engine immediately and try to find the cause. If an engine is run without oil, even for a short period of time, severe damage will occur.

2

Chapter 3
Fuel and exhaust systems

Contents

Degrees of difficulty

| Easy, suitable for novice with little experience | | Fairly easy, suitable for beginner with some experience | | Fairly difficult, suitable for competent DIY mechanic | | Difficult, suitable for experienced DIY mechanic | | Very difficult, suitable for expert DIY or professional | 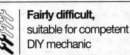 |

Specifications

Fuel

Grade
  Models with catalytic converter . . . . Premium grade unleaded petrol (gasoline) to German DIN 51607 standard or equivalent, minimum octane number 95 RON (Research Octane Number) or 85 MON (Motor Octane Number). **Do not** use leaded fuel.

  Models without catalytic converter . . . . Premium grade unleaded petrol (gasoline) to German DIN 51607 standard or equivalent, minimum octane number 95 RON (Research Octane Number) or 85 MON (Motor Octane Number) **or** Premium grade leaded petrol to German DIN 51600 standard or equivalent, minimum octane number 95 RON (Research Octane Number) or 85 MON (Motor Octane Number)

Tank capacity
  Full
    RS models . . . . 23 litres
    RT models . . . . 26 litres
    R models . . . . 21 litres
    GS models . . . . 25 litres
  Quantity remaining at fuel warning light illumination (approx.) . . . . . . 4 litres

Throttle bodies

Internal diameter . . . . 45 mm
Butterfly angle at idle
  RS and RT models . . . . 10°
  R and GS models . . . . 5°
Idle speed . . . . see Chapter 1

Torque settings

Fuel tank bolt . . . . 22 Nm
Intake manifold clamp screws . . . . 2 Nm
Silencer front mounting bolts . . . . 20 Nm
Silencer rear mounting bolt
  RS and RT models . . . . 35 Nm
  R and GS models . . . . 24 Nm
Silencer clamp bolt . . . . 50 Nm
Lambda sensor (where fitted) . . . . 55 Nm
Downpipe assembly nuts
  Plain nuts . . . . 22 Nm
  Domed nuts . . . . 18 Nm

3

1 General information and precautions

General information

The fuel supply system consists of the fuel tank, fuel filter, fuel pump, fuel hoses, fuel distributor rail, pressure regulator, throttle bodies, fuel injectors and control cables. For cold starting a choke lever on the left-hand handlebar opens both throttle bodies by a set amount. Information on fuel level is provided by a level sender in the tank which is linked to a low fuel level warning light in the instruments via a damping unit. Where fitted, a digital rider information display unit indicates fuel volume in the tank.

Fuel/air delivery to the engine, and ignition timing, are digitally controlled by the Bosch Motronic MA 2.2 engine management system, first fitted to BMW's 16-valve K-series bikes. Information on throttle angle, provided by the TPS (throttle position sensor) on the left-hand throttle body, is fed into the Motronic unit together with engine speed and TDC (top dead centre) information from the timing sensors on the front of the crankshaft. Additional information is provided by the air temperature sensor in the air filter housing lid, the oil temperature sensor, the CO potentiometer on the rear sub-frame (models without a catalytic converter), and the Lambda sensor in the exhaust (models with a catalytic converter).

All information supplied to the Motronic unit is compared with its pre-programmed data. The appropriate output signals are then sent to the fuel injectors, fuel pump and ignition HT coils and determine the amount of fuel required and the duration of the injection period.

The exhaust system is a two-into-one design, incorporating a catalytic converter and Lambda oxygen sensor on certain models.

Many of the fuel system service procedures are considered routine maintenance items and for that reason are included in Chapter 1.

Precautions

 Warning: Petrol (gasoline) is extremely flammable, so take extra precautions when you work on any part of the fuel system. Don't smoke or allow open flames or bare light bulbs near the work area, and don't work in a garage where a natural gas-type appliance is present. If you spill any fuel on your skin, rinse it off immediately with soap and water. When you perform any kind of work on the fuel system, wear safety glasses and have a fire extinguisher suitable for a class B type fire (flammable liquids) on hand.

Always perform fuel system procedures in a well-ventilated area to prevent a build-up of fumes.

Never work in a building containing a gas appliance with a pilot light, or any other form of naked flame. Ensure that there are no naked light bulbs or any sources of flame or sparks nearby.

Do not smoke (or allow anyone else to smoke) while in the vicinity of petrol (gasoline) or of components containing it. Remember the possible presence of vapour from these sources and move well clear before smoking.

Check all electrical equipment belonging to the house, garage or workshop where work is being undertaken (see the Safety first! section of this manual). Remember that certain electrical appliances such as drills, cutters etc. create sparks in the normal course of operation and must not be used near petrol (gasoline) or any component containing it. Again, remember the possible presence of fumes before using electrical equipment.

Always mop up any spilt fuel and safely dispose of the rag used.

Any stored fuel that is drained off during servicing work must be kept in sealed containers that are suitable for holding petrol (gasoline), and clearly marked as such; the containers themselves should be kept in a safe place. Note that this last point applies equally to the fuel tank if it is removed from the machine; also remember to keep its filler cap closed at all times.

Read the Safety first! section of this manual carefully before starting work.

Owners of machines used in the US, particularly California, should note that their machines must comply at all times with Federal or State legislation governing the permissible levels of noise and of pollutants such as unburnt hydrocarbons, carbon monoxide etc. that can be emitted by those machines. All vehicles offered for sale must comply with legislation in force at the date of manufacture and must not subsequently be altered in any way which will affect their emission of noise or of pollutants.

In practice, this means that adjustments may not be made to any part of the fuel, ignition or exhaust systems by anyone who is not authorised or mechanically qualified to do so, or who does not have the tools, equipment and data necessary to properly carry out the task. Also if any part of these systems is to be replaced it must be replaced with only genuine BMW components or by components which are approved under the relevant legislation. The machine must never be used with any part of these systems removed, modified or damaged.

 Warning: The very high output of the Motronic engine management system means that it can be very dangerous or even fatal to touch live components or terminals of any part of the system while in operation. Take care not to touch any part of the system when the engine is running, or even with it stopped and the ignition ON. Before working on an electrical component, make sure that the ignition switch is OFF, then disconnect the battery negative lead (-ve) and insulate it away from the battery terminal.

2 Fuel tank – removal and installation

 Warning: Refer to the precautions given in Section 1 before starting work, particularly the warnings relating to fuel and the Motronic system.

Removal

1 Remove the seat (see Chapter 8). Make sure the fuel filler cap is secure.
2 On RS and RT models, remove the fairing side panels, and on R and GS models remove the fuel tank trim (see Chapter 7). On RS models, also remove the screws securing the cockpit panels to the front of the tank.
3 Using hose clamps, clamp each fuel hose, then slacken the screw clamps securing the hoses to the pipes and detach the hoses, being prepared to catch any residue fuel **(see illustration)**.

HAYNES HiNT *Refer to Tools and Workshop Tips in the Reference section for hose clamping methods*

4 Disconnect the fuel pump and fuel level sender wiring connector **(see illustration)**.

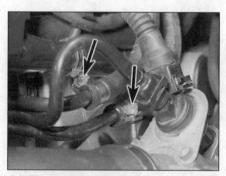

2.3 Slacken the clamps (arrowed) and pull the fuel hoses off the pipes

2.4 Disconnect the wiring connector ...

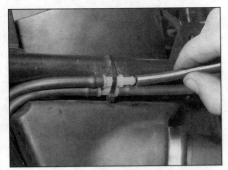

2.5 . . . and pull the breather hoses off the unions

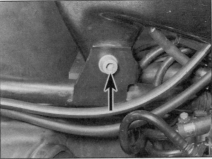

2.6a Unscrew the bolt (arrowed) . . .

2.6b . . . and remove the tank

5 Detach the breather hoses from the union on the rear sub-frame (see illustration).
6 Unscrew the bolt securing the rear of the right-hand side of the fuel tank to the frame (see illustration). Check that all hoses and wiring have been disconnected, then carefully lift the tank away from the machine (see illustration).
7 Inspect the tank mounting rubbers for signs of damage or deterioration and replace them if necessary.

Installation

8 Check that the tank mounting rubbers are fitted, then carefully lower the fuel tank into position (see illustration 2.6b). Install the bolt and tighten it to the torque setting specified at the beginning of the Chapter (see illustration).
9 Attach the fuel hoses to the pipes and the breather hoses to the unions, making sure they are secure (see illustrations 2.3 and 2.5).
10 Connect the fuel pump and fuel level sender wiring connector, making sure it is secure and correctly routed and cannot be trapped between the tank and the frame (see illustration 2.4).
11 Start the engine and check that there is no sign of fuel leakage, particularly around the clamps securing the fuel supply hoses to the unions on the tank and to the fuel pipes, then shut if off.
12 On RS and RT models install the fairing side panels, and on GS and R models install the fuel tank trim (see Chapter 7). On RS models, also install the screws securing the cockpit panels to the front of the tank.
13 Install the seat (see Chapter 7).

3 Fuel tank – cleaning and repair

1 All repairs to the plastic fuel tank should be carried out by a professional who has experience in this critical and potentially dangerous work. Even after cleaning and flushing of the fuel system, explosive fumes can remain and ignite during repair of the tank. If sending the tank for repair, first remove the fuel pump and filter assembly from inside the tank.
2 If the fuel tank is removed from the bike, it should not be placed in an area where sparks or open flames could ignite the fumes coming out of the tank. Be especially careful inside garages where a natural gas-type appliance is located, because the pilot light could cause an explosion.

4 Throttle bodies – removal, inspection and installation

⚠ **Warning:** *Refer to the precautions given in Section 1 before starting work, particularly the warnings relating to fuel and the Motronic system.*

Removal

1 On RS and RT models, remove the fairing side panels (see Chapter 7). Where fitted, remove the throttle body cover.

2.8 Install the mounting bolt and tighten it to the specified torque setting

2 Unscrew the bolt securing the earth (ground) cable to the bottom of the left-hand throttle body and detach the cable (see illustration).
3 Press in the clip on the throttle position sensor wiring connector on the left-hand throttle body and detach the connector (see illustration).
Caution: Do not remove the throttle position sensor from the throttle body as its position is pre-set. If it is disturbed and is not installed in exactly the original position, the information provided to the Motronic unit could be inaccurate, affecting engine performance.
4 Press in the clip on the injector wiring connector and detach the connector (see illustration).
5 Remove the fuel pipe clip on the top of the injector and detach the fuel pipe, being

4.2 Remove the bolt and detach the earth cable

4.3 Press in the clip and detach the TPS wiring connector

4.4 Press in the clip and detach the injector wiring connector

4.5a Remove the clip . . .

4.5b . . . and detach the fuel pipe from the injector

4.6a Slacken the clamp screws (arrowed) . . .

4.6b . . . and slide the duct back into the airbox

4.6c Slacken the clamp screw (arrowed) . . .

4.6d . . . and pull the throttle body out of the manifold

prepared to catch any residue fuel **(see illustrations)**. If required, remove the injector (see Section 5).

6 Slacken the clamp screws securing the air duct to the air box and the throttle body **(see illustration)**. Carefully pull the air duct off the throttle body and slide it back into the air box **(see illustration)**. Slacken the clamp screw securing the throttle body to the intake manifold on the cylinder head and carefully pull the throttle body out of the manifold **(see illustrations)**.

7 Detach the throttle cables and the choke cable from the throttle bodies (see Sections 6 and 7).

Inspection

8 Check the throttle bodies and air ducts for

cracks or any other damage which may result in air leakage.

9 Check that the throttle butterfly moves smoothly and freely in the body, and make sure that the inside of the body is completely clean.

10 Check that the throttle and choke cable cams move smoothly and freely, taking into account spring pressure. Clean any grit and dirt from around the cams.

11 Check the condition of all the O-rings and replace them if required **(see illustrations)**. It is advisable to replace the fuel pipe O-ring as a matter of course.

Installation

12 Installation is the reverse of removal, noting the following.

a) *Make sure the throttle bodies are fully engaged with the intake manifolds on the cylinder heads and with the air ducts. Make sure the tab on the throttle body locates in the notch in the manifold (see illustration).*

b) *Make sure the throttle body and air duct retaining clamps are secure, but take care not to overtighten them. If the tools are available, tighten them to the torque setting specified at the beginning of the Chapter.*

c) *Check the operation of the choke and throttle cables and adjust them as necessary (see Chapter 1).*

d) *Check idle speed and throttle synchronisation and adjust as necessary (see Chapter 1).*

4.11a Check the throttle body O-ring . . .

4.11b . . . and the fuel pipe O-ring and replace them if required

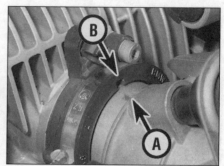

4.12 Locate the tab (A) in the notch (B)

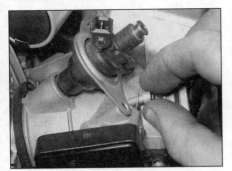

5.4a Remove the two screws . . .

5.4b . . . and lift off the bracket, manoeuvring it to clear the wiring socket

5.4c Gently pull the injector out of the throttle body

5 Injectors – removal and installation

> **Warning:** Refer to the precautions given in Section 1 before starting work, particularly the warnings relating to fuel and the Motronic system.

Removal

1 On RS and RT models, remove the fairing side panels (see Chapter 7). Where fitted, remove the throttle body cover.

2 Remove the fuel pipe clip on the top of the injector and detach the fuel pipe, being prepared to catch any residue fuel (see illustrations 4.5a and b).

3 Press in the clip on the injector wiring connector and detach the connector (see illustration 4.4).

4 Unscrew the two bolts securing the injector bracket and remove the bracket (see illustrations). Carefully pull the injector out of the throttle body (see illustration).

5 Modern fuels contain detergents which should keep the injectors clean and free of gum or varnish from residue fuel, additionally, the fuel filter should catch any dirt which may be present in the fuel or in the fuel tank. If either injector is suspected of being blocked, do not resort to cleaning it manually; instead use one of the aftermarket fuel injector cleaning agents which can be mixed with the fuel in the tank and run through the system to clean the injectors – make sure the cleaning agent manufacturer's instructions are followed.

Installation

6 Installation is the reverse of removal. Check the condition of the fuel pipe and injector O-rings and replace them if they are in any way damaged or deteriorated. It is advisable to replace them as a matter of course.

6 Throttle cables – removal and installation

> **Warning:** Refer to the precautions given in Section 1 before starting work, particularly the warnings relating to fuel and the Motronic system.

Note: Refer to the following procedure according to model year; note that 1993 to 1995 models have the throttle and choke cable adjusters located on the left-hand throttle body bracket, whereas 1996-on models have adjusters at the handlebar ends of the cables.

1993 to 1995 models

Removal

1 Remove the screw securing the cable elbow retainer and remove the retainer, noting how it fits (see illustration). Remove the elbow from its channel and free the cable nipple from the throttle pulley (see illustration).

2 Displace the throttle bodies (see Section 4, Steps 1 and 6). There is no need to disconnect the fuel hoses or the injector wiring connectors.

3 Slacken the locknuts on the main and joining cable adjusters on the left-hand throttle body bracket (see illustration). Thread the main cable adjuster out of the bracket, then free the inner cable nipple from the throttle cam (see illustrations).

6.1a Remove the screw securing the elbow retainer

6.1b Free the elbow and detach the throttle cable from the pulley

3

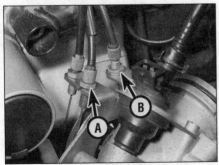

6.3a Main cable adjuster and locknut (A), joining cable adjuster and locknut (B)

6.3b Thread the main adjuster out of the bracket . . .

6.3c . . . and slip the inner cable and nipple out of the cam

6.8a Fit the inner cable into the pulley and the elbow into the channel . . .

6.8b . . . then fit the retainer and its screw

Now thread the joining cable adjuster out of the bracket. Slacken the locknut on the joining cable adjuster on the right-hand throttle body. Thread the adjuster out of the bracket and free the inner cable nipple from the throttle cam.

4 Withdraw the cable from the machine noting the correct routing.

Installation

5 Feed the cables through to the throttle bodies, making sure they are correctly routed and not kinked at any point. The cables must not interfere with any other component and should not be kinked or bent sharply.

6 Lubricate the joining cable nipple with grease and fit it into the right-hand throttle body cam, then thread the adjuster into the bracket, but do not yet tighten the locknut. Ensure that the outer cable end fits inside the head of the adjuster.

7 Lubricate the main cable nipple with grease. Thread the joining cable adjuster into the bracket on the left-hand throttle body, then feed the inner cable into its channel and the nipple into its socket in the throttle cam **(see illustration 6.3c)**. Thread the main cable adjuster into the bracket, but do not yet tighten the locknut **(see illustration 6.3b)**. Ensure that the outer cable end fits inside the head of the adjuster.

8 Lubricate the upper nipple with grease, then fit it into the throttle pulley and position the elbow in the channel **(see illustration)**. Fit the elbow retainer and secure it with the screw **(see illustration)**.

9 Install the throttle bodies (see Section 4).

6.13 Cable distributor unit location

10 Operate the throttle to check that it opens and closes freely. Turn the handlebars back and forth to make sure the cable doesn't cause the steering to bind.

11 Referring to Chapter 1, Section 8, pre-set cable freeplay and carry out throttle synchronisation.

12 Check that the idle speed does not rise as the handlebars are turned. If it does, the throttle cable is routed incorrectly. Correct the problem before riding the motorcycle.

1996-on models

Removal

13 The throttle, choke and throttle body link cables are all connected to a distributor unit located under the battery holder/ABS modulator **(see illustration)**. To replace the main throttle cable or either of the throttle body link cables, all four cables must be disconnected. Removal of the fairing side panels (RS and RT), fuel tank trim (R and GS) and fuel tank (all models) is advised for access (see Chapters 7 and 3).

14 Remove the screw securing the throttle cable elbow retainer and remove the retainer, noting how it fits **(see illustration 6.1a)**. Remove the elbow from its channel and free the cable nipple from the throttle pulley **(see illustration 6.1b)**.

15 Disconnect the choke cable from its operating lever (see Section 7, Step 1).

16 Displace the throttle bodies (see Section 4, Steps 1 and 6). There is no need to disconnect the fuel hoses or the injector wiring connectors. Slacken its locknut and screw in the adjuster on each throttle body link cable to create slack in the cable **(see illustration)**. Press in the tab on the throttle cam to release the cable trunnion from the cam.

17 Ease the cable distributor unit out of its housing under the battery holder/ABS modulator, towards the right-hand side. Make a sketch of the cable locations in the pulley, then release the spring clip and lift the distributor disc off its peg. Extract the cable required and make careful note of its routing before removing it from the machine.

Installation

18 Install the new cable so that it takes exactly the same route as the original. Make sure that all four cables are correctly seated in

the distributor disc and their guides, then install the disc retaining spring clip and slide the unit back into its housing **(see illustration)**.

19 Reconnect the cables in a reverse of their removal sequence, ensuring that the outer cable end fits inside the head of the adjuster on the throttle body bracket. Refit all disturbed components. Referring to Chapter 1, Section 8, pre-set cable freeplay and carry out throttle synchronisation.

20 Check that the idle speed does not rise as the handlebars are turned. If it does, the throttle cable is routed incorrectly. Correct the problem before riding the motorcycle.

7 Choke cable – removal and installation

> **Warning: Refer to the precautions given in Section 1 before starting work, particularly the warnings relating to fuel and the Motronic system.**

Note: Refer to the following procedure according to model year; note that 1993 to 1995 models have the throttle and choke cable adjusters located on the left-hand throttle body bracket, whereas 1996-on models have adjusters at the handlebar ends of the cables.

1993 to 1995 models

Removal

1 Carefully lever off the choke lever cap, then remove the screw and its washer **(see illustrations)**. Lift the lever out of the housing and pull the outer cable out of the housing, then free the inner cable nipple from its socket on the underside of the lever **(see illustrations)**.

2 Displace the left-hand throttle body (see Section 4, Steps 1 and 6); there is no need to disconnect the fuel hose or the injector wiring connector. Slacken the locknut on the choke cable adjuster, then thread the adjuster out of the bracket and slip the inner cable nipple out of the socket in the arm **(see illustration)**.

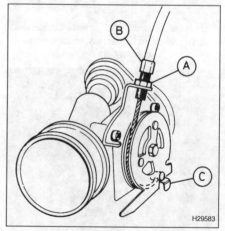

6.16 Throttle body link cable locknut (A), adjuster (B) and tab (C)

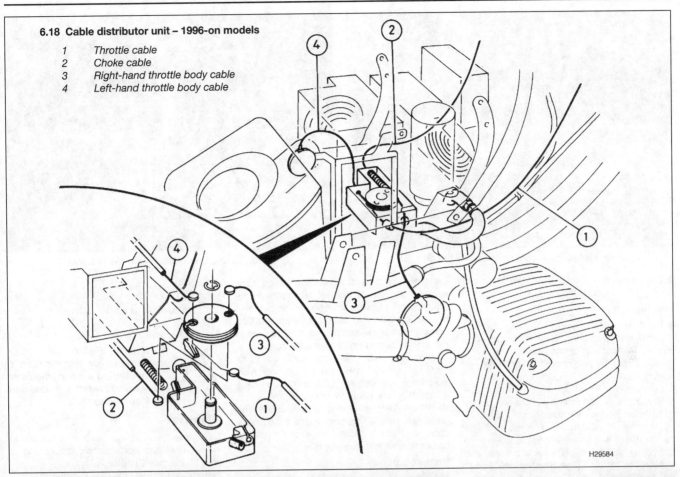

6.18 Cable distributor unit – 1996-on models

1   Throttle cable
2   Choke cable
3   Right-hand throttle body cable
4   Left-hand throttle body cable

H29584

7.1a Lever off the cap . . .

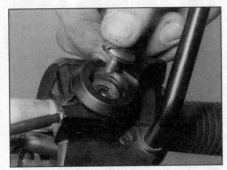

7.1b . . . then remove the screw . . .

7.1c . . . and its washer

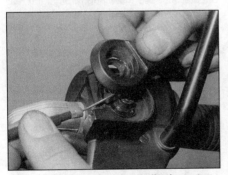

7.1d Lift off the lever and slip the outer cable out of the housing . . .

7.1e . . . then slip the inner cable nipple out of the socket

7.2 Slacken the locknut (A), unscrew the adjuster (B) and slip the nipple (C) out of the arm

3

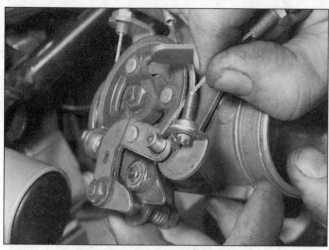

7.5a Fit the inner cable nipple into the arm . . .

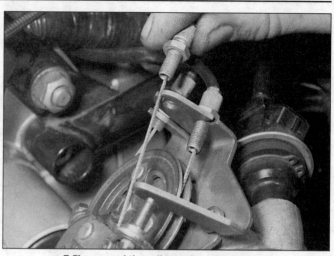

7.5b . . . and the adjuster into the bracket

3 Withdraw the cable from the machine noting the correct routing.

Installation

4 Feed the cable through to the left-hand throttle body, making sure it is correctly routed. The cable must not interfere with any other component and should not be kinked or bent sharply.
5 Lubricate the lower cable nipple with multi-purpose grease and attach it to the arm on the throttle body **(see illustration)**. Thread the outer cable adjuster into its bracket, setting it so that there is a small amount of freeplay in the cable before the choke is activated, and tighten the locknut **(see illustration)**. Install the throttle body.
6 Lubricate the upper cable nipple with multi-purpose grease and fit it into its socket on the underside of the choke lever **(see illustration 7.1e)**. Fit the end of the outer cable into the housing and position the lever on the housing, then fit the washer and tighten the screw **(see illustrations 7.1d, c and b)**. Fit the choke lever cap.
7 Check the operation and adjustment of the choke cable (see Chapter 1).

1996-on models

Removal

8 The choke, throttle and throttle body link cables are all connected to a distributor unit located under the battery holder/ABS modulator **(see illustration 6.13)**. To replace the main throttle cable or either of the throttle body link cables, all four cables must be disconnected. Removal of the fairing side panels (RS and RT), fuel tank trim (R and GS) and fuel tank (all models) is advised for access (see Chapters 7 and 3).
9 Remove the screw securing the throttle cable elbow retainer and remove the retainer, noting how it fits (see Section 6).
10 Disconnect the choke cable from its operating lever (see Step 1 above).
11 Displace the throttle bodies (see Section 4, Steps 1 and 6). There is no need to disconnect

the fuel hoses or the injector wiring connectors. Slacken its locknut and screw in the adjuster on each throttle body link cable to create slack in the cable **(see illustration 6.16)**. Press in the tab on the throttle cam to release the cable trunnion from the cam.
12 Ease the cable distributor unit out of its housing under the battery holder/ABS modulator, towards the right-hand side. Make a sketch of the cable locations in the pulley, then release the spring clip and lift the distributor disc off its peg. Extract the choke cable and make careful note of its routing before removing it from the machine.

Installation

13 Install the new choke cable, not forgetting to install the coil spring. Route the cable so that it takes exactly the same route as the original. Make sure that all four cables are correctly seated in the distributor disc and their guides, then install the disc retaining spring clip and slide the unit back into its housing **(see illustration 6.18)**.
14 Reconnect the cables in a reverse of their removal sequence. Refit all disturbed components. Referring to Chapter 1, Section 8, pre-set cable freeplay and carry out throttle synchronisation.
15 Check that the all cables operate smoothly before riding the motorcycle.

8.3 Unscrew the bolt (arrowed) and remove the duct

8 Airbox – removal and installation

Warning: Refer to the precautions given in Section 1 before starting work, particularly the warnings relating to fuel and the Motronic system.

Removal

1 Remove the seat and remove all the fairing side panels and body panels, according to your model (see Chapter 7).
2 Remove the fuel tank (see Chapter 3).
3 Remove the air filter (see Chapter 1). Unscrew the bolt securing the front of the air intake duct and lift the duct from the filter housing **(see illustration)**.
4 Remove the exhaust silencer (see Chapter 3).
5 Slacken the clamp screws securing the air ducts to the air box and the throttle bodies **(see illustration 4.6a)**. Carefully pull the air ducts off the throttle bodies and slide them back into the air box **(see illustration 4.6b)**.
6 Cut all the cable ties securing all the wiring to the rear sub-frame tubes. Remove the taillight unit and the rear turn signal units (see Chapter 8). Feed the wiring through the rear mudguard, noting its routing, and coil it on top of the fuse/relay box. All wiring must be clear of the rear sub-frame, as it is to be removed.
7 Unclip the fuse/relay box lid and remove it **(see illustration)**. Remove the four screws securing the fuse/relay holder section of the box to the wiring tray section, then remove the four screws securing the wiring tray to the rear mudguard and the frame **(see illustration)**. Free the wiring loom grommets from the cut-outs in the tray **(see illustrations)**. Disconnect the rear brake light switch wiring connector and, where fitted, the rear ABS sensor wiring connector. Lift the fuse/relay holder out of the frame and position it in the air filter housing so that it is out of the way. Working around the

8.7a Remove the fusebox lid . . .

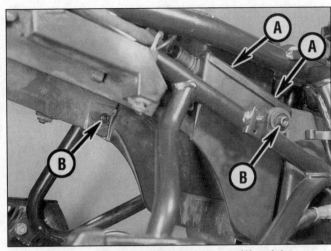

8.7b . . . followed by the holder screws (A) and the tray screws and bolts (B) on each side

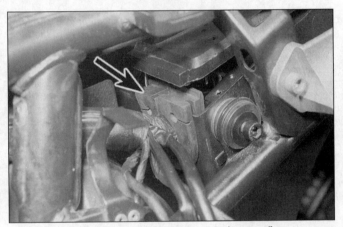

8.7c Pull the front wiring grommet (arrowed) . . .

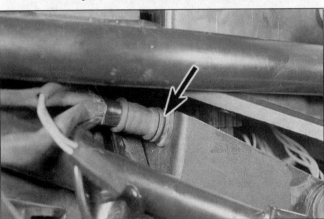

8.7d . . . and the rear grommet (arrowed) out of their cut-outs

3

rear frame, check that all wiring and electrical system components are clear of the rear frame. Trace the neutral switch, gear indicator, oil pressure switch and side stand wiring and disconnect them at the connectors. Feed the wiring back the components, noting its routing, and making sure it is free of any ties and clear of the frame.

8 Free the rear brake fluid reservoir from its clips and wrap it in a plastic bag to prevent any

fluid spillage **(see illustration)**. Also free the rear brake hose-to-pipe union grommet from the bracket on the frame **(see illustration)**.

9 On RT models, remove the bolts which secure the rear of each footrest mounting bracket to the rear sub-frame.

10 Place a small block of wood between the top of the gearbox casing and the swingarm to prevent them folding together when the rear shock absorber upper mounting bolt is

removed. Also place a jack under the gearbox to take the weight when it is separated from the engine. Check which side the rear shock upper mounting bolt is installed, then unscrew the two bolts and remove the seat support for that side **(see illustration)**. This allows clearance for the bolt to be withdrawn. Unscrew and remove the rear shock upper mounting bolt. Where fitted, also remove the bolt securing the remote pre-load adjuster.

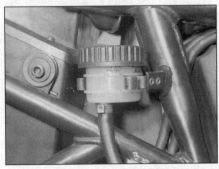

8.8a Release the reservoir from its clips . . .

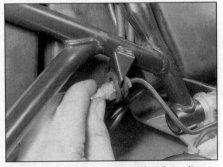

8.8b . . . and the pipe union from its grommet

8.10 Remove the seat support to provide clearance for the shock absorber bolt (arrowed)

8.11a Front frame strut/rear sub-frame through-bolt (A), rear sub-frame mounting bolts (B) (right-hand side)

8.11b Free the three pipes from the right-hand side . . .

11 Check around to verify that all cables, wiring, hoses, pipes, electrical components and anything else are all free from the rear sub-frame. Remove the screws securing the air box to the rear sub-frame. Unscrew the nut from the end of the through-bolt which secures the front upper rear sub-frame mountings and the strut side rails to the engine **(see illustration)**. Supporting the rear sub-frame, unscrew the four bolts which secure it to the engine and gearbox, then carefully manoeuvre it backwards off the engine and gearbox, making sure it is free from all cables, pipes and hoses, and remove it. Free the fuel supply pipes from the grommets at the front of the airbox **(see illustrations)**. Remove the air box.

Installation

12 Installation is the reverse of removal, noting the following points:

a) *When installing the rear sub-frame, install all four bolts and the through bolt that also secures the strut side rails finger-tight before tightening any of them to the specified torque setting. When torquing the bolts, tighten the right-hand rear bolt first, followed by the right-hand front bolt, then the left-hand front bolt and finally the left-hand rear bolt. Then tighten the through bolt to the specified torque.*

b) *Tighten all bolts to the torque settings specified either at the beginning of this Chapter, or at the beginning of the relevant Chapter elsewhere in the book.*
c) *Check the condition of all breather tubes, grommets and O-rings, and of all hose and pipe clamps before re-using them, and replace them if they are worn, damaged or deteriorated.*
d) *Make sure all wires, cables, pipes and hoses are correctly routed and connected, and secured by any clips or ties.*
e) *Check the operation of all electrics and lights before taking the bike on the road.*

> **9 Exhaust system –** 🔧
> removal and installation

⚠️ **Warning: Refer to the precautions given in Section 1 before starting work, particularly the warnings relating to fuel and the Motronic system.**
⚠️ **Warning: If the engine has been running the exhaust system will be very hot. Allow the system to cool before carrying out any**

work. This applies particularly to catalytic converter equipped models due to the high temperature of the converter.

Silencer

Note: *Whilst it is possible to remove the silencer separately from the downpipe assembly, the joint between the two components may well be seized, making them very difficult to separate. If this is the case, it is advisable to remove the silencer and downpipes as a complete assembly, thus avoiding the possibility of damaging anything.*

Removal

1 On models fitted with a catalytic converter, remove the fuel tank (see Section 2), then trace the wiring from the Lambda sensor on the silencer and disconnect it at the connector. Free the wiring from any clips or ties and feed it through to the silencer, noting its routing. If required, unscrew the sensor from the silencer after the silencer has been removed.
2 Slacken the bolt on the clamp securing the silencer to the downpipe assembly **(see illustration)**.

> **HAYNES HiNT** *The silencer clamp bolt can become corroded and seized. It is advisable to spray it with WD40 or a similar product before attempting to slacken it.*

3 Unscrew the bolts securing the silencer to the centre stand brackets **(see illustration)**.
4 Unscrew the bolt securing the silencer to the left-hand rear footrest bracket **(see illustration)**. On RS and RT models, the silencer is secured via a rubber mounting block. Release the silencer from the exhaust downpipe assembly using a slight twisting motion **(see illustration)**.

8.11c . . . and the single pipe from the left

9.2 Slacken the clamp bolt . . .

9.3 ... then unscrew the front mounting bolt (arrowed) on each side ...

9.4a ... and the rear mounting bolt (RS and RT models shown) ...

9.4b ... and remove the silencer

9.5a The rear mounting on RS and RT models is secured by a clip ...

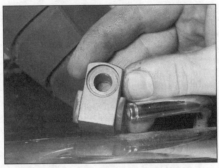

9.5b ... and slides off the bar

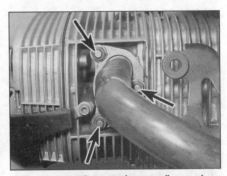

9.8 Unscrew the nuts (arrowed) securing the downpipe assembly to each head

5 Inspect the rubber mounting bushes for signs of damage and replace them if necessary. The front mountings are each secured by a nut. On RS and RT models, the rear is secured by a clip (see illustrations).

Installation

6 Installation is the reverse of removal. Apply some copper grease to the silencer clamp bolt threads. Tighten the silencer mounting bolt and clamp bolts, and the Lambda sensor where fitted, to the torque settings specified at the beginning of the Chapter. Run the engine and check the system for leaks.

Downpipe assembly

Note: *Whilst it is possible to remove the downpipe assembly separately from the silencer, the joint between the two may well be seized, making them very difficult to separate. If this is the case, it is advisable to remove the downpipes and silencer as a complete assembly, thus avoiding the possibility of damaging anything.*

Removal

7 Slacken the bolt on the clamp securing the downpipe assembly to the silencer (see *Haynes Hint* and illustration 9.2).

8 Unscrew the three nuts securing each side of the downpipe assembly to the cylinder heads (see illustration). Carefully manoeuvre the assembly off the heads and out of silencer and lower it to the floor.

9 Remove the gasket from each port in the cylinder heads and discard them as new ones must be fitted (see illustration).

Installation

10 Fit a new gasket into each of the cylinder head ports (see illustration). Apply a smear of grease to the gaskets to keep them in place whilst fitting the downpipe if necessary.

3

9.9 Remove the old gaskets ...

9.10 ... and fit new ones

9.11a Locate the downpipe assembly . . .

9.11b . . . then install the nuts . . .

9.11c . . . and tighten the nuts to the specified torque setting

11 Manoeuvre the assembly into position so that the head of each downpipe is located in its port in the cylinder head **(see illustration)**, and the rear of the assembly fits into the silencer **(see illustration 9.4b)**. Install the nuts and tighten them to the torque setting specified at the beginning of the Chapter **(see illustrations)**. Also tighten the silencer clamp bolt to the specified torque.
12 Run the engine and check the system for leaks.

---

10 Catalytic converter and lambda sensor –
general information

 Warning: Refer to the precautions given in Section 1 before starting work, particularly the warnings relating to fuel and the Motronic system.

Catalytic converter

1 US models and certain European models have a catalytic converter located in the exhaust system to minimise the amount of pollutants which escape into the atmosphere; the catalytic converter is also available as an option in markets where it is not fitted as standard. The catalytic converter operates within a three-way closed-loop system, with exhaust gas oxygen content information being fed back to the Motronic unit by the Lambda sensor.
2 The catalytic converter is simple in operation and requires no maintenance.

  a) *Always use unleaded fuel – the use of leaded fuel will destroy the converter.*
  b) *Do not use any fuel or oil additives.*
  c) *Keep the fuel and ignition systems in good order – if the fuel/air mixture is suspected of being incorrect have it checked on an exhaust gas analyser.*
  d) *If the catalytic converter is ever removed from the downpipes, handle it carefully and do not drop it.*

Lambda (oxygen) sensor

 Warning: The engine should never be run, nor should the ignition be switched on, with the Lambda sensor disconnected or removed from the exhaust, otherwise the Motronic fault memory will require resetting by a BMW dealer.

3 The Lambda sensor constantly measures exhaust gas oxygen content and relays this information back to the Motronic control unit. The Motronic unit can respond immediately to changes oxygen content and adjust the fuel/air mixture accordingly.
4 The sensor is threaded into the joint between the exhaust downpipe and silencer box on the right-hand side of the machine. To remove the sensor, first trace its wiring up to the block connector; this should be located under the fuel tank on the right-hand side of the frame – use the wiring diagram at the end of this manual for wire colour identification. Use an open-ended spanner to unscrew the sensor, noting that if there is not enough clearance to withdraw it fully, the silencer must be removed (see Section 9); do not risk damage to the sensor's tip if space is limited.
5 Before installing the Lambda sensor, clean the threads in the exhaust system and the sensor threads. Thread it into place and tighten it securely, to the specified torque setting if some means can be devised of applying the torque.
6 Re-route the sensor wiring up to the block connector and reconnect it. Install the fuel tank if this was removed for access to the connector. Secure the wiring with cable ties. Start the engine and check that there are no air leaks from the Lambda sensor for exhaust downpipe joint with the silencer.

Chapter 4
Ignition system

Contents

General information 1
Motronic control unit – removal and installation 5
Motronic control unit – testing 6
Ignition (main) switch – check, removal and installation ..see Chapter 8
Ignition HT coil – check, removal and installation 3
Ignition system – check 2
Ignition timing – checking 7
Timing sensors – removal and installation 4
Spark plugs – gap check and replacement ....see Chapter 1

Degrees of difficulty

Easy, suitable for novice with little experience	Fairly easy, suitable for beginner with some experience	Fairly difficult, suitable for competent DIY mechanic 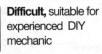	Difficult, suitable for experienced DIY mechanic	Very difficult, suitable for expert DIY or professional

Specifications

General information
Engine management system .... Motronic MA 2.2
Spark plugs and gaps .... See Chapter 1

Ignition timing
At idle .... 0° BTDC
Full advance .... 43° BTDC

Ignition HT coils
Primary winding resistance .... 0.5 ohms
Secondary winding resistance .... 13 K ohms

Torque settings
Alternator belt lower drive pulley bolt .... 50 Nm
Breather pipe
  8 mm bolt .... 20 Nm
  Banjo bolt .... 25 Nm

1 General information

Ignition is controlled electronically by the Bosch Motronic MA 2.2 engine management system, first fitted to BMW's 16-valve K-series bikes. Apart from spark plug maintenance and renewal, the engine management system is totally maintenance-free.

The system comprises a timing sensor for each cylinder, the Motronic control unit, ignition HT coil and the triple-electrode spark plugs.

Ignition timing and fuel/air delivery to the engine, are digitally controlled by the Bosch Motronic MA 2.2 engine management system. Information on throttle angle, provided by the TPS (throttle position sensor) on the left-hand throttle body, is fed into the Motronic unit together with engine speed and TDC (top dead centre) information from the timing sensors on the front of the crankshaft. Ignition advance is also controlled by the Motronic unit. Additional information is provided by the air temperature sensor in the air filter housing lid, the oil temperature sensor, the CO potentiometer on the rear sub-frame (models without a catalytic converter), and the Lambda sensor in the exhaust (models with a catalytic converter).

All information supplied to the Motronic unit is compared with its pre-programmed data. The appropriate output signals are then sent to the fuel injectors, fuel pump and ignition HT coils and determine the amount of fuel required and the duration of the injection period.

The system incorporates a safety interlock circuit which will cut the ignition if the side stand is put down whilst the engine is running and in gear, or if a gear is selected whilst the engine is running and the side stand is down.

Because of their nature, the individual ignition system components can be checked but not repaired. If ignition system troubles occur, and the faulty component can be isolated, the only cure for the problem is to replace the part with a new one. Keep in mind that most electrical parts, once purchased, cannot be returned. To avoid unnecessary expense, make very sure the faulty component has been positively identified before buying a replacement part.

Warning: The very high output of the Motronic engine management system means that it can be very dangerous or even fatal to touch live components or terminals of any part of the system while in operation. Take care not to touch any part of the system when the engine is running, or even with it stopped and the ignition ON. Before working on an electrical component, make sure that the ignition switch is OFF, then disconnect the battery negative lead (-ve) and insulate it away from the battery terminal. Additionally, if a system component is disconnected whilst the ignition is switched on a fault will be indicated in the Motronic fault memory, necessitating resetting on the Bosch diagnostic tester.

3.4 The HT coil is secured by two bolts (arrowed)

2 Ignition system – check

⚠ **Warning: The very high output of the Motronic engine management system means that it can be very dangerous or even fatal to touch live components or terminals of any part of the system while in operation. Take care not to touch any part of the system when the engine is running, or even with it stopped and the ignition ON.**

1 As no means of adjustment is available, any failure of the system can be traced to failure of a system component or a simple wiring fault. Of the two possibilities, the latter is by far the most likely. In the event of failure, check the system in a logical fashion, as described below. Before making any tests, check that the battery is fully charged and that all fuses are in good condition.

Checking for a spark

2 Place the machine on its centre stand. Remove the spark plug covers and disconnect the HT leads from the spark plugs (see Chapter 1). Connect each lead to a spare spark plug and lay each plug on the engine with the threads contacting the engine. Make sure the plug threads are in contact with the engine otherwise the Motronic system will be damaged. If the spark plug caps must be held, make sure you use an insulated tool – see **Warning** above.

3.5a Disconnect the primary circuit wiring connector (A), the earth wire (B) . . .

⚠ **Warning: Do not remove either of the spark plugs from the engine to perform this check – atomised fuel being pumped out of the open spark plug hole could ignite, causing severe injury! Also ensure that there is no fuel vapour present in the working area.**

3 Having observed the above precautions, check that the kill switch is in the RUN position, the gearbox is in neutral and the side stand is up, then turn the ignition switch ON and turn the engine over a couple of times on the starter motor. If the system is in good condition a regular, fat blue spark should be evident at each plug electrode. If the spark appears thin or yellowish, or is non-existent, further investigation will be necessary. Turn the ignition off, remove the test spark plugs and install the plug caps.

Ignition fault finding

4 Ignition faults can be divided into two categories, namely those where the ignition system has failed completely, and those which are due to a partial failure. The likely faults are listed below, starting with the most probable source of failure. Work through the list systematically, referring to the subsequent sections for full details of the necessary checks and tests, where information is available.

⚠ **Warning: Before working on an electrical component, make sure that the ignition switch is OFF, then disconnect the battery negative lead (-ve) and insulate it away from the battery terminal.**

a) Loose, corroded or damaged wiring connections, broken or shorted wiring between any of the component parts of the ignition system (see Chapter 8).
b) Faulty HT lead or spark plug cap, faulty spark plug, dirty, worn or corroded plug electrodes, or incorrect gap between electrodes.
c) Faulty ignition (main) switch or engine kill switch (see Chapter 8).
d) Faulty neutral or side stand switch (see Chapter 8).
e) Faulty timing sensor.
f) Faulty ignition HT coil.
g) Faulty Motronic control unit.

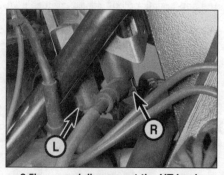

3.5b . . . and disconnect the HT leads (arrowed)

5 If the above checks don't reveal the cause of the problem, have the ignition system tested by a BMW dealer equipped with the diagnostic tester. The diagnostic tester permits full testing of the Motronic system components and reads the Motronic control unit fault code memory.

3 Ignition HT coil – check, removal and installation

⚠ **Warning: The very high output of the Motronic engine management system means that it can be very dangerous or even fatal to touch live components or terminals of any part of the system while in operation. Take care not to touch any part of the system when the engine is running, or even with it stopped and the ignition ON. Before working on an electrical component, make sure that the ignition switch is OFF, then disconnect the battery negative lead (-ve) and insulate it away from the battery terminal.**

Check

1 In order to determine conclusively that the ignition HT coil is defective, it should be tested by a BMW dealer equipped with the Bosch diagnostic tester.
2 However, the coil can be checked visually (for cracks and other damage) and the primary and secondary coil resistances can be measured with a multimeter. If the coil is undamaged, and if the resistance readings are as specified at the beginning of the Chapter, it is probably capable of proper operation.
3 Remove the fuel tank (see Chapter 3). Disconnect the battery negative (-ve) lead. On R and GS models, displace the Motronic control unit to access the coils.
4 The coil is mounted across the frame behind the steering head and is secured by two bolts **(see illustration)**. Unscrew the bolts and remove the coil.
5 Disconnect the primary circuit electrical connector, the earth (ground) wire and both HT leads from the coil **(see illustrations)**. Label each lead and its terminal on the coil as a means of correct location on refitting.
6 To check the condition of the primary windings, set the meter to the ohms x 1 scale and connect its probes across the two primary wire terminals on the coil, marked 15 and 1. This will give a resistance reading of the primary windings of the coil and should be consistent with the value given in the Specifications at the beginning of the Chapter.
7 To check the condition of the secondary windings, set the meter to the K ohms scale and connect its probes across the two HT lead terminals on the coil, marked 4a and 4b. This will give a resistance reading of the

4.4a Unscrew the pulley bolt (arrowed) and remove the pulley

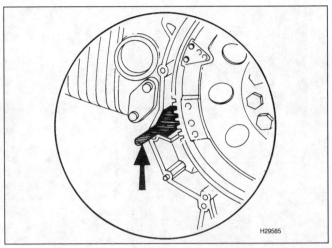

4.4b BMW tool No. 11 5 640 used to lock engine via flywheel

secondary windings of the coil and should be consistent with the value given in the Specifications at the beginning of the Chapter.

8 Should any of the above checks not produce the expected result, have your findings confirmed by a BMW dealer with the diagnostic tester (see Step 1). If the coil is confirmed to be faulty, it must be replaced; the coil is a sealed unit and cannot therefore be repaired.

9 The spark plug leads and caps may breakdown over a period of time and produce a mysterious ignition fault. To check the leads, disconnect them from the coil, one at a time, and carry out a continuity check from one end of the lead to the other. Although the cap and length of lead have a resistance value, if infinite resistance is shown (open-circuit) the lead/cap should be replaced – the cap is integral with the lead.

Removal

10 Remove the fuel tank (see Chapter 3). Disconnect the battery negative (-ve) lead. On R and GS models, displace the Motronic control unit to access the coil.

11 The coil is mounted across the frame behind the steering head and is secured by two bolts **(see illustration 3.4)**. Unscrew the bolts and lift the coil upwards to access the wire connections.

12 Disconnect the primary circuit electrical connector, the earth (ground) wire and the HT leads from the coil **(see illustrations 3.5a and b)**. Mark the locations of all wires and leads before disconnecting them, especially the HT leads.

13 Note the routing of the HT leads and remove them if required.

Installation

14 Installation is the reverse of removal. Make sure the wiring connectors and HT leads are correctly routed and fully inserted in their correct locations on the coil.

---

4 Timing sensors –
check, removal
and installation

> **Warning: The very high output of the Motronic engine management system means that it can be very dangerous or even fatal to touch live components or terminals of any part of the system while in operation. Take care not to touch any part of the system when the engine is running, or even with it stopped and the ignition ON. Before working on an electrical component, make sure that the ignition switch is OFF, then disconnect the battery negative lead (-ve) and insulate it away from the battery terminal.**

Check

1 BMW provide no resistance specification for the timing sensors, or Hall sensors as they are sometimes known. If they are thought to be faulty, first check that this is not due to damaged or broken wiring between the sensors and Motronic control unit, or a corroded wiring connector. If checking the wiring for continuity, first ensure that the ignition is OFF and the battery negative (-ve) lead is disconnected. Other than a wiring check, operation of the sensors can only be checked by a BMW dealer equipped with the Bosch diagnostic tester.

Removal

Caution: Do not remove the timing sensor plate from the inner cover unless absolutely necessary, as the ignition timing could be upset. If the plate is removed, make some accurate alignment marks between it and the cover to ensure it is correctly installed.

2 Remove the fuel tank (see Chapter 3). Disconnect the battery negative (-ve) lead. Remove the alternator drive belt (see Chapter 1, Section 18).

3 On all except RS models up to May 1994, unscrew the two bolts securing the breather pipe and remove the pipe. Discard the O-ring from the lower fitting and the sealing washers from the banjo bolt as new ones must be used.

4 Unscrew the bolt securing the alternator belt drive pulley and remove the pulley complete with timing rotor, noting how it locates in the notch in the end of the crankshaft **(see illustration)**. It will be necessary to lock the engine to prevent the crankshaft turning whilst slackening the bolt. This can be done in a number of ways:

a) *BMW provide a tool No. 11 5 640 (see illustration) which locks the flywheel, and can be fitted after removing the starter motor (see Chapter 8).*

b) *If the engine is in the frame, select a gear and have an assistant press on the rear brake pedal.*

c) *Use a TDC locking pin to stop the engine from rotating (see Haynes Hint).*

The engine can be locked in the TDC position by inserting a steel rod or bolt at least 100 mm long into the hole in the clutch housing on the left-hand side of the engine. There are corresponding holes in the flywheel and the rear of the crankcase. Locate the rod through all the holes, thereby locking the flywheel, and hence the engine, in the TDC position.

4

4.5a The sensor plate is secured by three bolts (arrowed)

4.5b Slacken the bolt (arrowed) to free the sensor wiring grommet

5 Make some accurate alignment marks between the sensor plate and the cover to ensure it is installed in exactly the same position. Unscrew the bolts securing the timing sensor plate and remove the plate **(see illustration)**. Note how the sensor wiring grommet locates in the cut-out and is secured by the tab – slacken the bolt to free it **(see illustration)**. Trace the sensor wiring and disconnect it at the connector inside the rubber boot **(see illustration)**.

Installation

6 Install the timing sensor plate, making sure the marks made on removal are exactly in

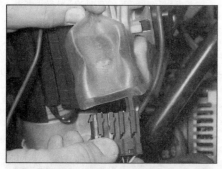

4.5c Disconnect the sensor wiring at the connector

alignment otherwise the ignition timing will be incorrect, then tighten its bolts securely **(see illustration 4.5a)**. Fit the grommet into its cut-out and secure it with the tab, then reconnect the wiring at the connector **(see illustrations 4.5b and c)**.

7 Install the rotor/pulley unit onto the end of the crankshaft, making sure the tab on the inside of the pulley locates in the notch in the end of the crankshaft **(see illustration)**. Note that on 1995-on models, the rotor is secured to the inside face of the pulley with Loctite instant adhesive – if the components require re-bonding, make sure both surfaces are degreased beforehand and apply the adhesive following its maker's instructions. Install the bolt and tighten it to the specified torque setting, using the method employed on removal to prevent the crankshaft turning **(see illustrations)**.

8 Install the breather pipe (except early RS models) using a new O-ring and sealing washers and tighten the bolts to the specified torque settings.

9 Install the alternator drive belt (see Chapter 1, Section 18). Check the drive belt along its entire length for splits, cracks, worn or damaged teeth, frays and any other damage or deterioration. Be careful not to bend the belt excessively or to get oil or grease on it.

Replace the belt if it is in any way worn, damaged or deteriorated, or if it is due for renewal under the maintenance schedule (see Chapter 1).

10 Reconnect the battery negative (-ve) lead and install the fuel tank (see Chapter 3).

Caution: If you are in any doubt as to the correct positioning of the timing sensor plate, check the timing as described in Section 7.

---

5 Motronic control unit –
    removal and installation

⚠️ *Warning: The very high output of the Motronic engine management system means that it can be very dangerous or even fatal to touch live components or terminals of any part of the system while in operation. Take care not to touch any part of the system when the engine is running, or even with it stopped and the ignition ON. Before working on an electrical component, make sure that the ignition switch is OFF, then disconnect the battery negative lead (-ve) and insulate it away from the battery terminal.*

4.7a Locate the tab (A) in the notch (B) . . .

4.7b . . . then install the bolt . . .

4.7c . . . and tighten it to the specified torque setting

5.2 The Motronic unit is secured by two screws on each side. The left-hand screws also secure two earth leads (arrowed)

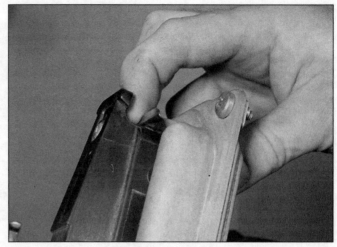

5.3a Press in the clip . . .

Removal

1 Remove the fuel tank (see Chapter 3). Disconnect the battery negative (-ve) lead.

2 Remove the four screws securing the Motronic unit in its holder, noting the two earth (ground) leads secured by the screws on the left-hand side (see illustration). Lift the unit out of the holder.

3 Press in the clip and disconnect the wiring connector from the Motronic unit (see illustrations). Handle the control unit carefully – it is a delicate and extremely expensive component.

Installation

4 Installation is the reverse of removal. Make sure the wiring connector is correctly and securely connected.

6 Motronic control unit – testing

1 The preceding sections of this chapter outline the checks which can be made by the DIY mechanic using home workshop equipment. If a fault is evident and the following systems are in good order, the machine must be taken to a BMW dealer

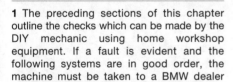

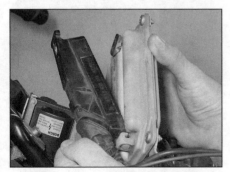

5.3b . . . and detach the connector

equipped with the Bosch diagnostic tester.

a) Make sure the battery is fully-charged.

b) Make sure all fuses are in good condition.

c) Check the spark plugs, caps and HT leads.

d) Check that the earth (ground) connection to the frame and all component earths are sound.

e) Check all wiring between the system components, checking that it is not trapped, pinched or badly routed.

f) Scrape off any corrosion from electrical connections and spray them with electrical contact cleaner.

2 This diagnostic tester permits full testing of the Motronic system components and reads the Motronic control unit fault code memory. The tester connects into the machine wiring harness via the diagnostic connector socket. A code displayed on the tester indicates which component is at fault and the tester allows the fault to be erased from the unit's memory once it has been traced and remedied.

7 Ignition timing – checking

1 Setting the ignition timing requires the use of a BMW TDC locking pin No. 11 2 650 (or the improvised tool – see *Haynes Hint* in Section 4) and a device which will indicate precisely when the timing sensor sends a pulse to the Motronic control unit. BMW recommend a service tool 12 3 650 which is used with a test lead 12 3 652; with this equipment an LED lights when the timing sensor reaches the firing point. Since it is expensive and not likely to be required very often, this equipment places the task beyond the scope of most owners.

2 Accordingly, owners are advised to take their machine to an authorised BMW dealer for the work to be carried out. For those who have the equipment, proceed as follows.

3 Remove the fuel tank (see Chapter 3). Disconnect the battery negative (-ve) lead. Trace the timing sensor wiring connector and disconnect it (see illustration 4.5c). Connect up the test lead and ignition tester to the timing sensor side of the connector.

4 Unscrew the four bolts securing the engine front cover and remove the cover (see illustration).

5 Remove the timing inspection plug from behind the right-hand cylinder (see illustration). Turn the engine until the "OT" mark on the flywheel, visible via the timing inspection hole, aligns with the middle of the hole (see illustration). The engine can be turned by selecting a high gear and rotating the rear wheel by hand in its normal direction of rotation. Alternatively, remove the engine front cover and turn the engine using a 17 mm spanner or socket on the belt drive pulley bolt, turning it in a clockwise direction only (see illustration).

Caution: Be sure to turn the engine in its normal direction of rotation only.

6 At this point, insert the TDC locking pin in the hole in the clutch housing on the left-hand side of the engine. There is a corresponding hole in the flywheel and the rear of the crankcase. Locate the rod through all the holes, thereby locking the flywheel, and hence the engine, in the TDC position.

7.4 Remove the four bolts (arrowed) securing the engine front cover

4

7.5a Remove the inspection hole blanking plug (arrowed)

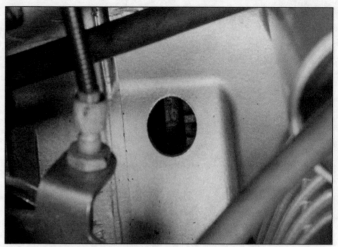

7.5b Align the "OT" mark with the middle of the hole

7.5c Turn the engine using a spanner or socket as shown

7 Slacken the three bolts which retain the timing sensor plate – these are just accessible with the belt in place – and turn the plate until the tester LED goes out. At this precise point, tighten the bolts.

8 Disconnect the test equipment and remake the timing sensor connector. Remove the TDC locking pin.

9 Refit all disturbed components. Reconnect the battery negative lead and install the fuel tank. If the ignition timing is still suspected of being inaccurate, the fault may lie in one of the Motronic system components or the Motronic control unit itself; have the system checked by a BMW dealer equipped with the Bosch diagnostic tester.

Chapter 5
Frame, suspension and final drive

Contents

Degrees of difficulty

Easy, suitable for novice with little experience	Fairly easy, suitable for beginner with some experience	Fairly difficult, suitable for competent DIY mechanic 🔧	Difficult, suitable for experienced DIY mechanic 🔧	Very difficult, suitable for expert DIY or professional 🔧

Specifications

Front suspension

Type	BMW Telelever
Fork oil type	BMW telescopic fork oil (5W or 10W)
Fork oil capacity	0.47 litre
Fork tube runout limit	0.4 mm

Rear suspension

Type	BMW Paralever

Final drive

Final drive oil type	Hypoid gear oil, API class GL5
Final drive oil viscosity	
Above 5°C	SAE 90
Below 5°C	SAE 80
All conditions	SAE 80W 90
Final drive oil capacity	
Oil change	0.23 litre
Following overhaul	0.25 litre
Final drive gear backlash	0.07 to 0.16 mm

Torque settings

Brake pedal pivot bolt	37 Nm
Gearchange lever pivot bolt	
R, RS and GS models	35 Nm
RT models	18 Nm
Centre stand pivots/pivot bolts	21 Nm
Side stand bolt	42 Nm
Handlebar mounting bolts	
RS models	
Handlebar bracket to rubber mounts	40 Nm
Rubber mounts to upper fork bridge	40 Nm
Handlebar end-weights	7 Nm
R and RT models	
Handlebars to upper fork bridge	21 Nm
Handlebar end-weights	20 Nm
GS models	
Handlebar clamp bolts	21 Nm
Handlebar end-weights	20 Nm

5

Torque settings

Brake lever pivot bolt	8 Nm
Forks to lower fork bridge	22 Nm
Fork clamp bolts (RS models)	22 Nm
Fork top bolt nut (R, RT and GS models)	45 Nm
ABS sensor bolts	4 Nm
Front shock absorber	
Upper mounting nut	47 Nm
Lower mounting bolt nut	
RS models with 8.8 bolt (number marked on bolt head)	43 Nm
RS models with 10.9 bolt (number marked on bolt head)	50 Nm
R, RT and GS models	50 Nm
Ball joint in lower fork bridge	230 Nm
Telelever nut-to-ball joint	130 Nm
Telelever pivot shaft bolt	73 Nm
Telelever threaded cap	42 Nm
Steering head ball joint-to-strut (RS models)	230 Nm
Steering head ball joint nut (RS models)	130 Nm
Steering head bearing bolt (R, RT and GS models)	130 Nm
Steering damper mounting bolts	20 Nm
Steering damper bracket to fork bridge	9 Nm
Steering damper bracket to Telelever	20 Nm
Rear shock absorber mounting bolts	
RS models with 8.8 bolt (number marked on bolt head)	43 Nm
RS models with 10.9 bolt (number marked on bolt head)	50 Nm
R, RT and GS models	50 Nm
Remote pre-load adjuster bolt (RT and GS models)	22 Nm
Swingarm right-hand pivot bolt	150 Nm
Swingarm left-hand pivot bolt	7 Nm
Swingarm left-hand pivot bolt locknut	105 Nm
Final drive housing right-hand pivot bolt	150 Nm
Final drive housing left-hand pivot bolt	7 Nm
Final drive housing left-hand pivot bolt locknut	105 Nm
Torque arm-to-final drive housing	43 Nm
Footrest holder mounting bolts (RT models)	
6 mm bolts	6 Nm
8 mm bolts	21 Nm
10 mm bolts	42 Nm

1 General information

There is no frame in the traditional sense because all components mount directly to the engine/transmission unit. Front and rear suspension systems pivot on the engine casing or swingarm at one end and to separate sub-frames at the other. The fuel tank, fairing, battery, ABS modulator and all electrical and electronic components mount directly to sub-frames, which are in turn bolted to the engine/transmission unit.

Front suspension and steering are managed separately by BMW's Telelever system. Telelever uses a combination of telescopic forks to support the front wheel and provide steering, and a swingarm and shock absorber arrangement to provide suspension.

Unlike conventional front forks, the telescopic forks contain neither damping mechanism nor springs, the fork tubes locate in nylon bushes set inside the fork sliders, and the oil contained in the fork slider is there merely to lubricate the fork tube and bush surfaces. There is a large overlap between the fork tubes and fork sliders, making the forks particularly stiff and stable. At the top, the fork tubes are held in a fork bridge, which is mounted to the suspension strut (or frame), and at the bottom, the fork sliders are mounted to a fork bridge which pivots on the Telelever ball joint.

The Telelever control arm, pivots at the engine unit on needle-roller bearings, and is fixed directly to the shock absorber lower mounting and ball joint on the fork lower bridge. The Showa shock absorber handles all springing and damping tasks.

Rear suspension is controlled by BMW's Paralever system, first seen on the 2-valve engined Boxers. The aluminium single-sided swingarm pivots at the transmission casing and at the final drive housing, allowing the drive housing a certain amount of movement in the same plane as the suspension. By linking the drive housing to the transmission unit with a torque arm, a parallelogram-shaped linkage is formed. A Showa shock absorber is mounted between the swingarm and rear sub-frame.

The drive to the rear wheel is by shaft, housed inside the swingarm. The final drive housing turns the drive through 90° to the rear wheel.

2 Frame – general information

1 In the case of these BMW models, there is little resembling a conventional frame, the front and rear suspension pivoting directly on the engine/transmission unit. The only framework on the machine consists of tubular steel sub-frames at the rear and fuel tank underside, and a cast aluminium strut at the front of the machine.

2 The rear sub-frame supports the seat, rear mudguard, rear lighting and signals, passenger footrests, and provides the top mounting for the shock absorber. The sub-frame bolts directly to the engine's crankcase and gearbox casing.

3 The front strut is bolted to the engine casting and provides the top mounting for the front shock absorber and the mounting for the fork top bridge and handlebar assembly.

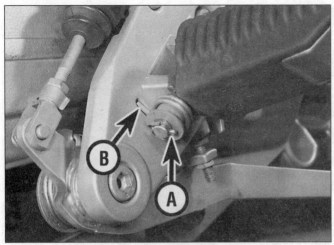

3.1a Remove the split pin (A) to free the pivot pin.
Note how the spring ends locate against the bracket (B) . . .

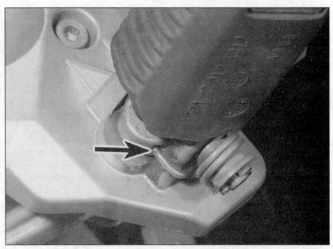

3.1b . . . and in the footrest (arrow)

3 Footrests, brake pedal and gearchange lever – removal and installation

Footrests

Removal

1 To remove the riders footrests, remove the split pin from the bottom of the footrest pivot pin, then withdraw the pivot pin and remove the return spring sleeve, the spring and the footrest (see illustrations). Note how the spring ends locate in the footrest and on the bracket.

2 To remove the passenger footrests, remove the split pin and washer from the bottom of the footrest pivot pin, then withdraw the pivot pin and remove the footrest.

3 If they are worn, the footrest rubbers can be drawn off the footrest and replaced.

Installation

4 Installation is the reverse of removal.

3.5a Release the clip . . .

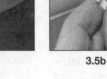

3.5b . . . and remove the pin

Always use new split pins to secure the pivot pins.

Brake pedal

Removal

5 Release the spring clip from around the base of the master cylinder pushrod, then swing it down and withdraw the pin (see illustrations). Separate the pushrod from the brake pedal.

6 Unscrew the nut on the inside of the pedal and withdraw the pivot bolt, then remove the pedal, noting how the return spring ends locate on the pedal and the lug on the inside of the bracket (see illustrations).

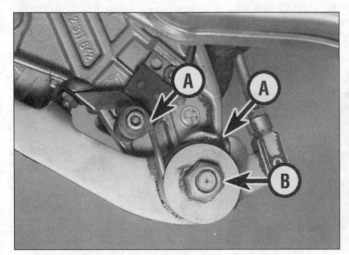

3.6a Note how the spring ends (A) locate, then unscrew the nut (B) . . .

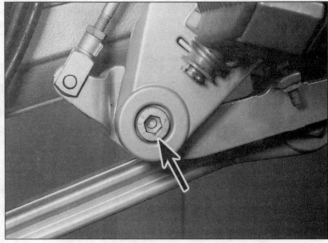

3.6b . . . and remove the pivot bolt (arrowed)

5

3.9a Remove the bolt (arrowed) and slide the arm off the shaft

Installation

7 Installation is the reverse of removal. Remove the bush from the lever, clean it and re-grease it on the inside and the outside. Make sure the spring ends are correctly positioned. Check the amount of brake pedal freeplay (see Chapter 1, Section 20).

Gearchange lever

Removal

8 On RT models, remove the left-hand fairing side panel.
9 To remove the lever with its linkage, unscrew the pinch bolt securing the linkage arm on the gearchange shaft on the gearbox, then slide the arm off the shaft, noting its alignment so that it can be installed in the same position **(see illustration)**. Now unscrew the pivot bolt securing the lever to the bracket and remove the lever with the linkage **(see illustration)**.

4.2 Unhook the centre stand springs

4.4a Unscrew the pivot bolts (arrowed) . . .

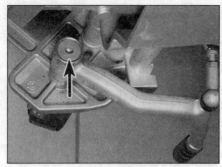

3.9b Unscrew the pivot bolt (arrowed) and remove the lever

10 To remove the lever by itself, release the clip securing the linkage rod to the lever, then prise the linkage rod socket off the ball joint on the lever **(see illustrations)**. Now unscrew the pivot bolt securing the lever to the bracket and remove the lever **(see illustration 3.9b)**.

Installation

11 Installation is the reverse of removal. Remove the sleeve from the lever, clean it and re-grease it on the inside and the outside.

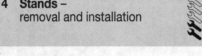

| 4 | Stands – removal and installation |

Centre stand

1 The centre stand is secured to the engine. Support the bike on its side stand.
2 On RT models, remove the lower fairing section. On GS models, remove the sump guard (see Chapter 7). On all models, unhook the springs **(see illustration)**.
3 On RT, R and GS models, and RS models from 4/95, remove the pivot caps, then unscrew the pivots and remove the stand.
4 On RS models up to 4/95, unscrew the pivot bolts, then withdraw the bushes and remove the stand **(see illustrations)**.
5 Inspect the stand, pivots and bushes for signs of wear and replace them if necessary. Apply a smear of grease to all moving parts and fit the stand back on the bike. Clean all pivot and bolt threads and apply Loctite 2701 or equivalent to them **(see illustration)**.

4.4b . . . and remove the bushes

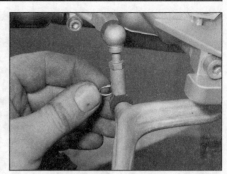

3.10a Release the clip . . .

3.10b . . . and prise the rod off the lever

Tighten the pivots or pivot bolts to the torque setting specified at the beginning of the Chapter.
6 Make sure the springs are in good condition and capable of holding the stand up when not in use. A broken or weak spring is an obvious safety hazard. Reconnect the springs.

Side stand

7 The side stand is attached to a pivot bracket on the engine. Springs anchored to the bracket ensure that the stand is held in the retracted or extended position.
8 Support the bike on its centre stand. Unhook the stand springs **(see illustration)**.
9 Unscrew the side stand bolt and remove the stand from its bracket **(see illustrations)**. Remove the bush from the bracket and clean it. On installation apply grease to the bush and Loctite 2701 to the threaded section of the side stand pivot **(see illustrations)**.

4.5 Apply threadlock to the pivot bolt threads

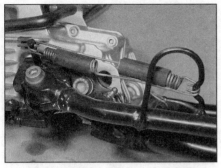

4.8 Unhook the side stand springs

4.9a Unscrew the bolt (arrowed) . . .

4.9b . . . and remove the side stand

4.9c Grease the bush . . .

4.9d . . . and apply threadlock to the threads

Tighten the bolt to the torque setting specified at the beginning of the Chapter. Reconnect the side stand springs and check that the return spring holds the stand securely up when not in use – an accident is almost certain to occur if the stand extends while the machine is in motion.

10 Refer to Chapter 8 for information on the side stand switch.

5 Handlebars and levers – removal and installation

Handlebars

Removal

Note: *The handlebars fitted on all R, RS and RT models can be removed individually. If only one side is being removed, ignore the steps*

5.2 Pull back the grip to access the screw (arrowed)

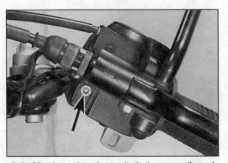

5.5 Slacken the clamp bolt (arrowed) and slide the clutch lever assembly off the handlebar

which apply to the other handlebar. Note that RS models available in certain markets may have one-piece handlebars..

1 On models with standard (unheated) handlebar grips, first remove the weight from the handlebar end. Carefully cut through the grips with a sharp knife and remove them from the bars; this is necessary in order to slide the brake master cylinder and clutch lever bracket assemblies off the bars. It may be possible to remove the grips without cutting them, but it is likely to prove difficult as they are glued in place. Obtain new grips for installation.

2 On models with heated handlebar grips, first remove the weight from the handlebar end, then fold back the inner end of the grip and remove the screw **(see illustration)**. Disconnect the wiring connector and release the cable shoes, then carefully pull the wiring through and remove the grip.

5.6 Slacken the master cylinder clamp bolt (arrowed)

3 Detach the choke cable (see Chapter 3) and the clutch cable (see Chapter 2) from their levers, and the throttle cable from the pulley (see Chapter 3).

4 Remove the left-hand and right-hand switch assemblies (see Chapter 8). There is no need to disconnect the wiring from its connector, but free it from any ties on the handlebar.

5 Slacken the clutch lever bracket clamp bolt and slide the bracket off the end of the handlebar **(see illustration)**.

6 Slacken the brake master cylinder clamp bolt, but do not yet attempt to slide it off the handlebar **(see illustration)**.

7 Unscrew the handlebar mounting bolt(s) and remove the bars **(see illustrations)**. On RS models, the handlebars can be removed individually by removing the bolt on the adjuster assembly **(see illustration)**, or together by removing the bolts securing the

5

5.7a Handlebar mounting bolts (arrowed) – R and RT models

5.7b Handlebar mounting bolts (arrowed) – GS models

5.7c Individual handlebar mounting bolt – RS models

5.7d Handlebar assembly mounting bolts (arrowed) – RS models

5.7e Remove the cover to access the bolts

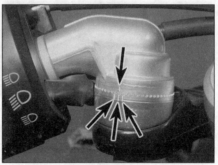

5.7f Note the alignment of the index marks before removing the bars

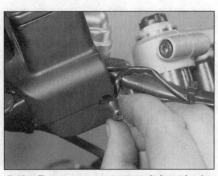

5.9 Check the small grommets between the brackets for wear

bracket to the rubber mounts on the upper fork bridge **(see illustration)**. Remove the cover to access the rubber mounting bolts **(see illustration)**. If removing the handlebars individually, note the position of the bars in the adjusters before removal **(see illustration)**. Slide the brake master cylinder off the end of the handlebar and support it in an upright position so that no strain is placed on the hose.

8 If necessary, unscrew the handlebar end-weight retaining screws, then remove the weights from the end of the handlebars.

Installation

9 Installation is the reverse of removal, noting the following.
  a) *Fit the brake master cylinder and clutch lever bracket so that the clamp mating surfaces align with the punch mark on the underside of the handlebar.*

  b) *On RS models, set the handlebar positions as required, but make sure the position is the same on each side. Check the condition of all the rubber mounting components and replace any that are worn, damaged or deteriorated* **(see illustration)**.
  c) *On GS models, make sure the handlebars are central in the clamps.*
  d) *Tighten the handlebar mounting bolts to the torque setting specified at the beginning of the Chapter.*
  e) *Tighten the brake master cylinder clamp bolt and the clutch lever bracket clamp bolt to the torque setting specified at the beginning of the Chapter.*
  f) *On models with heated handlebar grips, take care to locate the pin on the heater unit in the slot (early models) or hole (later models) in the handlebar end. Apply Loctite 2701 or equivalent to the*

  handlebar end-weight retaining screws.
  g) *On models without heated grips, secure the new grips to the handlebar with a suitable adhesive such as Loctite 638. Apply Loctite 2701 or equivalent to the handlebar end-weight retaining screws.*

Levers

Removal

10 To remove the front brake lever first remove the cover from the front of the front brake master cylinder; it is retained by a single screw **(see illustration)**. Unscrew the lever pivot bolt from the underside of the master cylinder **(see illustration)** and carefully withdraw the lever complete with the brake operating pushrod.

11 To remove the clutch lever, unscrew the pivot screw locknut and remove the wave washer, then withdraw the pivot screw and

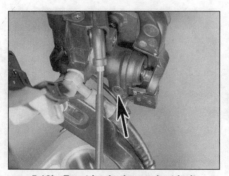

5.10a Remove cover to permit front brake lever removal

5.10b Front brake lever pivot bolt (arrowed)

5.11a Unscrew the locknut (arrowed) . . .

remove the lever, noting how the cable nipple fits into the socket in the lever (**see illustrations**). Take care not to lose the nipple as it is not attached to the inner cable.

Installation

12 Install the clutch lever in a reverse of the removal procedure. Apply grease to the pivot bolt shank, the bush and the contact areas between the lever and its bracket.

13 Before installing the front brake lever, fully unscrew the pushrod in the lever and remove all traces of old grease and threadlock. Apply Loctite 270 or equivalent to the pushrod threads and screw the pushrod into the lever, but only a little way in at this stage. Apply a smear of Optimoly MP3 grease to the pushrod plain section and tip. Install the lever, guiding the pushrod into the master cylinder boot. Apply a smear of grease to the pivot bolt shank and a drop of non-permanent thread locking compound (BMW recommend Tuflok blue) to its threads. Screw the pivot bolt into place to secure the lever, tightening it to the specified torque setting.

14 Carefully, screw in the pushrod screw until all freeplay in the lever is just taken up, then tighten it a further 1/2 turn (**see illustration**). BMW advise that the pushrod head is sealed with a dab or paint following correct adjustment. Install the master cylinder cover.

Caution: Do not tighten the pushrod screw any more than 1/2 a turn, otherwise the brakes could lock on or fail.

5.11b ... and remove the pivot screw (arrowed) to free clutch lever

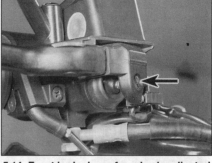

5.14 Front brake lever freeplay is adjusted at pushrod screw

6 Front suspension – removal, inspection and installation

Forks

Removal

1 Position the bike on its centre stand. Work can be made easier by raising the machine to a suitable working height on an hydraulic ramp or a suitable platform. Make sure the motorcycle is secure and will not topple over (see *Tools and Workshop Tips* in the Reference section). On RS and RT models, though not essential, it is advisable to remove the fairing side panels to avoid the possibility of damage should a tool slip.

2 Remove the front wheel (see Chapter 6).

3 On ABS models, unscrew the two bolts securing the ABS sensor to the left-hand fork slider and withdraw the sensor, taking care not to lose any of the shims (**see illustration**). Release the wiring from its clips on the inside of the fork leg and secure it with the brake caliper. It is advisable to secure all the shims to the sensor using a piece of wire through the bolt holes.

4 On RS models, slacken the fork clamp bolts in the upper fork bridge (**see illustration**). Remove the fork caps and press in the valve on the fork to release any air pressure (**see illustration**). Keeping the valve depressed, push the fork tube down through the bridge and fully into the slider.

5 On R and RT models, unscrew the handlebar mounting bolts and displace the bars to access the fork tops (**see illustration 5.7a**). There is no need to detach any wiring, hoses or cables. On R, RT and GS models, remove the fork cap, then counter-hold the fork top bolt via the hex on its body and unscrew the nut (**see illustration**). Push the fork tube down through the bridge and fully into the slider.

6 On R and RT models, unscrew the brake line union nut and position the line clear, making sure no strain is placed on the hoses or pipe (**see illustration**).

7 Unscrew the bolts securing the fork to the lower fork bridge and remove the fork (**see illustration**).

6.3a Unscrew the sensor bolts ...

6.3b ... and remove the sensor and all its shims

6.4a Slacken the clamp bolt (arrowed) and remove the fork cap ...

6.4b ... then depress the valve and slide the tube down through the bridge

6.5 Counter-hold top bolt hex (arrowed) whilst nut is removed

5

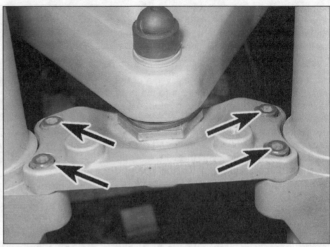

6.6 Remove the union mounting nut

6.7 Unscrew the bolts (arrowed) securing the sliders to the bridge and remove the forks

Inspection

8 Draw the tube out of the slider **(see illustration)**, releasing any suction effect on RS models by pressing in the valve **(see illustration 6.4b)** or on all other models, by slackening the bleed screw (all other models) on the fork top bolt **(see illustration 6.16)**. If required, invert the slider and drain the oil, or alternatively, remove the drain screw from the base of the fork slider and allow the oil to drain.

9 Check the fork tube for score marks, scratches, flaking of the chrome finish and excessive or abnormal wear. Look for dents in the tube and replace the tube in both forks if any are found. Check the fork tube for runout

6.8 Draw the fork tube out of the slider

using V-blocks and a dial gauge **(see illustration)**. If the amount of runout exceeds the service limit specified, the tube should be replaced. Note that each fork slider contains four nylon bushes which support the fork tube and reduce fork noise; these bushes are available as replacement parts should renewal be necessary.

 Warning: If either fork tube is bent, it should not be straightened; replace both fork tubes with new ones.

10 On R, RT and GS models, check the fork top joints in the upper bridge. If they are worn or damaged, remove the top circlip and drift the joint out from the underside of the bridge using a suitable drift or socket. Drive the new joint into place and fit the circlip.

11 Inspect the area above the dust seal on the fork tubes for signs of oil leakage, then carefully lever off the dust seal using a flat-bladed screwdriver and inspect the area around the fork seal **(see illustration)**. If leakage is evident, discard the dust seal and replace the oil seal as follows.

12 Carefully remove the retaining clip, taking care not to scratch the surface of the tube **(see illustration)**.

13 Carefully lever out the oil seal using a large flat-bladed screwdriver, taking care not

to damage the rim of the slider, then remove the washer. If the seal is difficult to remove, heat the top of the slider with a hot-air gun. Check the fork seal seat in the slider for nicks, gouges and scratches. If damage is evident, leaks will occur.

14 Install the oil drain screw (with a new O-ring) in the base of the fork slider.

15 Refill the slider with the specified type and amount of oil.

16 Oil the fork tube with the specified fork oil and insert it into the slider. Push the tube fully into the slider releasing any pressure effect by pressing in the valve (RS models) or slackening the bleed screw (all other models) on the fork top bolt **(see illustration)**.

17 Install the oil seal washer, then install the new oil seal **(see illustrations)**. Smear the seal's lips with fork oil and slide it over the tube so that its markings face upwards and drive the seal into place using a suitable piece of tubing or drift; the tubing must be slightly larger in diameter than the fork tube and slightly smaller in diameter than the seal recess in the slider. Take care not to scratch the fork tube during this operation; it is best to make sure that the fork tube is pushed fully into the slider so that any accidental scratching is confined to the area above the oil seal. Make sure the retaining clip groove is visible above the seal.

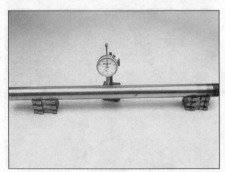

6.9 Check fork tube runout using V-blocks and a dial gauge

6.11 Prise out the dust seal using a flat-bladed screwdriver

6.12 Prise out the retaining clip using a flat-bladed screwdriver

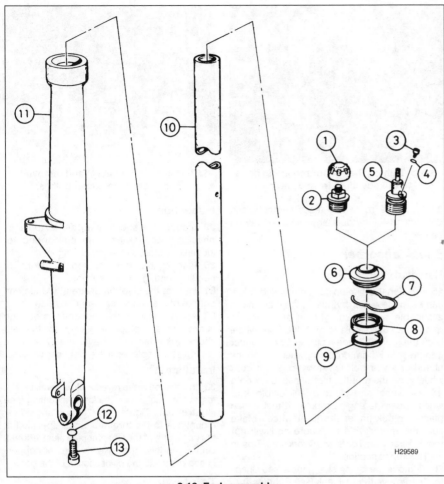

6.17a Install the washer . . .

6.17b . . . followed by the oil seal, making sure it is the correct way up

6.16 Fork assembly

1	Cap – RS models	4	Bleed screw O-ring
2	Top bolt and air valve –	5	Top bolt – R, RT, GS
	RS models		models
3	Bleed screw – R, RT,	6	Dust seal
	GS models	7	Retaining clip
		8	Oil seal

9	Washer
10	Fork tube
11	Slider
12	O-ring
13	Oil drain screw

6.18 Fit the retaining clip into its groove . . .

18 Once the seal is correctly seated, fit the retaining clip, making sure it is correctly located in its groove **(see illustration)**.
19 Lubricate the lips of the new dust seal then slide it down the fork tube and press it into position **(see illustration)**.

Installation

20 Install the forks onto the lower fork bridge **(see illustration 6.7)**. Apply Loctite 243 or equivalent to the bolt threads and tighten them to the torque setting specified at the beginning of the Chapter.

21 On RS models set the fork tubes in the upper bridge so that they project 4.5 to 5.5 mm above the bridge; release any suction effect by pressing in the valve on the fork top bolt **(see illustration)**. **Note:** *The forks must be bled of air in the no-load (extended) position.* Tighten

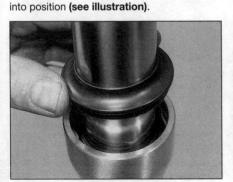

6.19 . . . then press the dust seal into the top of the slider

6.21a Set the projection of each fork tube above the bridge equally and as specified

6.21b On RS models, tighten the clamp bolt to the specified torque setting

5

6.23a Secure the wiring in the clips (arrowed) and route the sensor as shown

6.23b Tighten the sensor mounting bolts to the specified torque setting

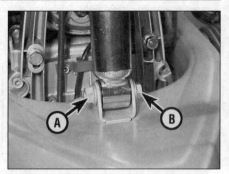

6.27 Unscrew the nut (A) and remove the shock absorber lower bolt (B)

the clamp bolts to the specified torque setting **(see illustration)**.

22 On R, RT and GS models, insert the forks into the joints set in the upper bridge, releasing any suction effect by slackening the bleed screw on the fork top bolt. **Note:** *The forks must be bled of air in the no-load (extended) position.* Counter-hold the top bolt hex **(see illustration 6.5)** and tighten the nut to the specified torque setting. Install the fork top caps. On R and RT models, mount the handlebars and tighten the bolts to the specified torque setting.

23 On ABS models, route the sensor through the bracket on the left-hand fork slider and fit the wiring in the clips **(see illustration)**. Install the sensor with all the shims previously fitted **(see illustration 6.3b)** and tighten the bolts to the specified torque setting **(see illustration)**. Check the sensor air gap (see Chapter 6).

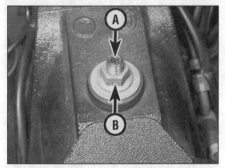

6.28 Counter-hold the stud (A) and unscrew the nut (B)

24 Install the front wheel (see Chapter 6), and on RS and RT models any removed fairing panels.

Shock absorber

Removal

25 Position the bike on its centre stand. Work can be made easier by raising the machine to a suitable working height on an hydraulic ramp or a suitable platform. If the front wheel is off the ground, place a block of wood under it to prevent it from dropping when the shock absorber lower bolt is removed. If the front wheel is on the ground, tilt the bike back onto its rear wheel and place a jack or block of wood under the engine to keep it there, then place a block under the front wheel. Make sure the motorcycle is secure and will not topple over (see *Tools and Workshop Tips* in the Reference section).

26 Remove the fairing side panels (see Chapter 7). Remove the fuel tank (see Chapter 3).

27 Unscrew the nut and withdraw the bolt from the shock absorber lower mounting **(see illustration)**.

28 Counter-hold the stud on the top of the shock using an Allen key and unscrew the nut **(see illustration)**. Remove the washer and the upper rubber bush, noting how it fits. Remove the block from under the front tyre to lower the Telelever, thereby freeing the shock from its mount, and collect the washers. Draw the shock down and out of the strut; free the lower rubber bush and spacer from the upper mounting to allow the shock to be angled to clear the Telelever.

Inspection

29 Inspect the shock absorber for obvious physical damage and the coil spring for looseness, cracks or signs of fatigue.

30 Inspect the damper rod for signs of bending, pitting and oil leakage.

31 Inspect the pivot hardware at the top and bottom of the shock for wear or damage.

32 Apart from the shock mounting components, replacement parts for the rear shock absorber are not available. The entire unit must be replaced if it is worn or damaged.

Installation

33 Installation is the reverse of removal. Do not forget the washers on the lower mounting **(see illustration)**. Tighten the upper and lower mountings to the torque settings specified at the beginning of the Chapter **(see illustrations)**. On RS models, apply Never Seez or copper-based grease to the pivot section of the bolt.

Telelever

Removal

34 Position the bike on its centre stand. Work can be made easier by raising the machine to a suitable working height on an hydraulic ramp or a suitable platform. Make sure the motorcycle is secure and will not topple over (see *Tools and Workshop Tips* in the Reference section).

35 Remove the forks (see above).

36 On RS and GS models, unscrew the bolt securing the front brake hose union to the underside of the lower fork bridge, then position the calipers clear of the Telelever,

6.33a Fit a washer on each side of the shock ...

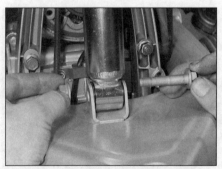

6.33b ... then raise the Telelever and install the bolt and nut ...

6.33c ... and tighten the nut to the specified torque

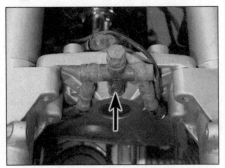

6.36 Unscrew the bolt (arrowed) securing the brake hose union

6.39a Remove the plastic caps . . .

making sure no strain is placed on the hoses (see illustration).

37 On R models, if the lower fork bridge is to be separated from the Telelever, remove the steering damper (see Section 8). Otherwise it can be left in place.

38 Unscrew the shock absorber lower mounting bolt nut, but do not yet remove the bolt (see illustration 6.27).

39 Remove the plastic cap from each end of the Telelever pivot (see illustration). Unscrew the threaded cap on the inside of the left-hand pivot (see illustration). Remove the circlip and withdraw the inner cap from the right-hand pivot (see illustration). Unscrew the pivot shaft bolt from the right-hand end of the shaft (see

illustration), then drift the shaft out and withdraw it from the left (see illustration 6.45a).

40 Supporting the Telelever, withdraw the shock absorber lower bolt and draw the Telelever forward off the engine. Take care not to lose the washers from the shock mount.

Inspection

41 Thoroughly clean all components, removing all traces of dirt, corrosion and grease. Inspect all components closely, looking for obvious signs of wear such as heavy scoring, and cracks or distortion due to accident damage. Any damaged or worn component must be replaced.

42 Inspect the ball joint on the fork bridge for

signs of wear or damage. Hold the Telelever and move the fork bridge around – there should be no noticeable play, and the joint should move smoothly and freely with no signs of roughness or notchiness. Note that on R models, the steering damper must be removed to make this check (see Section 8). If any wear or damage is evident, the joint must be replaced. Remove the joint cap (see illustration). Counter-hold the joint stud and unscrew the nut securing the Telelever to the fork bridge and separate them (see illustration); use a screwdriver as a lever if necessary, but take care not to scratch or gouge the surfaces. Note: *The ball joint nut has been known to work loose on some models, and this nut has therefore been replaced by a locknut. If a locknut is not fitted on your model, obtain one from a BMW dealer.*

> **HAYNES HiNT** *If the nut is difficult to remove, heat it using a hot-air gun. This softens the threadlock and will make it easier to remove the nut. This will work on the particular threadlock that BMW use on assembly, however if the nut has since been tightened using a different type of threadlock, heat may actually make it more difficult to remove. If this is the case, allow the parts to cool and try again.*

6.39b . . . then unscrew the threaded cap in the left-hand pivot (arrow)

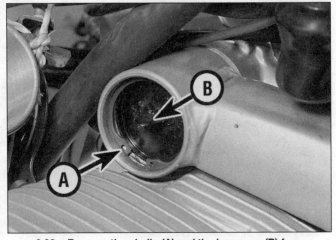

6.39c Remove the circlip (A) and the inner cap (B) from the right-hand pivot . . .

6.39d . . . then unscrew the pivot shaft bolt (arrowed)

6.42a Prise off the cap . . .

6.42b . . . then counter-hold the stud and unscrew the nut

6.43a Apply threadlock to the stud . . .

6.43b . . . then fit a new nut

6.43c Tighten the nut to the specified torque

43 Clamp the bridge in a soft-jawed vice, using rags to protect it, and unscrew the ball joint. Install the new joint and tighten it to the torque setting specified at the beginning of the Chapter. Fit the bridge onto the Telelever, then apply Loctite 2701 or equivalent to the ball joint stud

threads and install a new nut (the nut is of the self-locking type and can only be used once); tighten the nut whilst counter-holding the stud **(see illustrations)**. The nut must be tightened to the specified torque, so the final tightening has to be done without counter-holding the joint.

Make a reference mark across the nut and stud to check that the joint does not turn with the nut and tighten it to the specified torque **(see illustration)**. Mark the nut with a dab of paint to indicate that the later type self-locking nut is fitted. Fit the joint cap **(see illustration)**.

6.43d Fit the cap onto the ball joint

6.44a Lever out the spacer . . .

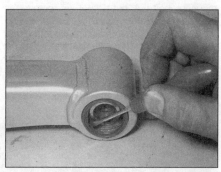

6.44b . . . and the seal

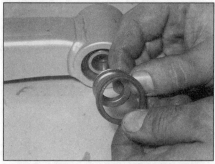

6.44c Fit the new seal onto the spacer

6.45a Install the Telelever pivot shaft . . .

6.45b . . . and its bolt . . .

6.45c . . . and tighten the bolt to the specified torque

6.45d Fit the inner cap . . .

6.45e . . . and secure it with the circlip

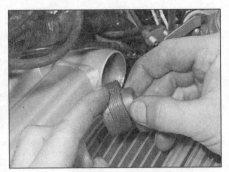

6.46a Grease the threaded cap threads . . .

6.46b . . . then fit it into the left-hand pivot . . .

6.46c . . . and tighten it to the specified torque

44 Ease out the spacer and seal from the inner side of the Telelever pivots **(see illustrations)**. Wipe all old grease off the four bearings in the Telelever (two on each side). Referring to *Tools and Workshop Tips (Sections 5 and 6)* in the Reference Section, inspect the bearings for wear or damage. If the bearings do not run smoothly and freely or if there is excessive freeplay, they must be replaced. Lubricate the bearings with fresh grease. Check the condition of the bearing seals and replace them if necessary. Lubricate the lips of the new seals with grease and install them over the shoulder on the spacers **(see illustration)**.

Installation

45 On 1993 to 1995 RS models, apply grease to the pivot section of the shock absorber bolt. Fit the Telelever onto its engine mountings and loosely install the shock absorber bolt to keep it in place, not forgetting the washers **(see illustration 6.33a)**. Lubricate the pivot shaft with grease and slide it through from the left **(see illustration)**. Install the shaft bolt and tighten it to the torque setting specified at the beginning of the Chapter **(see illustrations)**. Fit the inner cap over the bolt and secure it with the circlip, making sure it fits properly in its groove **(see illustrations)**.

46 Apply a smear of Never Seez or copper-based grease to the threads of the threaded cap, then install it in the left-hand side of the Telelever and tighten it to the specified torque setting **(see illustrations)**. Fit the plastic caps to each side, making sure their O-rings are in good condition **(see illustration)**. Replace them if they are worn, damaged or deteriorated.

47 Thread the nut onto the shock absorber bolt and tighten it to the specified torque setting **(see illustration 6.33c)**.

48 On R models, if removed, install the steering damper (see Section 8).

49 On RS and GS models, fit the front brake hose union onto the underside of the fork bridge and tighten the bolt securely **(see illustration)**.

50 Install the forks (see above).

6.46d Replace the cap O-rings if necessary

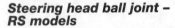

7 Steering head ball joint (RS) or bearing (R, RT and GS) – removal and installation

Steering head ball joint – RS models

Removal

1 Place the motorcycle on its centre stand and support it so that the front wheel is raised off the ground. Prise off the ball joint cap **(see illustration)**. Counter-hold the ball joint stud and unscrew the nut (see *Haynes Hint* – Section 6) **(see illustration)**.

2 Slacken the fork clamp bolts in the upper fork bridge **(see illustration 6.4a)**. Remove the fork caps and press in the valve on the fork to release any air pressure **(see illustration 6.4b)**. Keeping the valve depressed, push the

7.1a Prise off the cap . . .

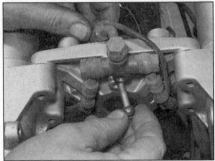

6.49 Fit the brake hose union onto the fork bridge

top of the fork down through the bridge and fully into the slider. Disconnect the ignition (main) switch wiring at the connector. Remove the bolts which secure the handlebar bracket to the upper bridge (see Section 5).

3 Lift off the upper bridge and place it clear of the ball joint. Use a screwdriver as a lever if necessary, but take care not to scratch or gouge the surfaces. Make sure there is no strain placed on the cables, hose and wiring.

Inspection

4 Inspect the ball joint for signs of wear or damage. Move the joint around – there should be no noticeable play, and the joint should move smoothly and freely with no signs of roughness or notchiness. If any wear or damage is evident, unscrew the joint and discard it.

5 Install the new joint and tighten it to the torque setting specified at the beginning of the Chapter.

5

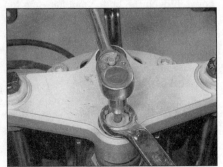

7.1b . . . then counter-hold the stud and remove the nut

7.8 Fit the cap

Installation

6 Fit the bridge onto the strut. Set the fork tubes in the upper bridge so that they project 4.5 to 5.5 mm above the bridge, and tighten the clamp bolts to the specified torque setting **(see illustrations 6.21a and b)**. Press in the valve on the top of the forks as you set them in the bridge to release any suction effect.

7 Apply Loctite 2701 or equivalent to the ball joint stud threads and tighten the nut whilst counter-holding the stud using an Allen key **(see illustration 7.1b)**. The nut must be tightened to the specified torque, so the final tightening has to be done without counter-holding the joint. Make a reference mark across the stud and bridge to check that the joint does not turn with the nut as it is tightened.

8 Fit the ball joint cap **(see illustration)**.

Steering head bearing – R, RT and GS models

Removal

9 Remove the fuel tank (see Chapter 3). Trace the ignition switch wiring and disconnect it at the connector. On R models without a headlight cowl the switch panel extends over the fork bridge – remove the instrument cluster.

10 Unscrew the handlebar mounting bolts and displace the bars **(see illustrations 5.7a and b)**. There is no need to detach any wiring, hoses or cables.

11 Remove the fork cap, then counter-hold the fork top bolt hex **(see illustration 6.5)** and unscrew the nut. Push the top of the fork down through the bridge and fully into the slider.

12 Remove the steering head cap **(see illustration 7.1a)**, then unscrew the bolt and remove the fork bridge. The bolt is a press fit in the bearing and will remain in it.

Inspection

13 Inspect the bearing set in the upper bridge. If the bearing requires removal, it must be pressed out of the bridge and then the bolt pressed out of the bearing. Commence by removing the ignition (main) switch from the upper bridge (see Chapter 8) and then removing the retaining circlip from the bearing housing. If the upper bridge will be heated, it

is advisable to also remove the two fork joints; each joint can be pressed out using a drawbolt arrangement after removal of the circlip on each side.

14 Note that BMW advise heating the upper bridge to 100°C (212°F) to aid bearing removal. Support the upper bridge upside down on wood blocks, and use a tubular drift to drive the bearing out towards the upper face of the bridge. Once free of the upper bridge, support the edges of the bearing and drive the bolt out.

15 Press the bolt into the new bearing. Heat the upper bridge to 100°C (212°F) and support it upright on wood blocks placed around the bearing housing. Using a tubular drift, such as a socket, which bears only on the bearing's outer race, drive the bearing into its housing from the upper side.

16 Secure it with the circlip, making sure it seats properly in its groove. Install the ignition (main) switch (see Chapter 8). Use a drawbolt tool to install the fork joints and retain them with a circlip on each side.

Installation

17 Clean the threads of the bolt and apply Loctite 243 or equivalent to them. Mount the bridge onto the frame strut and tighten the bolt to the torque setting specified at the beginning of the Chapter. Fit the steering head cap **(see illustration 7.8)**.

18 Insert the fork tubes back into the upper bridge joints. Counter-hold the fork top bolt hex whilst tightening the nut to the specified torque setting. Install the fork top caps.

19 Mount the handlebars and tighten the bolts to the specified torque settings.

20 Where applicable on R models, install the instrument cluster.

21 Connect the ignition switch wiring connector, then install the fuel tank (see Chapter 3).

---

8 Steering damper (R models) – check, removal and installation

Check

1 With the bike on its centre stand, turn the steering from lock to lock and check that

8.3 Each bracket is secured by two bolts (arrowed)

there is no freeplay between the mountings and the mounting brackets. If any freeplay is evident, check that the mounting bolts are tightened to the correct torque settings. If they are, then the joint links between the damper and the brackets are worn and the worn mounting components and/or the steering damper must be replaced.

2 Check that the damper rod moves smoothly in and out of the damper with no signs of roughness or notchiness, or freeplay. Check the rod for pitting and corrosion, and around each end of the damper body for signs of fluid leakage. Replace the damper if any of the above are evident.

Removal and installation

3 Unscrew the two bolts securing each steering damper mounting bracket and remove the damper with its brackets attached (see *Haynes Hint* (Section 6) **(see illustration)**.

4 If required, unscrew the damper mounting bolts and remove the brackets, carefully noting the order and position of the various sleeves, washers and spacers.

5 Reassemble the brackets on the damper, making sure all sleeves, washers and spacers are correctly positioned. Clean the damper mounting bolt threads, then apply Loctite 2701 or equivalent and tighten them to the torque setting specified at the beginning of the Chapter.

6 Clean the bolt threads for the bracket on the fork bridge, then apply Loctite 2701 or equivalent and tighten them to the torque setting specified at the beginning of the Chapter. The bolts for the bracket on the Telelever require no threadlock, but tighten them to the specified torque.

---

9 Rear shock absorber – removal, inspection and installation

Removal

1 Remove the seat (see Chapter 7).

2 On RT models, remove the side panels (see Chapter 7).

3 Remove the rear wheel (see Chapter 6).

4 On RT and GS models with the remote spring pre-load adjuster, unscrew the bolt securing the adjuster.

5 Unscrew the nut and remove the bolt securing the bottom of the shock absorber to the swingarm **(see illustration)**.

6 On some models, there may not be enough clearance for the shock absorber upper mounting bolt to be withdrawn because of the seat mounting. If this is the case on your model, check from which side the upper mounting bolt is installed, then unscrew the two bolts and remove the seat support for that side **(see illustration)**. Unscrew and remove the upper mounting bolt and manoeuvre the shock out of the frame **(see illustrations)**.

9.5 Unscrew the nut and remove the bolt

9.6a If necessary, unscrew the seat support bolts (arrowed) and remove the support

9.6b Unscrew the nut . . .

9.6c . . . remove the bolt . . .

9.6d . . . and manoeuvre the shock down and out

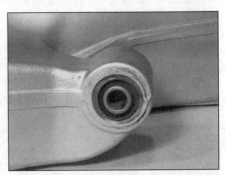

9.9 Check the condition of the bush in the swingarm

Inspection

7 Inspect the shock absorber for obvious physical damage and the coil spring for looseness, cracks or signs of fatigue.
8 Inspect the damper rod for signs of bending, pitting and oil leakage.
9 Inspect the pivot hardware at the top and bottom of the shock for wear or damage (see illustration).
10 Individual components for the rear shock absorber are not available. The entire unit must be replaced if it is worn or damaged.

Installation

11 Installation is the reverse of removal, noting the following.

a) Grease the sleeve in the upper mounting (see illustration).

b) Counter-hold the bolts and tighten the nuts to the torque settings specified at the beginning of the Chapter (see illustration).
c) Adjust the suspension as required (see Section 10).

10 Suspension – adjustments

Front forks

1 The front suspension is non-adjustable on R, RS and RT models.
2 On GS models (and certain RS models), the front shock is adjustable for spring pre-load. Adjustment is made using a suitable C-spanner (one is provided in the toolkit) to

turn the spring seat on the base of the shock absorber. There are five positions. Position 1 is the softest setting, position 5 is the hardest. Align the setting required with the adjustment stopper.

Rear shock absorber

3 On all models the rear shock absorber is adjustable for spring pre-load and rebound damping.
4 On R and RS models, pre-load adjustment is made using a suitable C-spanner (one is provided in the toolkit) to turn the spring seat on the base of the shock absorber (see illustration). There are seven positions. Position 1 is the softest setting, position 7 is the hardest. Align the setting required with the stopper. To increase the pre-load, turn the spring seat clockwise. To decrease the pre-load, turn the spring seat anti-clockwise.

5

9.11a Withdraw the sleeve and grease it

9.11b Counter-hold the bolts and tighten the nuts

10.4 Pre-load adjuster on R and RS models – align the setting required with the stopper (arrowed)

5 On RT and GS models, pre-load adjustment is made via the remote adjuster mounted on the frame (see illustration). Set the adjuster as required – there are no pre-set positions. To increase the pre-load, turn the adjuster knob clockwise. To decrease the pre-load, turn the adjuster knob anti-clockwise.

6 Rebound damping adjustment is made on all models by turning the adjuster on the bottom of the shock absorber using a screwdriver (see illustration). To set the standard position, turn the adjuster clockwise until it stops, then turn it anti-clockwise 1/2 turn. To increase the damping from the standard position, turn the adjuster clockwise. To decrease the damping, turn the adjuster anti-clockwise.

11 Final drive housing –
removal, inspection and installation

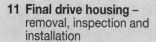

Note: *This procedure covers removal/ installation of the drive housing and disconnection of the driveshaft and driveshaft joint. Further dismantling and set-up of the drive housing are beyond the scope of this manual.*

Removal

1 If kept upright the final drive housing need not be drained of oil. If desired, however, the oil can be drained as described in Chapter 1.

2 On R, RT and GS models equipped with ABS, unscrew the bolt securing the sensor in

11.5 Unscrew the nut on the torque arm bolt (arrowed)

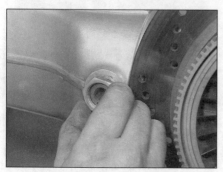

11.8a Unscrew the locknut . . .

10.5 Remote pre-load adjuster – RT and GS models

the final drive housing and remove the sensor, making sure all the shims are kept with it. It is advisable to secure all the shims to the sensor using wire through the bolt sockets.

3 Remove the rear brake caliper (see Chapter 6). There is no need to detach the brake hose, or the ABS sensor on RS models (where fitted), but secure the caliper clear of the drive housing, making sure there is no strain placed on the hose or wiring (ABS models). Release the wiring from any ties if necessary.

4 Remove the rear wheel (see Chapter 6).

5 Unscrew the nut on the bolt securing the torque arm to the drive housing, but do not yet remove the bolt (see illustration).

6 Slacken the rubber gaiter front clamp screw and draw the gaiter off the end of the swingarm (Paralever housing) (see illustration). Support the final drive housing using a block of wood.

11.6 Slacken the clamp screw and draw the gaiter off the swingarm

11.8b . . . and the left-hand pivot bolt . . .

10.6 Rebound damping adjuster screw (arrowed)

7 Using a hot air gun, heat the final drive housing pivots in the swingarm. This softens the threadlock and will make it easier to remove the pivots.

Caution: The application of heat will work on the particular threadlock that BMW use on assembly, however if the pivots have subsequently been assembled using a different type of threadlock, heat may actually make them more difficult to remove. If this is the case, allow the parts to cool and try again.

8 Counter-hold the left-hand pivot bolt and unscrew the locknut, then unscrew and remove both left-and right-hand pivots (see illustrations).

Caution: It has been known for the pivots to strip out the threads in the swingarm when removing them, so take great care and do not force them. If the pivots start to become tight after unscrewing them a little, screw them in and out to clean the threadlock off the threads until they become looser.

9 Support the final drive housing and withdraw the torque arm bolt, then draw the housing back off the swingarm and driveshaft (see illustration 11.18a).

10 Before removing the driveshaft joint, rotate the universal joint by hand, checking that the drive through the housing is smooth and free from wear or notchiness. Also inspect the shaft splines for wear. Note that there may be wear in the universal joint. To check the joint lever it out of the front of the final drive housing (see illustration). The universal joint has a spring clip which locates

11.8c . . . followed by the right-hand pivot bolt

11.10 Lever the driveshaft joint off the housing shaft

11.11 Check for any signs of freeplay or roughness in the universal joint

11.12 Check the bearings in the housing

in a groove in the housing shaft – the clip will expand and slip out of the groove as the joint is levered out, allowing the joint to be drawn off the shaft.

Inspection

11 Inspect the universal joint for signs of wear or damage. There should be no noticeable play in the bearings, and the joint should move smoothly and freely with no signs of roughness or notchiness **(see illustration)**. If any wear or damage is evident, the joint must be replaced.

12 Refer to *Tools and Workshop* Tips in the Reference Section and also check the pivot bearings in the swingarm and replace them if necessary **(see illustration)**.

13 Inspect the rubber bush set in the final drive housing torque arm mount **(see illustration)**. If the rubber has deteriorated, creating play in the joint, the bush should be

replaced with a new one. Use a drawbolt arrangement to press the old bush out and the new bush in (see *Tools and Workshop Tips* in the reference section).

14 Where applicable, remove the brake disc and/or ABS pulse ring from the final drive housing (see Chapter 6). Check the housing for any evidence of oil leakage from the large O-ring around the final drive housing cover and the oil seal at the drive output flange. Renewal of either O-ring or oil seal is a task for a BMW dealer. Also check the rubber gaiter for cracks and deterioration and replace it if necessary **(see illustration)**.

15 If the final drive housing gears, bearings, O-ring or oil seal require attention, the complete unit should be taken to a BMW dealer who will have the facilities and expertise to overhaul the assembly and correctly set up gear lash and shimming.

Installation

16 Apply molybdenum grease to the universal joint splines. Push the driveshaft joint onto the shaft in the final drive housing until the spring clip is felt to locate in its groove in the shaft – tap it with a hammer to make sure it is fully home **(see illustrations)**. Make sure the joint is secure on the shaft.

17 Apply molybdenum grease to the driveshaft joint splines **(see illustration)**. Also grease the bearings in the final drive housing **(see illustration)**.

18 Install the final drive housing, making sure the driveshaft joint engages correctly into the driveshaft **(see illustration)**. Support the housing with a block of wood or have an assistant hold it. Apply Loctite 2701 or equivalent to the pivot bolt threads, making sure the entire length of the threads are covered on the left-hand pivot, and screw them

11.13 Torque arm bush is a press-fit in housing

11.14 Check the gaiter and replace it if worn or damaged

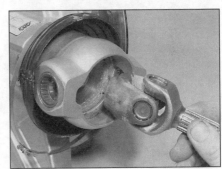

11.16a Slide the driveshaft joint into the housing . . .

11.16b . . . and tap it with a soft-faced hammer to locate the spring clip in its groove

11.17a Grease the driveshaft joint splines . . .

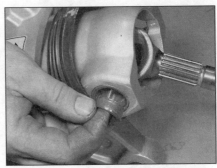

11.17b . . . and the pivot bearings

5

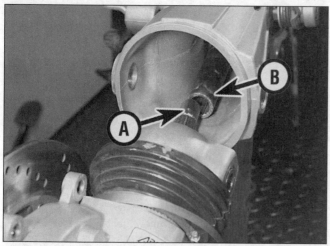

11.18a Slide the joint shaft (A) into the driveshaft (B)

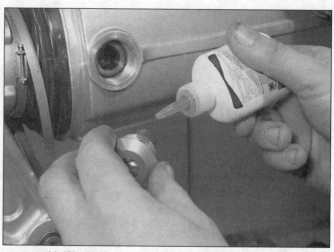

11.18b Apply the threadlock to the pivot threads

11.18c Tighten the right-hand . . .

11.18d . . . and left-hand pivot bolts to the specified torque settings

into the swingarm, making sure they align with the bearings in the housing (see illustration and 11.8c and b). Tighten the right-hand pivot bolt to the torque setting specified at the beginning of the Chapter, then tighten the left-hand pivot to the specified torque (see illustrations). Thread the locknut onto the left-hand pivot, then counter-hold the pivot and tighten the locknut. The nut must be tightened to the specified torque, so the final tightening has to be done without counter-holding the pivot. Make a reference mark across the pivot and swingarm to check that the pivot does not turn with the nut as it is tightened to the specified torque (see illustrations).

Caution: Do not delay between applying the Loctite and tightening the pivots – the Loctite sets quickly and will affect the torque setting if the pivots are not tightened immediately.

19 Locate the torque arm onto the final drive housing, then install the bolt and tighten the nut to the specified torque (see illustrations).

20 Fit the rubber gaiter around the end of the swingarm and tighten the clamp screw (see illustration).

21 If removed, install the rear brake disc and/or ABS pulse ring (see Chapter 6). Install the rear wheel and the brake caliper (see Chapter 6).

11.18e Make a reference mark between the pivot and the swingarm . . .

11.18f . . . and tighten the locknut

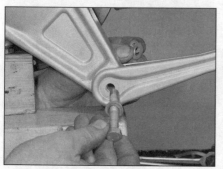

11.19a Fit the torque arm . . .

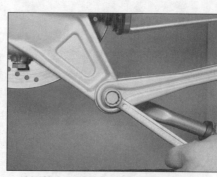

11.19b . . . and tighten the nut to the specified torque setting

11.20 Fit the gaiter over the swingarm and tighten the clip

12.4 Use a hot air gun to heat the swingarm pivots

12.5a Unscrew the locknut . . .

12.5b . . . and the left-hand . . .

22 On R, RT and GS models equipped with ABS, install the sensor with its shims and tighten the bolt. If any wiring ties were cut on removal, replace them with new ones.

23 If drained, fill the final drive housing with the correct grade and quantity of oil (see Chapter 1).

12 Swingarm (Paralever) and driveshaft – removal and installation

Removal

1 Remove the final drive housing (Section 11).

2 Place a small block of wood between the underside of the swingarm and the gearbox casing to prevent the swingarm dropping when the lower shock absorber bolt is removed. Unscrew the nut and remove the bolt securing the bottom of the shock absorber to the swingarm (see illustration 9.5).

3 On RT models, remove the fairing side panels (see Chapter 7), then remove the gearchange lever (see Section 3) and the rear brake master cylinder (see Chapter 6) – there

is no need to detach the brake lines. Unscrew the bolts securing both footrest holders and remove them. On the left-hand holder, disconnect the auxiliary power supply socket wiring connector, and on the right-hand holder disconnect the rear brake light switch wiring connector.

4 Using a hot air gun, heat the swingarm pivots in the gearbox casing (see illustration). This softens the threadlock and will make it easier to remove the pivots.

Caution: The application of heat will work on the particular threadlock that BMW use on assembly, however if the pivots have since been tightened using a different type of threadlock, heat may actually make them more difficult to remove. If this is the case, allow the parts to cool and try again.

5 Counter-hold the left-hand pivot bolt and slacken its locknut, then unscrew and remove the left-and right-hand pivots (see illustrations).

Caution: It has been known for the pivots to strip out the threads in the gearbox housing when removing them, so take great care and do not force them. If the pivots start to become tight after unscrewing them a little, screw them in

12.5c . . . and right-hand pivots

and out to clean the threadlock off the threads until they become looser.

6 Draw the swingarm back off the gearbox, supporting the driveshaft as you do to prevent damage to all surfaces (see illustration). Note the rubber gaiter fitted around the front of the swingarm.

7 Lever the driveshaft off the gearbox output shaft (see illustration). The driveshaft has a spring clip which locates in a groove in the output shaft – the clip will expand and slip out of the groove as the driveshaft is levered out, allowing it to be drawn off the output shaft.

5

12.6 Draw the swingarm back out of the casing . . .

12.7 . . . then lever the driveshaft off the output shaft

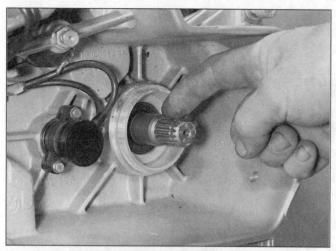

12.9a Lubricate the shaft splines . . .

12.9b . . . then slide the driveshaft onto the output shaft . . .

12.9c . . . and tap it with a soft-faced hammer to locate the spring clip in its groove

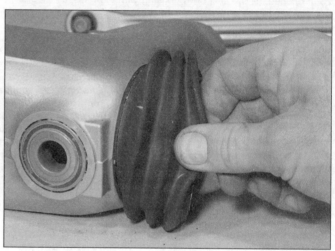

12.10 Check the gaiter and replace it if worn or damaged

8 Inspect all components for wear or damage as described in Section 13.

Installation

9 Lubricate the gearbox output shaft splines with molybdenum disulphide grease **(see illustration)**. Push the driveshaft onto the output shaft until the spring clip is felt to locate in its groove in the output shaft – tap it with a hammer to make sure it is fully home **(see illustrations)**. Make sure the driveshaft is secure on the output shaft.

10 Check the condition of the rubber gaiter and replace it if it is cracked, damaged or deteriorated **(see illustration)**. If removed, fit the rubber gaiter around the front of the swingarm. Grease the inside of the front of the gaiter so that it slides easily onto the output shaft neck. Check that the bearings are in place in the swingarm.

11 Manoeuvre the swingarm into position over the driveshaft and into the gearbox casing, making sure the gaiter locates correctly around the output shaft neck **(see illustration 12.6)**. Feed the gaiter on using a blunt screwdriver if required, then check that it is located by pulling back slightly on the swingarm. Install the shock absorber bolt, but do not yet tighten the nut. Place the small block of wood used on removal between the underside of the swingarm and the gearbox casing to support the swingarm.

12 Apply Loctite 2701 or equivalent to the pivot threads, making sure the entire length of the threads are covered on the left-hand pivot bolt, and screw them into the gearbox casing, making sure they align with the bearings in the swingarm **(see illustration and 12.5c and b)**. Tighten the right-hand pivot bolt to the torque setting specified at the beginning of the Chapter **(see illustration)**, then tighten the left-hand

12.12a Apply the threadlock . . .

12.12b . . . then tighten the right-hand pivot to the specified torque

12.12c Tighten the left-hand pivot as described, then to the specified torque

12.12d Make a reference mark across the pivot and casing (arrows) . . .

pivot to the specified torque. When tightening the left-hand pivot, tighten it without a torque wrench until all lateral freeplay in the swingarm is just taken up, then tighten it using the torque wrench to the specified setting **(see illustration)**. This removes the possibility of the friction caused by the threadlock giving a false torque reading before the freeplay is taken up. Remove the shock absorber bolt and the block of wood and move the swingarm up and down several times to settle the bearings, then replace the bolt and check that the left-hand pivot is still tightened to the torque setting specified and adjust if necessary. Thread the locknut onto the left-hand pivot, then counter-hold the pivot and tighten the locknut. The nut must be tightened to the specified torque, so the final tightening has to be done without counter-holding the pivot. Make a reference mark across the pivot and casing to check that the pivot does not turn with the nut as it is tightened to the specified torque **(see illustrations)**.

Caution: Do not delay between applying the Loctite and tightening the pivots – the Loctite sets quickly and will affect the torque setting if the pivots are not tightened immediately.

13 Fit the shock absorber bolt nut and tighten it to the specified torque setting **(see illustration 9.11b)**.

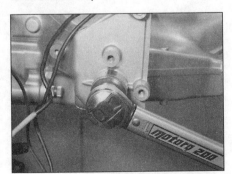

12.12e . . . and tighten the locknut to the specified torque

14 Install the final drive housing onto the swingarm (see Section 11).
15 On RT models, install the footrest holders and tighten the bolts to the specified torque settings. Connect the auxiliary power supply and brake light switch wiring connectors. Install the gearchange lever (see Section 3) and the rear brake master cylinder (see Chapter 6) and the fairing side panels (see Chapter 7).
16 Check the operation of the rear suspension and driveshaft before taking the machine on the road.

13 Swingarm (Paralever) and driveshaft – inspection and bearing replacement

Inspection

1 Thoroughly clean all components, removing all traces of dirt, corrosion and grease.
2 Inspect all components closely, looking for obvious signs of wear such as heavy scoring, and cracks or distortion due to accident damage. Any damaged or worn component must be replaced.
3 Inspect the driveshaft and gearbox output shaft splines for wear or damage. If wear is evident and there is excessive clearance between the output shaft and the driveshaft, either or both components must be replaced as required.

4 Inspect the universal joint for signs of wear or damage. There should be no noticeable play in the bearings, and the joint should move smoothly and freely with no signs of roughness or notchiness **(see illustration 11.11)**. If any wear or damage is evident, the universal joint must be replaced.
5 Rotate the tapered roller bearing in each side of the swingarm **(see illustration)**. If the bearings do not run smoothly in their outer races, or feel rough or gritty, they must be replaced.

Bearing replacement

6 Working on one side of the swingarm at a time, slip the tapered roller bearing out of its outer race. The outer race is a press-fit in the swingarm and will require a knife-edged bearing puller with slide-hammer attachment for removal. Refer to *Tools and Workshop Tips* in the *Reference* Section for details of using a puller.
7 BMW advise that the swingarm is heated to 80°(176°F) to ease bearing outer race installation. Use a drift which bears only on the race's outer edge, not its working face, to drive it into the swingarm. Lubricate the races and the bearings using a multi-purpose lithium-based grease and fit the bearings into their races.

13.5 Swingarm bearings are of the tapered roller type

5

Chapter 6
Brakes, wheels and tyres

Contents

Degrees of difficulty

Easy, suitable for novice with little experience	Fairly easy, suitable for beginner with some experience	Fairly difficult, suitable for competent DIY mechanic	Difficult, suitable for experienced DIY mechanic	Very difficult, suitable for expert DIY or professional

Specifications

Front brake

Brake fluid type	DOT 4
Caliper piston diameter	32 and 34 mm
Master cylinder piston diameter	20 mm
Disc diameter	305 mm
Disc thickness	
Standard	4.9 to 5.1 mm
Service limit (min)	4.5 mm
ABS sensor air gap	0.45 to 0.55 mm

Rear brake

Brake fluid type	DOT 4
Caliper piston diameter	
RS models	38 mm
R, RT and GS models	26 and 28 mm
Master cylinder piston diameter	
R, RS and RT models	12 mm
GS models	13 mm
Disc diameter	
RS models	285 mm
R, RT and GS models	276 mm
Disc thickness	
Standard	5.0 mm
Service limit (min)	
RS models	4.6 mm
R, RT and GS models	4.5 mm
ABS sensor air gap	0.45 to 0.55 mm

Wheels

Maximum wheel runout (axial and radial)	
Cast wheels	
Front	0.5 mm
Rear	0.3 mm
Spoke wheels (front and rear)	1.3 mm

Tyres

Tyre pressures	see *Daily (pre-ride) checks*
Tyre sizes*	
R models	
Front (cast wheel)	120/70 ZR17
Front (spoke wheel)	110/80 ZR18
Rear (cast wheel)	160/60 ZR18
Rear (spoke wheel)	150/70 ZR17
RS and RT models	
Front	120/70 ZR17
Rear	160/60 ZR18
GS models	
Front	110/80 R17 59H
Rear	150/70 R17 69H

Also refer to your owners handbook or the tyre information label under the bike's seat.

Torque settings

Front brake caliper mounting bolts	40 Nm
Front brake hose banjo bolt	15 Nm
Front brake caliper bleed valve	7 Nm
Front ABS sensor bolts	4 Nm (hand-tight)
Front brake disc bolts	
RS and RT models, and R models with cast wheels	21 Nm
R models with spoke wheels and GS models	24 Nm
Rear brake caliper mounting bolts	40 Nm
Rear brake hose banjo bolt	15 Nm
Rear ABS sensor bolts	4 Nm (hand-tight)
Rear brake caliper bleed valve	
RS models	7 Nm
R, RT and GS models	4 Nm
Rear brake disc screws	21 Nm
Front axle bolt	30 Nm
Front axle clamp bolts	22 Nm
Rear wheel bolts	
Initial setting	50 Nm
Final setting	105 Nm
Silencer rear mounting bolt (RS models)	35 Nm

1 General information

All models covered in this manual are fitted with wheels designed for tubeless tyres only. On RS and RT models, the wheels are cast alloy, on R models the wheels are either cast alloy or spoked, and on GS models the wheels are spoked. The design of spoked wheel used is somewhat unusual, the spokes being attached to the rim outboard of the tyre's sealing area, and thus permitting the use of tubeless tyres.

Both front and rear brakes are hydraulically operated disc brakes. On all models, the front calipers have four opposed pistons. On RS models, the rear caliper has two opposed pistons. On R, RT and GS models, the rear caliper has twin pistons on a sliding caliper. All calipers are manufactured by Brembo.

The FAG-Kuggelfischer anti-lock braking system (ABS) prevents the wheels from locking up under hard braking. The system is managed by a highly complex system of electronics and hydraulics. Such is the nature of the system that no attempt is being made to cover it fully in this manual. A functional and operational description of the system is given later in this Chapter. Note that the models covered in this manual use BMW's ABS II system; this differs from the system fitted to the 2-valve engined models in that a single pressure modulator is fitted and the system is thus considerably lighter.

Caution: Disc brake components rarely require disassembly. Do not disassemble components unless absolutely necessary. If a hydraulic brake line is loosened, the entire system must be disassembled, drained, cleaned and then properly filled and bled upon reassembly. Do not use solvents on internal brake components. Solvents will cause the seals to swell and distort. Use only clean brake fluid or denatured alcohol for cleaning. Use care when working with brake fluid as it can injure your eyes and it will damage painted surfaces and plastic parts.

2 Brake pads – replacement

Warning: The dust created by the brake system may contain asbestos, which is harmful to your health. Never blow it out with compressed air and don't inhale any of it. An approved filtering mask should be worn when working on the brakes.

Front brake pads

1 Remove the brake caliper (see Section 3). There is no need to detach the brake hose.
2 Drive the pad retaining pin out from the back of the caliper and withdraw it from the front **(see illustrations)**. Withdraw the pads from the bottom of the caliper **(see illustration)**. Note that on certain calipers, the pad retaining pin may be retained by a spring clip **(see illustration)**.

2.2a Drive the pin out from the back . . .

2.2b . . . and withdraw it from the caliper . . .

2.2c . . . then remove the pads

2.2d Pad retaining pin may have a spring clip (arrowed) on certain caliper types

3 Inspect the surface of each pad for contamination and check that the friction material has not reached the wear indicator cut-out or worn down to the minimum thickness – refer to Chapter 1, Section 5 for details. If either pad is worn, fouled with oil or grease, or heavily scored or damaged by dirt and debris, both pads must be replaced as a set. Note that it is not possible to degrease the friction material; if the pads are contaminated in any way they must be replaced. In the case of the front brake, always replace the pads in both front brake calipers at the same time.

4 If the pads are in good condition clean them carefully, using a fine wire brush which is completely free of oil and grease to remove all traces of road dirt and corrosion. Use a pointed instrument to dig out any embedded particles of foreign matter. Any areas of glazing may be removed using emery cloth.

5 Check the condition of the brake disc (see Section 4).

6 Remove all traces of corrosion from the pad pin. Inspect the pin for signs of damage and replace if necessary.

7 Push the pistons as far back into the caliper as possible using hand pressure. Due to the increased friction material thickness of new pads, it may be necessary to remove the master cylinder reservoir cover and diaphragm and siphon out some fluid. Note that if difficulty is experienced in pushing the pistons back, especially if the increased thickness of new pads has to be accommodated, the BMW piston resetting tool No. 34 1 500 can be used.

8 Smear the backs of the pads and the shank of the pad pin with copper-based grease, making sure that none gets on the front or sides of the pads. Make sure that the anti-rattle spring clips are in place around the pad pin hole on each pad.

9 Installation of the pads is the reverse of removal. Insert the pads into the caliper so that the friction material will face the disc, then slide the pad retaining pin through **(see illustration 2.2b)**. Make sure the pin passes through the hole in each pad **(see illustration)**. Drive the retaining pin home until the olive on the pin is located in the caliper **(see illustration)**. On calipers which have a spring clip to retain the pad retaining pin, make sure the pad retaining pin is fitted with the clip hole uppermost, then slip the spring clip through the hole **(see illustration 2.2d)**.

10 Top up the master cylinder reservoir if necessary (see *Daily (pre-ride) checks*), and replace the diaphragm and reservoir cover.

11 Operate the brake lever several times to bring the pads into contact with the disc. Check the operation of the brake before riding the motorcycle.

Rear brake pads

12 On RS models, lever off the caliper cover, then drive the pad retaining pins out from the back of the caliper and withdraw them from the front **(see illustrations)**. Remove the spring and spring pin, then withdraw the pads from the top of the caliper **(see illustrations)**.

13 On R, RT and GS models, remove the brake caliper (see Section 6). There is no need to detach the brake hose. Remove the R-pin or clip securing the pad retaining pin, then drive the pin out towards the wheel side of the caliper **(see illustration)**. Remove the pads.

2.9a Install the pin . . .

2.9b . . . and tap it fully home

6

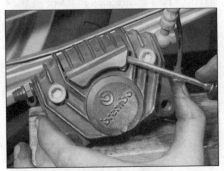

2.12a Lever off the cover . . .

2.12b . . . then drive the pins out from the back . . .

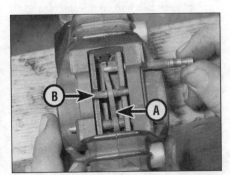

2.12c . . . and withdraw them from the front. Remove the spring (A) and the spring pin (B) . . .

2.12d . . . and withdraw the pads

14 Inspect the surface of each pad for contamination and check that the friction material has not reached the wear indicator cut-out or worn down to the minimum thickness – refer to Chapter 1, Section 5 for details. If either pad is worn down to, or beyond, the service limit, fouled with oil or grease, or heavily scored or damaged by dirt and debris, both pads must be replaced as a set. Note that it is not possible to degrease the friction material; if the pads are contaminated in any way they must be replaced.

15 If the pads are in good condition clean them carefully, using a fine wire brush which is completely free of oil and grease to remove all traces of road dirt and corrosion. Using a pointed instrument to dig out any embedded particles of foreign matter. Any areas of glazing may be removed using emery cloth.

16 Check the condition of the brake disc (see Section 7).

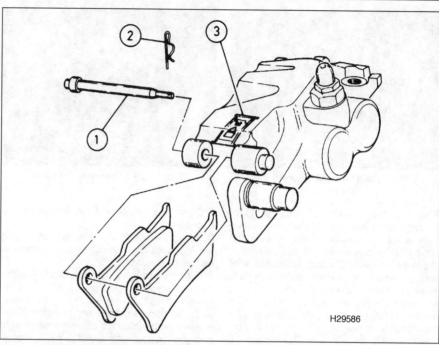

2.13 Brake pad retaining pin (1), R-pin or spring clip (2) and pad spring (3)

17 Remove all traces of corrosion from the pad pin(s). Inspect the pin(s) for signs of damage and replace if necessary.

18 Push the pistons as far back into the caliper as possible using hand pressure. Due to the increased friction material thickness of new pads, it may be necessary to remove the master cylinder reservoir cover and diaphragm and siphon out some fluid. Note that if difficulty is experienced in pushing the pistons back on the opposed piston caliper fitted to RS models, the BMW piston resetting tool No. 34 1 500 can be used.

19 Smear the backs of the pads and the shank(s) of the pad pin(s) with copper-based grease, making sure that none gets on the front or sides of the pads (see illustration).

20 On RS models, insert the pads into the caliper so that the friction material faces the disc, then slide the rear pad retaining pin part-way through and fit the spring onto it, then slide it all the way through (see illustrations). Fit the spring pin between the pads, then slide the front retaining pin through, making sure the spring locates under it, and drive the retaining pins fully home (see illustrations). Make sure the pins pass through the holes in each pad and the spring and spring pin are correctly located in the grooves in the pins and pads (see illustration). Fit the cover back

2.19 Smear copper grease onto the back of each pad and on the pad pins

2.20a Install the pads . . .

2.20b . . . then slide the rear pin half way through and fit the spring

2.20c Locate the spring pin across the pads, making sure it fits in the grooves . . .

2.20d . . . then fit the front pin and drive them both fully home

2.20e The installed pad assembly should be as shown

2.20f Fit the cover back onto the caliper

onto the caliper, making sure it clips into place and is secure (see illustration).

21 On R, RT and GS models, first check that the pad spring is in place in the caliper body; it should be positioned so that the arrowhead on its top surface points towards the front of the bike (see illustration 2.13). Insert the inner pad into the caliper so that the friction material faces the disc, then slide the pad retaining pin part-way through to keep the pad in place whilst installing the outer pad. Make sure the pin passes through the hole in each pad. Drive the retaining pin home and secure it with the R-pin or clip. Install the caliper (see Section 6).

22 Top up the master cylinder reservoir if necessary (see *Daily (pre-ride) checks*), and replace the diaphragm and reservoir cover.

23 Operate the brake pedal several times to bring the pads into contact with the disc. Check the operation of the brake before riding the motorcycle.

3.1 Note the alignment of the hose before removing the banjo bolt (arrowed)

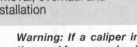

3 Front brake calipers –
removal, overhaul and installation

⚠ **Warning: If a caliper indicates the need for an overhaul (usually due to leaking fluid or sticky operation), all old brake fluid should be flushed from the system. Also, the dust created by the brake system may contain asbestos, which is harmful to your health. Never blow it out with compressed air and don't inhale any of it. An approved filtering mask should be worn when working on the brakes. Do not, under any circumstances, use petroleum-based solvents to clean brake parts. Use clean brake fluid, brake cleaner or denatured alcohol only.**

Removal

1 Note that disconnection of the brake hose is only necessary if the caliper is being overhauled – it can remain attached if pad renewal only is being carried out. Pad the area around the brake hose union bolt with rag to catch fluid spills, then remove the union bolt, noting its alignment on the caliper and separate the hose from the caliper (see illustration).

2 Plug the hose end or wrap a plastic bag tightly around it to minimise fluid loss and prevent dirt entering the system. Discard the sealing washers as new ones must be used on installation. **Note:** *If you are planning to overhaul the caliper and don't have a source of compressed air to blow out the pistons, just*

loosen the banjo bolt at this stage and retighten it lightly. The bike's hydraulic system can then be used to force the pistons out of the body once the pads have been removed. Disconnect the hose once the pistons have been sufficiently displaced.

> **HAYNES HINT** *Stick some protective tape on the wheel rim adjacent to the caliper to protect the surface – there is very little clearance between the wheel and caliper when the caliper is drawn off the disc.*

3 Unscrew the caliper mounting bolts, and slide the caliper off the disc, angling it to provide the best clearance, then twist it out so that the top of the caliper clears the wheel – the pressure of the disc on the pads as you twist the caliper will part the pads to provide the necessary clearance, but do not twist too much or the pads could be damaged (see illustrations).

4 Remove the brake pads (see Section 2).

Overhaul

Note: *There is no clear information as to the availability of caliper rebuild kits for the models covered. Before overhauling the caliper, check with a BMW dealer or Brembo brake specialist, as to the availability of seals and pistons for your model. Otherwise an entire new caliper must be installed. Always install new seals in both front calipers at the same time.*

5 Clean the exterior of the caliper with denatured alcohol or brake system cleaner.

6 Remove the pistons from the caliper body, either by pumping them out by operating the front brake lever until the pistons are displaced, or by forcing them out using compressed air. Mark each piston head and caliper body with a felt marker to ensure that the pistons can be matched to their original bores on reassembly. If the compressed air method is used, place a wad of rag between the pistons and the caliper to act as a cushion, then use compressed air directed into the fluid inlet to force the pistons out of the body. Use only low pressure to ease the pistons out and make sure both pistons are displaced at the same time. If the air pressure is too high and the pistons are forced out, the caliper and/or pistons may be damaged.

6

3.3a Unscrew the caliper mounting bolts (arrowed) . . .

3.3b . . . then slide the caliper off the disc . . .

3.3c . . . and twist it away from the wheel rim

3.16a Install the bolts . . .

3.16b . . . and tighten them to the specified torque setting

⚠️ *Warning: Never place your fingers in front of the pistons in an attempt to catch or protect them when applying compressed air, as serious injury could result.*

7 Using a wooden or plastic tool, remove the dust seals from the caliper bores. Discard them as new ones must be used on installation. If a metal tool is being used, take great care not to damage the caliper bores.

8 Remove and discard the piston seals in the same way.

9 Clean the pistons and bores with denatured alcohol, clean brake fluid or brake system cleaner. If compressed air is available, use it to dry the parts thoroughly (make sure it's filtered and unlubricated).

10 Inspect the caliper bores and pistons for signs of corrosion, nicks and burrs and loss of plating. If surface defects are present, the caliper

assembly must be replaced. If the caliper is in bad shape the master cylinder should also be checked (see Section 5).

Caution: Do not attempt to separate the caliper halves.

11 Lubricate the new piston seals with clean brake fluid and install them in their grooves in the caliper bores. Note that different sizes of bore and piston are used (see Specifications), and care must therefore be taken to ensure that the correct size seals are fitted to the correct bores. The same applies when fitting the new dust seals and pistons.

12 Lubricate the new dust seals with clean brake fluid and install them in their grooves in the caliper bores.

13 Lubricate the pistons with clean brake fluid and install them closed-end first into the caliper bores. Using your thumbs, push the pistons all the way in, making sure they enter the bore squarely.

Installation

14 Install the brake pads (see Section 2).

15 Manoeuvre the caliper on to the brake disc (see illustrations 3.3c and b). Make sure the pads sit squarely either side of the disc.

16 Install the caliper mounting bolts, and tighten them to the torque setting specified at the beginning of this Chapter (see illustrations).

17 Connect the brake hose to the caliper, using new sealing washers on each side of the fitting. Align the hose as noted on removal (see illustration 3.1). Tighten the banjo bolt to the torque setting specified at the beginning of the Chapter.

18 Fill the master cylinder reservoir with DOT 4 brake fluid (see *Daily (pre-ride) checks*) and bleed the hydraulic system as described in Section 10.

19 Check for leaks and thoroughly test the operation of the front brake before riding the motorcycle.

4 Front brake discs – inspection, removal and installation

Inspection

1 Visually inspect the surface of the disc for score marks and other damage. Light scratches are normal after use and won't affect brake operation, but deep grooves and heavy score marks will reduce braking efficiency and accelerate pad wear. If a disc is badly grooved it must be machined or replaced.

2 To check disc runout, position the bike on its centre stand and support it so that the front wheel is raised off the ground. Mount a dial gauge to the fork slider, according to wheel, with its tip touching the surface of the disc about 10 mm (1/2 in) from the outer edge (see illustration). Rotate the wheel and watch the gauge needle. BMW provide no figures, but if the runout is excessive, check the wheel bearings for play (see Chapter 1). If the bearings are worn, replace them (see Section 15) and repeat this check. If the disc runout is still excessive, it will have to be replaced, although machining by an engineer may be possible.

3 The disc must not be machined or allowed to wear down to a thickness less than the service limit as listed in this Chapter's Specifications and as marked on the disc itself (see illustration). The thickness of the disc can be checked with a micrometer (see illustration). If the thickness of the disc is less than the service limit, it must be replaced.

Removal

4 Remove the wheel (see Section 13).

Caution: Do not lay the wheel down and allow it to rest on the disc – the disc could become warped. Set the wheel on wood blocks so the disc doesn't support the weight of the wheel.

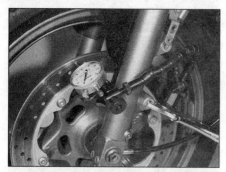

4.2 Set up a dial gauge with its tip contacting the brake disc, then rotate the wheel to check for runout

4.3a The minimum disc thickness is marked on each disc

4.3b Using a micrometer to measure disc thickness

4.5 Each disc is secured by six bolts (arrowed)

5 Mark the relationship of the disc to the wheel, so it can be installed in the same position, and mark the disc itself to indicate left-or right-hand side, and which is the outer face. Unscrew the disc retaining bolts, loosening them a little at a time in a criss-cross pattern to avoid distorting the disc, then remove the disc from the wheel (see illustration).

Installation

6 Install the disc on the wheel and align the previously applied matchmarks (if you're reinstalling the original disc). If a new disc is being fitted, note that the drillings in the disc surface should run in the normal direction of wheel rotation, eg the disc shown in illustration 4.5 is fitted to the left-hand side of the wheel. Additionally, the left-hand disc carries the ABS pulse ring, where applicable.

7 On models with spoke wheels, apply Loctite 243 or equivalent to the disc bolt threads.

8 Install the bolts and tighten them in a criss-cross pattern evenly and progressively to the torque setting specified at the beginning of the Chapter. Clean off all grease from the brake disc(s) using acetone or brake system cleaner. If a new brake disc has been installed, remove any protective coating from its working surfaces.

9 Install the wheel (see Section 13).

10 Operate the brake lever several times to bring the pads into contact with the disc. Check the operation of the brakes carefully before riding the bike. On ABS models, check the sensor air gap (see Section 17). Note: On ABS models if the left-hand disc or pulse ring have been renewed, the wheel speed sensor-to-pulse ring alignment must be reset as described in Section 17.

5	Front brake master cylinder – removal, overhaul and installation

1 If the master cylinder is leaking fluid, or if the lever does not produce a firm feel when the brake is applied, and bleeding the brakes does not help (see Section 10), and the hydraulic hoses are all in good condition, then master cylinder overhaul is recommended.

2 Before disassembling the master cylinder, read through the entire procedure and make sure that you have the correct rebuild kit, a fresh supply of DOT 4 brake fluid and some clean rags. Note: To prevent damage to the paint from spilled brake fluid, always cover the fuel tank when working on the master cylinder. Caution: Disassembly, overhaul and reassembly of the brake master cylinder must be done in a spotlessly clean work area to avoid contamination and possible failure of the brake hydraulic system components. Caution: Do not, under any circumstances, use a petroleum-based solvent to clean brake parts.

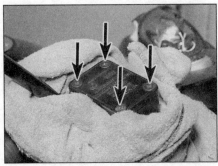

5.3 Slacken the screws (arrowed) securing the reservoir cover

Removal

3 Slacken, but do not remove, the screws holding the reservoir cover in place (see illustration).

4 Remove the brake light switch (Chapter 8).

5 Remove the front brake lever (Chapter 5).

6 Unscrew the brake hose banjo bolt and separate the hose from the master cylinder, noting its alignment (see illustration). Discard the two sealing washers as they must be replaced with new ones. Wrap the end of the hose in a clean rag and suspend it in an upright position or bend it down carefully and place the open end in a clean container. The objective is to prevent excessive loss of brake fluid, fluid spills and system contamination.

7 Note that the master cylinder can be overhauled in situ on the handlebar. If removal is required, refer to Chapter 5, Section 5, ignoring the Steps which do not apply.

Caution: Do not tip the master cylinder upside down or brake fluid will run out.

Overhaul

Note: There is no clear information as to the availability of master cylinder rebuild kits for the models covered. Before overhauling the master cylinder, check with a BMW dealer or

5.6 Note the alignment of the brake hose before removing the union bolt

Brembo brake specialist as to the availability of a kit for your model. Otherwise an entire new master cylinder must be installed.

8 Remove the reservoir cover retaining screws and lift off the cover, then remove the rubber diaphragm. Drain the brake fluid from the reservoir into a container. Wipe any remaining fluid out of the reservoir with a clean rag.

9 Carefully remove the dust boot from the end of the piston. Check for signs of brake fluid leakage as the dust boot is removed. If fluid has leaked past the piston seals, renewal of the piston/seal assembly (or complete master cylinder if individual parts cannot be obtained) will be necessary. Note that early RS models were fitted with a modified piston/seal assembly – master cylinders containing the modified assembly can be identified by the number 368 or higher, stamped in the body of the master cylinder, next to the Magura emblem above the hose connection. Master cylinder sealing was further improved in 1996 – the identification number of 663 indicating the later type.

10 Press the piston in slightly and slacken off the stop screw until the piston is free – take care that the piston does not spring out once the stop screw is backed off (see illustration).

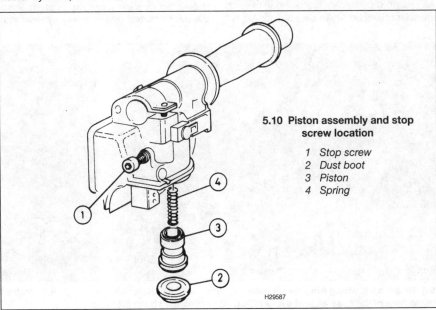

5.10 Piston assembly and stop screw location

1 Stop screw
2 Dust boot
3 Piston
4 Spring

H29587

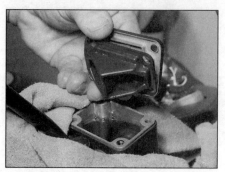

5.23 Fit the diaphragm and cover

11 Slide out the piston assembly and the spring, noting how they fit. Lay the parts out in the proper order to prevent confusion during reassembly.

12 Clean all parts with clean brake fluid or denatured alcohol. If compressed air is available, use it to dry the parts thoroughly (make sure it's filtered and unlubricated).

13 Check the master cylinder bore for corrosion, scratches, nicks and score marks. If damage or wear is evident, the master cylinder must be replaced with a new one. If the master cylinder is in poor condition, then the calipers should be checked as well. Check that the fluid inlet and outlet ports in the master cylinder are clear.

14 Use all of the new parts provided in the rebuild kit, regardless of the apparent condition of the old ones. If new seals are supplied separately from the piston, make sure the seals are installed so that their sealing lips face towards the fluid chamber.

15 Install the spring in the master cylinder in the same direction as was noted on dismantling.

16 Lubricate the piston assembly with clean brake fluid. Install the piston assembly into the master cylinder, making sure it is the correct way round with the seal lips facing in towards the fluid chamber. Make sure the lips do not

turn inside out when the assembly is slipped into the bore. Depress the piston and tighten the stop screw by the amount previously slackened.

17 Install the rubber dust boot, making sure the lip is seated correctly in the piston groove.

Installation

18 If the master cylinder was removed from the handlebar, refer to Chapter 5, Section 5 for the installation procedure, ignoring the Steps which do not apply.

19 Connect the brake hose to the master cylinder, using new sealing washers on each side of the union, and aligning the hose as noted on removal **(see illustration 5.6)**. Tighten the banjo bolt to the torque setting specified at the beginning of this Chapter.

20 Install the brake lever, noting that the pushrod freeplay must be correctly set (see Chapter 5).

21 Install the brake light switch (Chapter 8).

22 Fill the fluid reservoir with new DOT 4 brake fluid as described in *Daily (pre-ride) checks*. Refer to Section 10 of this Chapter and bleed the air from the system.

23 Inspect the reservoir cover rubber diaphragm and replace if damaged or deteriorated. Fit the rubber diaphragm, making sure it is correctly seated, the diaphragm plate and the cover onto the master cylinder reservoir **(see illustration)**.

24 Check the operation of the front brake before riding the motorcycle.

6 Rear brake caliper – removal, overhaul and installation

⚠ **Warning: If a caliper indicates the need for an overhaul (usually due to leaking fluid or sticky operation), all old brake fluid should be flushed from the system. Also, the dust created by the brake system may**

contain asbestos, which is harmful to your health. Never blow it out with compressed air and don't inhale any of it. An approved filtering mask should be worn when working on the brakes. Do not, under any circumstances, use petroleum-based solvents to clean brake parts. Use clean brake fluid, brake cleaner or denatured alcohol only.

Removal

1 On R, RT and GS models equipped with ABS, release the sensor wiring from the grommet and cut the first cable tie to free it from the brake hose **(see illustration)**.

2 On RS models, remove the brake pads (see Section 2).

3 Remove the brake hose banjo bolt, noting its alignment on the caliper and separate the hose from the caliper **(see illustration and 6.1)**. Plug the hose end or wrap a plastic bag tightly around it to minimise fluid loss and prevent dirt entering the system. Discard the sealing washers as new ones must be used on installation. **Note:** *If you are planning to overhaul the caliper and don't have a source of compressed air to blow out the pistons, just loosen the banjo bolt at this stage and retighten it lightly. The bike's hydraulic system can then be used to force the pistons out of the body once the pads have been removed. Disconnect the hose once the pistons have been sufficiently displaced.*

4 Unscrew the caliper mounting bolts, and slide the caliper off the disc, angling it to provide the best clearance **(see illustrations)**.

5 On RS models with ABS, unscrew the two bolts securing the ABS sensor to the caliper and withdraw the sensor, taking care not to lose any of the shims **(see illustration)**. It is advisable to secure all the shims to the sensor using wire through the bolt holes. If a Torx key is not available to remove the sensor bolts, trace the wiring and disconnect it at the connector, then free it from any ties.

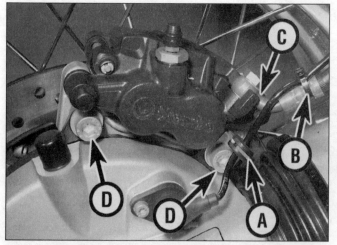

6.1 Release the wiring from the grommet (A) and the tie (B). Note the alignment of the hose against the lug (C). Caliper mounting bolts (D)

6.3 Brake hose banjo bolt (arrowed) – RS models

6.4a Unscrew the caliper mounting bolts (arrowed) . . .

6.4b . . . and slide the caliper off the disc

6.5 The ABS sensor is secured by two Torx bolts (arrowed) – RS models

6 On R, RT and GS models, remove the brake pads (see Section 2). Remove the pad spring, noting how it fits.

Overhaul

Note: *There is no clear information as to the availability of caliper rebuild kits for the models covered. Before overhauling the caliper, check with a BMW dealer or Brembo brake specialist as to the availability of seals and pistons for your model. Otherwise an entire new caliper must be installed.*

7 Clean the exterior of the caliper with denatured alcohol or brake system cleaner.
8 Remove the pistons from the caliper body, either by pumping them out by operating the front brake lever until the pistons are displaced, or by forcing them out using compressed air. Mark each piston head and caliper body with a felt marker to ensure that the pistons can be matched to their original bores on reassembly – this is essential on RS models because the pistons are the same diameter, whereas their diameter's differ by 2 mm on R, RT and GS models. If the compressed air method is used, place a wad of rag between the pistons and the caliper to act as a cushion, then use compressed air directed into the fluid inlet to force the pistons out of the body. Use only low pressure to ease the pistons out and make sure both pistons are displaced at the same time. If the air pressure is too high and the pistons are forced out, the caliper and/or pistons may be damaged.

Warning: Never place your fingers in front of the pistons in an attempt to catch or protect them when applying comp-ressed air, as serious injury could result.

9 Using a wooden or plastic tool, remove the dust seals from the caliper bores. Discard them as new ones must be used on installation. If a metal tool is being used, take great care not to damage the caliper bores.
10 Remove and discard the piston seals in the same way.
11 Clean the pistons and bores with denatured alcohol, clean brake fluid or brake system cleaner. If compressed air is available, use it to dry the parts thoroughly (make sure it's filtered and unlubricated).
12 Inspect the caliper bores and pistons for signs of corrosion, nicks and burrs and loss of

plating. If surface defects are present, the caliper assembly must be replaced. If the caliper is in bad shape the master cylinder should also be checked.
Caution: Do not attempt to separate the caliper halves on the opposed piston caliper fitted to RS models.
13 On R, RT and GS models, check that the caliper body is able to slide freely on the slider pins. If seized due to corrosion, separate the caliper and bracket and clean off all traces of corrosion and hardened grease. Apply a smear of Shell Retinax A grease to the slider pins and reassemble the two components. Replace the rubber boots if they are damaged or deteriorated.
14 Lubricate the new piston seals with clean brake fluid and install them in their grooves in the caliper bores. Note that on R, RT and GS models, different sizes of bore and piston are used (see Specifications), and care must therefore be taken to ensure that the correct size seals are fitted to the correct bores. The same applies when fitting the new dust seals and pistons.
15 Lubricate the new dust seals with clean brake fluid and install them in their grooves in the caliper bores.
16 Lubricate the pistons with clean brake fluid and install them closed-end first into the caliper bores. Using your thumbs, push the pistons all the way in, making sure they enter the bore squarely. Check that the each piston is returned to its original bore on RS models; on R, RT and GS models the piston diameter's differ by 2 mm thus preventing transposition of components.

Installation

17 On R, RT and GS models, fit the pad spring, then install the brake pads (see Section 2).
18 On RS models with ABS, if removed install the sensor with all the shims previously fitted and tighten the bolts to the specified torque setting **(see illustration 6.5)**.
19 Manoeuvre the caliper on to the brake disc **(see illustration 6.4b)**. Make sure the pads sit squarely either side of the disc.
20 Install the caliper mounting bolts, and tighten them to the torque setting specified at the beginning of this Chapter **(see illustrations)**.
21 Connect the brake hose to the caliper, using new sealing washers on each side of the fitting. Align the hose as noted on removal **(see illustrations 6.1 or 6.3)**. Tighten the banjo bolt to the torque setting specified at the beginning of the Chapter.
22 On RS models, install the brake pads (see Section 2). On RS models with ABS, check the sensor air gap (see Section 17). Connect the wiring at the connector if it was previously disconnected, and secure the wiring with ties.
23 On R, RT and GS models equipped with ABS, fit the sensor wiring into the grommet and secure the wire to the brake hose using a new cable tie **(see illustration 6.1)**.
24 Fill the master cylinder reservoir with DOT 4 brake fluid (see *Daily (pre-ride) checks*) and bleed the hydraulic system as described in Section 10.
25 Check for leaks and thoroughly test the operation of the brake before riding the motorcycle.

 6

6.20a Mount the caliper and install the bolts . . .

6.20b . . . and tighten them to the specified torque

7.4 Heat the screws (arrowed), then unscrew them and remove the disc

7 Rear brake disc – inspection, removal and installation

Inspection

1 Refer to Section 4 of this Chapter, noting that the dial gauge should be attached to the Paralever housing. If runout is excessive, either the disc is warped or the final drive unit bearings are worn.

Removal

2 Unbolt the rear brake caliper, leaving the brake hose attached (see Section 6). Remove the rear wheel (see Section 14). On RS, RT and R models with cast wheels, the disc is mounted on the final drive housing. On GS and R models with spoke wheels, the disc is mounted on the wheel hub.
Caution: On spoked wheel models, do not

8.4a Release the clip . . .

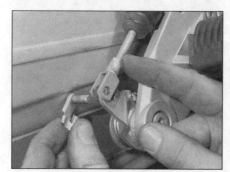

8.4b . . . and withdraw the pin

lay the wheel down and allow it to rest on the disc – the disc could become warped. Set the wheel on wood blocks so the disc doesn't support the weight of the wheel.
3 Mark the relationship of the disc to the housing or wheel hub, so it can be installed in the same position – this is particularly important on ABS-equipped models where the pulse ring is part of the brake disc, to retain the sensor and pulse ring alignment. On spoked wheel models, mark the disc itself to indicate which is the outer face.
4 Where the disc is mounted on the final drive housing, heat the two retaining screws using a hot air gun before removing them **(see illustration)**; this will soften the locking compound used on their threads and ease their removal.
5 Where the disc is mounted on the wheel hub, unscrew the disc retaining screws, loosening them a little at a time in a criss-cross pattern to avoid distorting the disc, then remove the disc.

Installation

6 Install the disc on the final drive housing or wheel hub (as applicable), aligning the previously applied matchmarks (if you're reinstalling the original disc).
7 Remove all traces of old locking compound from the disc retaining screws. If the screws were damaged on removal, replace them with new ones. If new screws are used, they should be of the micro-encapsulated type which contains a thread locking agent – inspection them carefully before use.
8 Where the disc is mounted on the wheel hub, apply Loctite 243 or equivalent to the screw threads, then install the screws and tighten them in a criss-cross pattern evenly and progressively to the torque setting specified at the beginning of the Chapter. Where the disc is mounted on the final drive housing, apply Loctite 273 or equivalent to the screw threads, then install the screws and tighten them progressively to the torque setting specified at the beginning of the Chapter. Clean off all grease from the brake disc using acetone or brake system cleaner. If a new brake disc has been installed, remove any protective coating from its working surfaces.
9 Install the wheel (see Section 14). Operate

the brake pedal several times to bring the pads into contact with the disc.
10 On all ABS-equipped models, check the sensor air gap (see Section 17). On ABS-equipped models which have the disc mounted to the final drive housing, if a new disc has been fitted, or care has not been taken to return the disc to its original position on the housing, the sensor-to-pulse ring alignment must be checked and reset if necessary (see Section 17).
11 Check the operation of the brakes carefully before riding the bike.

8 Rear brake master cylinder – removal, overhaul and installation

1 If the master cylinder is leaking fluid, or if the lever does not produce a firm feel when the brake is applied, and bleeding the brakes does not help (see Section 10), and the hydraulic hoses are all in good condition, then master cylinder overhaul is recommended.
2 Before disassembling the master cylinder, read through the entire procedure and make sure that you have the correct rebuild kit, a new supply of DOT 4 brake fluid and some clean rags. **Note:** *To prevent damage to the paint from spilled brake fluid, always cover the surrounding components when working on the master cylinder.*
Caution: Disassembly, overhaul and reassembly of the brake master cylinder must be done in a spotlessly clean work area to avoid contamination and possible failure of the brake hydraulic system components. Do not, under any circumstances, use a petroleum-based solvent to clean brake parts.

Removal

3 On RS and RT models, remove the right-hand fairing side panel (see Chapter 7).
4 Release the spring clip from around the base of the master cylinder pushrod, then swing it down and withdraw the pin **(see illustrations)**. Separate the pushrod from the brake pedal.
5 Release the reservoir from its clips on the frame, then remove the reservoir cover and pour the fluid into a container **(see illustration)**.

8.5a Pull the reservoir out of its clips

8.5b Release the clamp (arrowed) and detach the hose

8.6 Unscrew the nut (arrowed) and detach the pipe

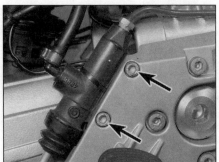

8.7 The master cylinder is secured by two bolts (arrowed)

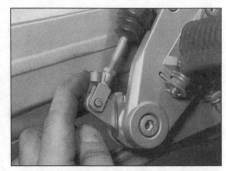

8.17 Press the clip onto the pushrod to secure the pin

Separate the fluid reservoir hose from the elbow on the master cylinder by releasing the hose clamp **(see illustration)**.

6 Unscrew the brake pipe nut and separate the pipe from the master cylinder **(see illustration)**. Wrap the end of the pipe in a clean rag or place the open end in a clean container. The objective is to prevent excessive loss of brake fluid, fluid spills and system contamination.

7 Unscrew the two bolts securing the master cylinder to the bracket **(see illustration)**.

Overhaul

Note: *There is no clear information as to the availability of master cylinder rebuild kits for the models covered. Before overhauling the master cylinder, check with a BMW dealer or Brembo brake specialist as to the availability of a kit for your model. Otherwise an entire new master cylinder must be installed.*

8 Dislodge the rubber dust boot from the base of the master cylinder to reveal the pushrod retaining circlip.

9 Depress the pushrod and, using circlip pliers, remove the circlip. Slide out the piston assembly and spring. If they are difficult to remove, apply low pressure compressed air to the fluid inlet. Lay the parts out in the proper order to prevent confusion during reassembly.

10 Clean all of the parts with clean brake fluid or denatured alcohol. If compressed air is available, use it to dry the parts thoroughly (make sure it's filtered and unlubricated).

11 Check the master cylinder bore for corrosion, scratches, nicks and score marks. If damage is evident, the master cylinder must be replaced with a new one. If the master cylinder is in poor condition, then the caliper should be checked as well.

12 Use all of the parts supplied in the rebuild kit. Install the spring in the master cylinder bore in the same direction as noted on dismantling.

13 Lubricate the piston assembly with clean hydraulic fluid and install the assembly into the master cylinder, making sure it is the correct way round with the seal lips facing in. Make sure the lips on the seals do not turn inside out when they are slipped into the bore.

14 Install and depress the pushrod, then install a new circlip, making sure it is properly seated in the groove.

15 Install the rubber dust boot, making sure the lip is seated properly in the groove.

Installation

16 Install the master cylinder onto the footrest bracket and tighten its mounting bolts to the torque setting specified at the beginning of the Chapter **(see illustration 8.7)**.

17 Align the brake pedal with the master cylinder pushrod, then slide in the pin **(see illustration 8.4b)** and secure its clip onto the pushrod **(see illustration)**.

18 Connect the brake pipe to the master cylinder, and tighten the nut securely **(see illustration 8.6)**.

19 Secure the fluid reservoir in its clips on the frame **(see illustration 8.5a)**. Ensure that the hose is correctly routed, then connect it to the union on the master cylinder and secure it with the clamp **(see illustration 8.5b)**. Check that the hose is secure and clamped at the reservoir end as well. If the clamps have weakened, use new ones.

20 Fill the fluid reservoir with new DOT 4 brake fluid (see *Daily (pre-ride) checks*) and bleed the system following the procedure in Section 10.

21 Check the relief clearance of the pushrod by measuring the gap between the brake pedal stop screw head and the cast stop on the footrest bracket **(see illustration)**. A gap of 0.2 mm should be present. To adjust, slacken the locknut on the pushrod and rotate the pushrod by hand until the gap is correct, then tighten the locknut **(see illustration)**.

22 On RS and RT models, install the right-hand fairing side panel (see Chapter 7).

23 Check the operation of the brake carefully before riding the motorcycle.

6

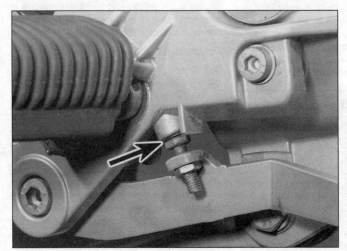

8.21a Check the clearance between the pedal stop screw and the stop (arrowed)

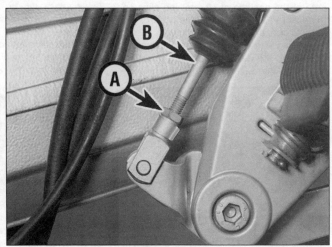

8.21b To adjust clearance, slacken nut (A) and turn pushrod (B) as required

9.2 Flex the brake hoses and check for cracks, bulges and leaking fluid

9.4 Remove the banjo bolt and separate the hose from the caliper; there is a sealing washer on each side of the fitting

9 Brake hoses, pipes and unions – inspection and replacement

Inspection

1 Brake hose condition should be checked regularly as described in Chapter 1.

2 Twist and flex the rubber hoses while looking for cracks, bulges and seeping fluid **(see illustration)**. Check extra carefully around the areas where the hoses connect with the banjo fittings, as these are common areas for hose failure.

3 On ABS models, metal brake pipes connect the pressure modulator to the flexible brake hoses from the master cylinders and calipers. Remove the fuel tank (see Chapter 3) and fairing side panels (where applicable) to inspect the metal brake pipes. Check that there is no sign of corrosion or fluid leakage.

Replacement

4 The brake hoses have banjo union fittings on each end and the brake pipes have joint nuts **(see illustration)**. Cover the surrounding area with plenty of rags and unscrew the banjo bolt or joint nut at each end of the hose or pipe, noting its alignment. Make careful note of the position of all retaining clips and the exact routing, then free the hose or pipe from its clips or guides and remove it. Discard the sealing washers on the hose banjo unions.

5 Position the new hose or pipe exactly as noted on removal, making sure it isn't twisted

or otherwise strained, and abut the tab on the hose union with the lug on the component casting, where present. Otherwise align the hose or pipe as noted on removal. Install the hose banjo bolts using new sealing washers on both sides of the unions. Tighten the banjo bolts and joint nuts to the torque settings specified at the beginning of this Chapter. Make sure the hoses and pipes are correctly aligned and routed clear of all moving components.

6 Flush the old brake fluid from the system, refill with new DOT 4 brake fluid (see *Daily (pre-ride) checks*) and bleed the air from the system (see Section 10). Check the operation of the brakes carefully before riding the motorcycle.

10 Brake system – bleeding

Note (ABS models): *The following procedure details the conventional method of brake bleeding. It may be found that due to the volume of fluid in the system and the small-bore metal pipes which connect the pressure modulator, that it is difficult to bleed the brake effectively. In this case, a pressure bleeder kit is advised, which attaches to the fluid reservoir and dispenses with hand operation of the lever/pedal. If use of pressure bleeder fails to disperse air from the system, seek the advise of a BMW dealer – it may be necessary to bleed air from the pressure modulator unit.*

1 Bleeding the brakes is simply the process of removing all the air bubbles from the brake fluid reservoirs, the hoses, the brake calipers, and on ABS-equipped models the pressure modulator and connecting pipes. Bleeding is necessary whenever a brake system hydraulic connection is loosened, when a component or hose is replaced, or when the master cylinder or caliper is overhauled. Leaks in the system may also allow air to enter, but leaking brake fluid will reveal their presence and warn you of the need for repair.

Preparation

2 To bleed the brakes, you will need some new DOT 4 brake fluid, a length of clear vinyl or plastic tubing, a small container partially filled with clean brake fluid, some rags and a spanner to fit the brake caliper bleed valves.

3 Note that BMW advise that bleeding of the front calipers is more effective if the calipers are detached from the brake discs, the pads removed, and the pistons pushed fully back into the caliper, thus reducing the fluid volume. The piston resetting service tool No. 34 1 500 can be used to retract the pistons in one caliper, then a spacer inserted No. 34 1 520 whilst the tool is removed and applied to the other front caliper. With the spacer and tool in place both front calipers are bled simultaneously. However, if new brake pads have recently been installed, the pistons will already be pushed well back in their bores and in this case the front brakes can be bled with the calipers in situ on the brake discs.

4 If bleeding the front brake, cover the fuel tank and other painted components to prevent damage in the event that brake fluid is spilled. If bleeding the rear brake on RS and RT models, remove the right-hand fairing side panel (see Chapter 7) for access to the fluid reservoir.

5 If bleeding the rear brake, note that on R and RT models, the caliper must be detached from the disc and positioned so that its bleed valve is at the highest point. Retract the pistons slightly in the caliper and insert a hardwood or metal spacer between them. The caliper can remain in situ on RS and GS models.

Procedure

6 Remove the reservoir cover, diaphragm plate (where fitted) and diaphragm and slowly pump the brake lever or pedal a few times, until no air bubbles can be seen floating up from the holes in the bottom of the reservoir. Doing this bleeds the air from the master cylinder end of the line. Loosely refit the reservoir cover.

7 Pull the dust cap off the bleed valve **(see illustration)**. Attach one end of the clear vinyl or plastic tubing to the bleed valve and submerge the other end in the brake fluid in the container **(see illustration)**. Remove the reservoir cover and check the fluid level. Do not allow the fluid level to drop below the lower mark during the bleeding process otherwise air will enter the system.

10.7a Brake caliper bleed valve (arrowed)

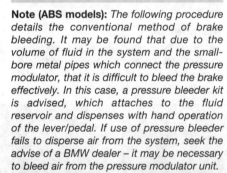

10.7b To bleed the brakes, you need a spanner, a short section of clear tubing, and a clear container half-filled with brake fluid

8 Carefully pump the brake lever or pedal three or four times and hold it in (front) or down (rear) while opening the caliper bleed valve. When the valve is opened, brake fluid will flow out of the caliper into the clear tubing and the lever will move toward the handlebar or the pedal will move down.

9 Retighten the bleed valve, then release the brake lever or pedal gradually. Repeat the process until no air bubbles are visible in the brake fluid leaving the caliper and the lever or pedal is firm when applied. On completion, disconnect the bleeding equipment, then tighten the bleed valve to the torque setting specified at the beginning of the chapter and install the dust cap.

10 Install the diaphragm and cover assembly, wipe up any spilled brake fluid and check the entire system for leaks.

> **HAYNES HiNT** *If it's not possible to produce a firm feel to the lever or pedal the fluid my be aerated. Let the brake fluid in the system stabilise for a few hours and then repeat the procedure when the tiny bubbles in the system have settled out.*

11 Wheels –
inspection and repair

1 In order to carry out a proper inspection of the wheels, it is necessary to support the bike upright so that the wheel being inspected is raised off the ground. Position the motorcycle on its centre stand or an auxiliary stand. Clean the wheels thoroughly to remove mud and dirt that may interfere with the inspection procedure or mask defects. Make a general check of the wheels (see Chapter 1) and tyres (see *Daily (pre-ride) checks*).

2 In order to accurately check wheel runout, the wheel must be removed from the machine and the tyre removed from the rim. This allows the runout measurement to be made against the inner, machined, surfaces of the wheel **(see illustration)**. To carry this out successfully, the wheel must be supported centrally on a jig to allow it to be rotated whilst the readings are taken. In view of the equipment required and carefully set-up, it is advised that wheel runout is checked by a BMW dealer or wheel building specialist.

3 On models with cast wheels, inspected for cracks, flat spots on the rim and other damage. Look very closely for dents in the area where the tyre bead contacts the rim. Dents in this area may prevent complete sealing of the tyre against the rim, which leads to deflation of the tyre over a period of time. If damage is evident the wheel will have to be replaced with a new one. Never attempt to repair a damaged cast alloy wheel.

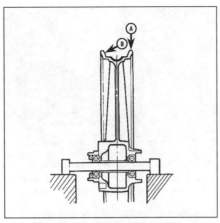

11.2 Check for radial (out-of-round) runout at point A and axial (side-to-side) runout at point B

4 On R and GS models with spoke wheels, wheel rebuilding or spoke replacement must be left to a BMW dealer or wheel building specialist. A great deal of skill and equipment is required, and given the potential for poor handling and machine instability that could result from a poorly-built wheel, it is essential that owners do not attempt repairs themselves. Additionally, due to the unique spoke and rim design used, replacement parts are only likely to be available through a BMW dealer.

12 Wheels –
alignment check

1 Misalignment of the wheels can cause strange and possibly serious handling problems. Due to the BMW's solid engine and running gear construction, the normal problem areas for wheel misalignment (distorted frame or cocked rear wheel) to not apply. If the wheels are out of alignment, this could be due to loose or damaged Telelever components at the front, or loose swingarm (Paralever) bearings at the rear. If accident damage has occurred, the machine should be taken to a BMW dealer for a thorough check of wheel alignment using the track alignment gauge and thorough inspection of the structural components.

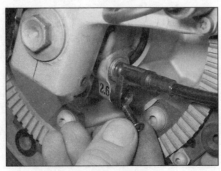

13.4a Remove the screw . . .

2 To check the alignment you will need an assistant, a length of string or a perfectly straight metal bar and a ruler.

3 In order to make a proper check of the wheels it is necessary to support the bike in an upright position, either on its centre stand or on an auxiliary stand. Measure the width of both tyres at their widest points. Subtract the smaller measurement from the larger measurement, then divide the difference by two. The result is the amount of tyre width offset that should exist between the front and rear tyres on both sides.

4 If a string is used, have your assistant hold one end of it about halfway between the floor and the rear axle, touching the rear sidewall of the rear tyre.

5 Run the other end of the string forward and pull it tight so that it is roughly parallel to the floor. Slowly bring the string into contact with the front sidewall of the rear tyre, then turn the front wheel until it is parallel with the string. Measure the distance from the front tyre sidewall to the string. Subtract the tyre width offset from the distance measured.

6 Repeat the procedure on the other side of the motorcycle.

7 The distance from the front tyre sidewall to the string should be equal on both sides of the bike and equal to the tyre width offset calculated previously. If the measurement is greater than the tyre width offset on either side of the bike, the wheels are out of alignment by this amount.

8 As was previously pointed out, a perfectly straight length of metal bar may be substituted for the string. The procedure is the same.

9 If the wheels are out of alignment, and the fault cannot be traced to the Telelever or Paralever assemblies, the bike should be taken to a BMW dealer for verification of your findings using the track alignment gauge. A maximum wheel offset of 9 mm is permissible.

13 Front wheel –
removal and installation

Removal

1 Position the motorcycle on its centre stand and support it under the crankcase so that the front wheel is off the ground. Always make sure the motorcycle is properly supported.

2 On RS and RT models, remove the front mudguard (see Chapter 7).

3 Remove the front brake calipers (see Section 3). There is no need to disconnect the hoses from the calipers. Support the calipers with a piece of wire or a bungee cord so that no strain is placed on their hydraulic hoses.

4 If required, remove the screw securing the speedometer cable to its drive housing on the left-hand side of the wheel hub, and detach the cable **(see illustrations)**. The cable can be left attached to the drive housing if the housing does not need to be removed separately.

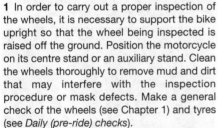

13.4b . . . and detach the speedometer cable

13.5 Unscrew the axle bolt (A) and slacken the clamp bolt (B) on each fork

13.6 Withdraw the axle and remove the wheel

5 Unscrew the axle bolt on the left-hand end of the axle, then slacken the axle clamp bolts on the bottom of each fork **(see illustration)**.
6 Support the wheel, then withdraw the axle from the right-hand side and carefully lower the wheel **(see illustration)**.
7 Remove the wheel spacer from the right-hand side of the wheel and the speedometer drive housing from the left-hand side **(see illustrations)**. **Note:** *Do not operate the front brake lever with the wheel removed.*
Caution: Don't lay the wheel down and allow it to rest on either disc – the disc could become warped. Set the wheel on wood blocks so neither disc supports the weight of the wheel.
8 Check the axle for straightness by rolling it on a flat surface such as a piece of plate glass (first wipe off all old grease and remove any corrosion using fine emery cloth). If the equipment is available, place the axle in

V-blocks and measure the runout using a dial gauge. No service limit is given by BMW for the amount of runout, but normally runout should not exceed 0.2 mm.
9 Check the condition of the wheel bearings (see Section 15).

Installation

10 Apply a smear of grease to the speedometer drive components. Fit the speedometer drive to the wheel's left-hand side, aligning its driven gear slots with the drive plate tabs **(see illustration 13.7b)**.
11 Apply a smear of grease to the inside of the wheel spacer, and also to the inner face where it fits into the wheel. Fit the spacer into the right-hand side of the wheel **(see illustration 13.7a)**.
12 Manoeuvre the wheel into position. Apply a thin coat of grease to the axle.
13 Lift the wheel into place between the fork

sliders, making sure the spacer and speedometer drive remain in position. Slide the axle in from the right-hand side **(see illustration 13.6)**. Align the speedometer drive housing so that the protrusion on the top of the housing locates against the front of the lug on the fork slider **(see illustration)**.
14 Install the axle bolt and tighten it to the torque setting specified at the beginning of the Chapter **(see illustrations)**. Move the motorcycle off its stand, apply the front brake and pump the front forks a few times to settle all components in position. Check that the bolt is still at the correct torque.
15 Tighten the axle clamp bolts to the specified torque setting **(see illustration)**.
16 Install the brake calipers (see Section 3).
17 If removed, connect the speedometer cable to the drive housing, aligning the slot in the cable end with the drive tab, and securely tighten its screw **(see illustrations 13.4b and a)**.

13.7a Remove the spacer from the right-hand side . . .

13.7b . . . and the speedometer drive housing from the left-hand side

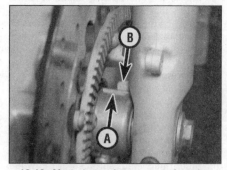

13.13 Abut protrusion on speedometer drive (A) against lug (B) on the fork slider

13.14a Install the axle bolt . . .

13.14b . . . and tighten it to the specified torque

13.15 Tighten the axle clamp bolts to the specified torque

14.2 Unscrew the mounting bolt to allow the silencer to be moved

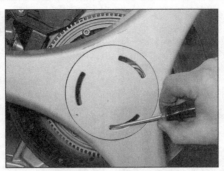
14.4 Remove the hub cover

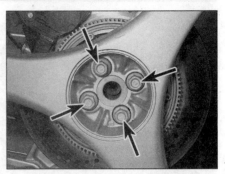
14.5 Unscrew the bolts (arrowed) . . .

18 On RS and RT models, install the front mudguard (see Chapter 7).
19 Apply the front brake a few times to bring the pads back into contact with the discs.
20 On ABS-equipped models, check the front wheel speed sensor air gap (see Section 17).
21 Check for correct operation of the front brake before riding the motorcycle.

14 Rear wheel – removal and installation

Removal

1 Position the motorcycle on its centre stand.
2 On RS models, remove the number plate carrier. On RS models equipped with ABS, unscrew the silencer rear mounting bolt to allow the silencer to be pressed away from

the bike to provide clearance for the wheel **(see illustration)**.
3 On spoked wheel models, remove the brake caliper (see Section 6). There is no need to disconnect the hose from the caliper. Support the caliper with a piece of wire or a bungee cord so that no strain is placed on its hydraulic hose.
4 On cast wheel models, lever off the hub cover **(see illustration)**.
5 Unscrew the four bolts securing the wheel to the final drive housing **(see illustration)**. Remove the bolts along with their tapered collars.
6 Grasp the wheel and draw it off the final drive housing **(see illustration)**. On GS models, carefully move the wheel guard aside – do not remove it.
7 On cast wheel models, remove the spacer plate from the inside of the wheel **(see illustration)**.

Caution: On spoked wheel models, do not lay the wheel down and allow it to rest on the disc – it could become warped. Set the wheel on wood blocks so the disc doesn't support the weight of the wheel. Do not operate the brake pedal with the wheel removed.

Installation

8 Manoeuvre the wheel into position. Make sure the contact faces between the wheel hub and the final drive, and on cast wheel models the spacer plate, are clean and free of grease.
9 On cast wheel models, fit the spacer plate onto the inside of the wheel, aligning the holes **(see illustration 14.7)**.
10 Clean the wheel bolt threads. Lift the wheel into position, making sure it engages correctly, and install the bolts hand-tight, making sure the tapered ends of the collars are facing into the wheel **(see illustration)**. On GS models, move the wheel guard aside to provide clearance for the wheel.
Caution: The rear wheel bolts are of a particular specification and have the code number 60 stamped into their heads. Do not use any other type of bolt.
11 Tighten the bolts evenly in a criss-cross pattern to the initial torque setting specified, then tighten them to the final toque specified **(see illustration)**.
12 On spoked wheel models, install the brake caliper (see Section 6).
13 On cast wheel models, fit the hub cover, aligning its tab with the wheel cutout **(see illustration)**.

14.6 . . . and remove the wheel

14.7 Remove the spacer plate from the inside of the wheel

14.10 Install the bolts . . .

14.11 . . . and tighten them as described to the specified torque

14.13 Fit the hub cover, aligning its tab with the cutout in the wheel

6

15.3a Lever out the grease seal . . .

15.3b . . . then remove the speedometer drive plate . . .

14 On RS models, install the number plate carrier. On RS models equipped with ABS, install the silencer rear mounting bolt and tighten it to the specified torque setting (see illustration 14.2).
15 Operate the brake pedal several times to bring the pads into contact with the disc. On spoked wheel models equipped with ABS, check the wheel speed sensor air gap (see Section 17).
16 Check the operation of the rear brake carefully before riding the bike.

15 Wheel bearings – removal, inspection and installation

Front wheel bearings

Note: *Always replace the wheel bearings in pairs. Never replace the bearings individually. Avoid using a high pressure cleaner on the wheel bearing area.*
1 Remove the wheel (see Section 13).
2 Set the wheel on blocks so as not to allow the weight of the wheel to rest on the brake discs and on ABS models, the wheel speed sensor rings.
3 Prise out the grease seal on the left-hand side of the wheel using a flat-bladed screwdriver – use a block of wood to protect the disc (see illustration). Discard the seal as a new one must be used. Remove the speedometer drive plate after removing the seal, then remove the bearing circlip (see illustrations). Where fitted, remove the headed spacer from the left-hand bearing.
4 Heat the wheel hub around the bearing housing to 80°C (176°F) and use a metal rod (preferably a brass drift punch) inserted through the centre of the upper bearing, to tap evenly around the inner race of the lower bearing to drive it from the hub (see illustrations). The bearing spacer will also come out.
5 Lay the wheel on its other side so that the remaining bearing faces down. Drive the bearing out of the wheel using the same technique as above.
6 If the bearings are of the unsealed type or are only sealed on one side, clean them with a high flash-point solvent (one which won't leave any residue) and blow them dry with compressed air (don't let the bearings spin as you dry them). Apply a few drops of oil to the bearing. **Note:** *If the bearing is sealed on both sides don't attempt to clean it.*
7 Hold the outer race of the bearing and rotate the inner race – if the bearing doesn't turn smoothly, has rough spots or is noisy, replace it with a new one.

HAYNES HiNT *Refer to Tools and Workshop Tips for more information about bearings.*

8 If the bearing is good and can be re-used, wash it in solvent once again and dry it, then pack the bearing with grease.
9 Thoroughly clean the hub area of the wheel. Heat the wheel hub around the bearing housing to 80°C (176°F) and install the left-hand side bearing into its recess in the hub, with the marked or sealed side facing outwards – note that it is a larger size than the right-hand bearing. Using the old bearing (if new ones are being fitted), a bearing driver or a socket large enough to contact the outer race of the bearing, drive it in until it's completely seated (see illustration).
10 Turn the wheel over and install the bearing spacer. Drive the right-hand side bearing into place as described above. Where fitted, install

15.3c . . . and the circlip

15.4a Use a drift to knock out the bearings . . .

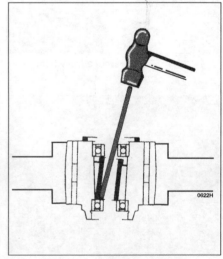

15.4b . . . locating it as shown

15.9 Drive the bearing in using a driver or suitable socket

15.11a Fit the grease seal . . .

15.11b . . . and drift it in using a block of wood

the headed bush into the hub left-hand bearing. Install the bearing circlip, making sure it fits properly in its groove **(see illustration 15.3c)**.

11 Fit the speedometer drive plate into the left-hand side of the wheel, locating its outer tabs in the slots in the hub **(see illustration 15.3b)**. Apply a smear of grease to the lips of the new grease seal, then install it in the wheel **(see illustration)**. Drive it into place using a seal or bearing driver, a suitable socket or a flat piece of wood **(see illustration)**.

12 Clean off all grease from the brake discs using acetone or brake system cleaner then install the wheel (see Section 13).

Rear wheel bearings

13 The rear wheel bearings are part of the final drive housing, the dismantling of which is beyond the scope of this manual.

---

16 Tyres –
general information
and fitting

General information

1 The wheels fitted to all models are designed to take tubeless tyres only. Tyre sizes are given in the Specifications at the beginning of this chapter.

2 Refer to the *Daily (pre-ride) checks* listed at the beginning of this manual for tyre maintenance.

Fitting new tyres

3 When selecting new tyres, refer to the tyre sizes given in the Specifications section of this Chapter, the tyre options listed in your owners handbook and the tyre information label under the bike's seat. Ensure that front and rear tyre types are compatible, the correct size and correct speed rating; if necessary seek advice from a BMW dealer or tyre fitting specialist **(see illustration)**.

4 It is recommended that tyres are fitted by a motorcycle tyre specialist rather than attempted

6

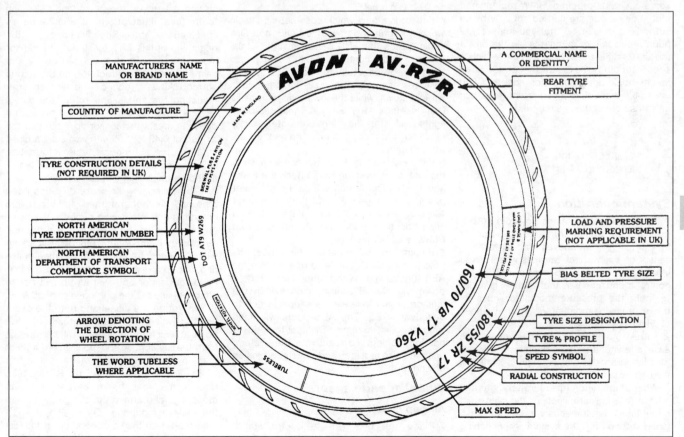

16.3 Common tyre sidewall markings

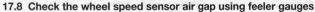

17.8 Check the wheel speed sensor air gap using feeler gauges

17.11a Unscrew the two bolts (arrowed) . . .

in the home workshop. This is particularly relevant in the case of tubeless tyres because the force required to break the seal between the wheel rim and tyre bead is substantial, and is usually beyond the capabilities of an individual working with normal tyre levers. Additionally, the specialist will be able to balance the wheels after tyre fitting.

5 Note that punctured tubeless tyres can in some cases be repaired. BMW recommend that such repairs are carried out only by an authorised dealer. A get-you-home type puncture repair kit is included in the bike's toolkit. If this is used, note that the vehicle's speed should not exceed 37 mph (60 kmh) and it should not be used for more than 250 miles (400 km) before a permanent repair can be carried out or the tyre renewed.

17 ABS – system operation and components

System operation

1 The anti-lock braking system (ABS) prevents the wheels from locking up under hard braking or on uneven road surfaces. A sensor on each wheel transmits wheel speed information to the ABS control unit. If the control module senses that a wheel is about to lock, the pressure modulator releases brake pressure momentarily to that wheel, preventing a skid.
2 The ABS system is self-checking, and is always switched on, although it will not function at speeds below 3 mph (5 kmh) or if the battery is flat.
3 When the ignition (main) switch is turned on, the ABS indicator lights on the instrument panel flash simultaneously, then extinguish when setting off if the system is functioning normally. If the indicator lights remain on, or start flashing alternately, the rider is alerted

that ABS is not functioning and that a fault is indicated in the system. If this occurs, stop the motorcycle and switch the ignition off. Switch it on again – if the lights flash simultaneously then the system is OK, but if the lights remain on or flash alternately, take the machine to a BMW dealer for testing. **Note:** *The light may flash when placing the machine on the centre stand with the ignition on. This is normal.*
4 If there is a fault which does not cure itself when the ignition is switched OFF and ON again, press the ABS cancel switch – this confirms to the control unit that the fault has been acknowledged by the rider. The system switches itself off. The upper indicator light will extinguish, while the lower one remains lit (not flashing). After every 4.5 minutes both lights will start flashing alternately as a reminder that there is a fault. Press the ABS switch again.
5 If the indicator lights do not come on when the ignition is switched on, check the fuses and the indicator bulbs in the instrument panel (see Chapter 8), and the control unit wiring connector (see Step 7). If the bulbs and connections are good, take the machine to a BMW dealer for testing.
6 If using an ABS-equipped GS model off-road, it may be desirable to ride without the ABS functioning. With the ignition off, hold down the cancel switch, then switch the ignition on and release the cancel switch to deactivate ABS. The ABS lights will flash simultaneously to remind the rider that ABS is not functioning. To reactivate the system, bring the bike to a stop and switch the ignition off and on.

Wheel speed sensors

Air gap check

7 Place the bike on its centre stand, supporting it so that the wheel is off the ground. Locate the sensor feeler gauge in the

bike's toolkit. Before checking the air gap, make sure that the pulse ring is clean. Rotate the wheel and locate the paint dot on the outer edge of the pulse ring – this point has been marked following a runout check of the pulse ring and determines the widest gap between the sensor and pulse ring. Align the paint dot with the sensor.
8 Using the feeler gauges, measure the clearance between the sensor and pulse ring tooth **(see illustration)**. It should be as specified in the Specifications at the beginning of this Chapter. If not, the sensor must be removed as described below and shims added or subtracted until the gap is correct. Always recheck the air gap after adjusting the shims and rotate the wheel several times to check that the pulse ring doesn't contact the sensor tip. Note that the air gap of 0.45 to 0.55 mm was standardised in 1997 for all models; this information supersedes that detailed on the fork leg or final drive unit stickers of earlier bikes.
9 If the pulse ring or brake disc has been renewed or not returned to the same mounting holes following removal, the pulse ring runout must be measured and the ring remarked. Remove the old paint marking where applicable. For the front wheel pulse ring, set up a dial gauge attached to the left-hand fork slider and rest its tip (with shoe attached) against the pulse ring toothed face. Slowly rotate the wheel and note the exact point at which the runout is greatest (ie the point at which the pulse ring will be furthest from the sensor). At this point make a paint mark on the outer edge of the pulse ring tooth. For the rear wheel on RS models, detach the rear brake caliper from its mountings and attach the dial gauge arm to the caliper mounting. On R, RT and GS models, detach the sensor and insert the dial gauge through its mounting hole in the final drive housing.

17.11b ... and withdraw the sensor with shims

17.11c Tighten the bolts to the specified torque

Removal and refitting – front wheel sensor

10 If the sensor is to be removed completely, trace the wiring from the sensor up to the wire connectors under the fuel tank right-hand side on R, RS and GS models, or under the fairing right-hand side on RT models. Use the wiring diagrams at the end of this manual for wire colour information. Free the sensor wiring from all retaining clips, having taken note of its exact routing. Where necessary for access, remove the front mudguard (see Chapter 7).

11 Remove the sensor mounting bolts and gently withdraw the sensor and shims from the fork slider mounting **(see illustrations)**. Clean off any dirt from the sensor's tip, install the shims on its body and refit it to the fork slider, tightening its bolts to the specified torque setting **(see illustration)**.

Removal and refitting – rear wheel sensor

12 If the sensor is to be removed completely, trace the wiring from the sensor up to the wire connectors under the right-hand side of the rear sub-frame. Use the wiring diagrams at the end of this manual for wire colour information. Free the sensor wiring from all retaining clips, having taken note of its exact routing.

13 On RS models, unbolt the rear brake caliper to access the sensor mountings. Remove the sensor mounting bolts and gently withdraw the sensor and shims from its mounting bracket **(see illustration 6.5)**.

14 On R, RT and GS models, remove the single screw which retains the sensor to the final drive unit and ease the sensor and shims out of its bore **(see illustration 6.1)**.

15 On all models, clean off any dirt from the sensor's tip, install the shims on its body and refit it, tightening its bolts to the specified torque setting.

Control unit and pressure modulator

17 The ABS control unit, pressure modulator and main ABS relay are mounted under the fuel tank **(see illustration)**. Remove the fuel tank for access (see Chapter 3). Work on these components is outside of the scope of this manual. If a problem is indicated, all that can be done is to check that the wiring connections to the unit are sound – check with the battery negative lead disconnected.

18 A protective cap should be fitted over the front brake hose union bolt where it connects to the metal modulator pipe on the right-hand side of the frame strut **(see illustration)**. The cap protects the brake line from damage should the right-hand fork tube contact it due to a damaged steering stop. If there isn't one on your bike, obtain one from a BMW dealer.

System testing

19 The ABS control unit is capable of storing faults in its memory which can be read out as codes and cleared using the Bosch diagnostic tester. The tester connects into a diagnostic plug on the machine. It follows that all testing of the ABS system must be carried out by a BMW dealer.

6

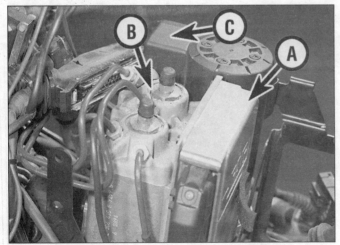

17.17 ABS control unit (A), pressure modulator (B) and main relay (C)

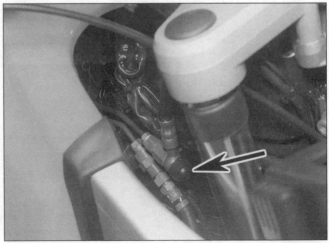

17.18 Protective cap (arrowed) fitted over brake hose union bolt

Chapter 7
Bodywork

Contents

Degrees of difficulty

| Easy, suitable for novice with little experience | | Fairly easy, suitable for beginner with some experience | | Fairly difficult, suitable for competent DIY mechanic | | Difficult, suitable for experienced DIY mechanic | | Very difficult, suitable for expert DIY or professional | 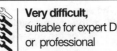 |

1 General information

This Chapter covers the procedures necessary to remove and install the body parts. Since many service and repair operations on these motorcycles require the removal of the body parts, the procedures are grouped here and referred to from other Chapters.

In the case of damage to the body parts, it is usually necessary to remove the broken component and replace it with a new (or used) one. The material that the body panels are composed of doesn't lend itself to conventional repair techniques. There are however some shops that specialise in 'plastic welding', so it may be worthwhile seeking the advice of one of these specialists before consigning an expensive component to the bin.

When attempting to remove any body panel, first study it closely, noting any fasteners and associated fittings, to be sure of returning everything to its correct place on installation. In some cases the aid of an assistant will be required when removing panels, to help avoid the risk of damage to paintwork. Once the evident fasteners have been removed, try to withdraw the panel as described but DO NOT FORCE IT – if it will not release, check that all fasteners have been removed and try again. Where a panel engages another by means of tabs, be careful not to break the tab or its mating slot or to damage the paintwork. Remember that a few moments of patience at this stage will save you a lot of money in replacing broken fairing panels!

When installing a body panel, first study it closely, noting any fasteners and associated fittings removed with it, to be sure of returning everything to its correct place. Check that all fasteners are in good condition, including all trim nuts or clips and damping/rubber mounts; any of these must be replaced if faulty before the panel is reassembled. Check also that all mounting brackets are straight and repair or replace them if necessary before attempting to install the panel. Where assistance was required to remove a panel, make sure your assistant is on hand to install it.

Tighten the fasteners securely, but be careful not to overtighten any of them or the panel may break (not always immediately) due to the uneven stress.

2 Bodywork – R models

Front mudguard

1 Remove the front wheel and pass the speedometer cable through its guide on the mudguard (see Chapter 6).
2 Remove the bolt and nut from each side which secure the mudguard to the fork sliders, then remove the bolt from the underside of the mudguard which secures it to the fork bridge. Unscrew the bolts securing the mudguard and draw it forward from between the forks.
3 Installation is the reverse of removal.

Fuel tank trim

4 Remove the seat (see below).
5 Remove the screws securing the various sections of trim to the fuel tank, then remove the sections, noting how they fit together **(see illustrations)**.
6 Installation is the reverse of removal.

7

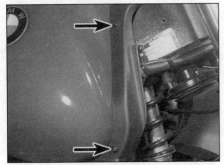

2.5a Remove the screws (arrowed) securing the front ...

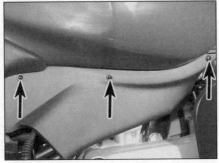

2.5b ... the middle ...

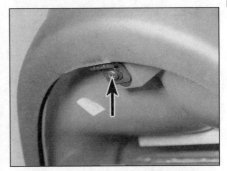

2.5c ... and the rear trim sections

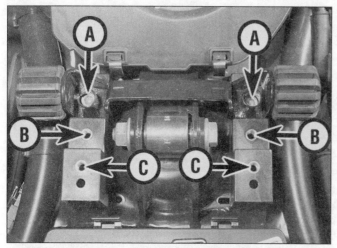

2.9 The seat mount is secured by two bolts (A). Position the mount in the lower position (shown), the middle position (B) or the upper position (C) as required

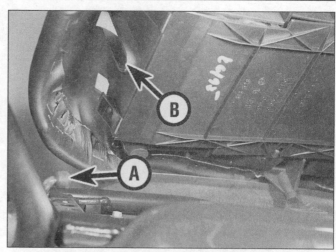

2.10 Locate the bar end (A) in the slot (B)

2.11 Locate the mounts (A) into the slots (B), and the pegs (C) in the holes (D)

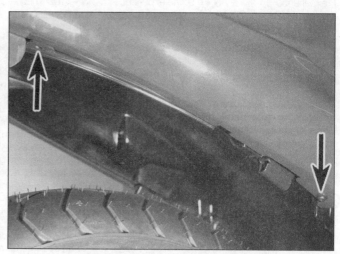

2.13a Remove the two fasteners (arrowed)

Seat

Note: *A modified seat catch was introduced in early 1996 to prevent the seat opening in cases where rider weight is thrust forward, such as on heavy application of the brakes. Refer to a BMW dealer for details.*

7 Insert the ignition key into the seat lock located on the left-hand side, and turn it clockwise to unlock the seat.

8 Keeping the key turned, remove the rider's section of the seat, followed by the passenger's section, noting how they locate.

9 The rider's seat height can be adjusted to suit individual needs. Remove the bolts securing the adjustable front mounts and position them as required, then tighten the bolts securely **(see illustration)**. There are three positions.

10 Install the passenger's seat first, locating the bar end in the slot in the back of the seat **(see illustration)**.

11 Locate the front of the rider's seat onto the front mounts, then locate the rear pegs in the holes and press down on the seat to engage the latches **(see illustration)**.

Side panels

12 Remove the seat (see above).

13 Release the two fasteners on the underside of the panel **(see illustration)**, then unscrew the upper grab-rail mounting bolt and slacken the lower one **(see illustration 2.15)**. Carefully pull the panel away to release the clip from the frame, then draw it forward to free its rear edge from between the grab-rail and the frame **(see illustrations)**.

14 Installation is the reverse of removal.

2.13b Release the clip from the frame . . .

2.13c . . . and draw the panel from between the grab-rail and the frame

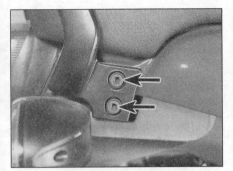

2.15 The passenger grab-rail is secured by two bolts (arrowed)

3.1 Remove the two bolts (arrowed) to free the side section

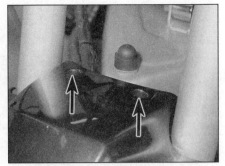

3.2a Remove the two screws (arrowed) . . .

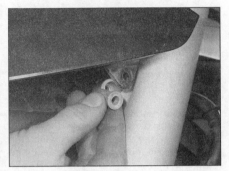

3.2b . . . noting how the collar fits between the mudguard and the fork

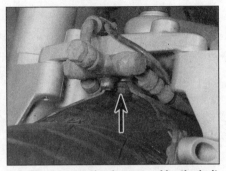

3.3 The rear section is secured by the bolt (arrowed)

3.5 The seat lock is below the headlight

Passenger grab-rail

15 Unscrew the two bolts securing each side of the grab-rail to the frame and remove the grab-rail **(see illustration)**.

16 Installation is the reverse of removal.

3 Bodywork – RS models

Front mudguard

1 Unscrew the two bolts securing each side section (where fitted) to the fork, noting that

the top bolts also secure the sides of the top and rear sections **(see illustration)**.

2 Remove the two screws securing the top section to the fork bridge and remove the top section, noting how the collar fits between it and the fork mounting **(see illustrations)**.

3 Unscrew the bolt securing the rear section to the underside of the fork bridge and remove the rear section, noting that the speedometer cable must first be freed from its guide **(see illustration)**.

4 Installation is the reverse of removal.

Seat

5 Insert the ignition key into the seat lock located below the tail light, and turn it

clockwise to unlock the seat **(see illustration)**.

6 Keeping the key turned, remove the passenger's section of the seat, followed by the rider's section, noting how they locate.

7 The rider's seat height can be adjusted to suit individual needs. When installing the seat, set it in the slots required in the mountings **(see illustration)**. There are three positions.

8 Install the rider's seat first, locating the mounting bar in the required slots in the mountings **(see illustration 3.7)**.

9 Locate the front of the passenger's seat, then locate the rear bar in the lock and press down on the seat to engage the latch **(see illustration)**.

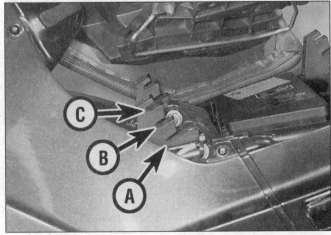

3.7 The seat has three positions, low (A), medium (B) and high (C)

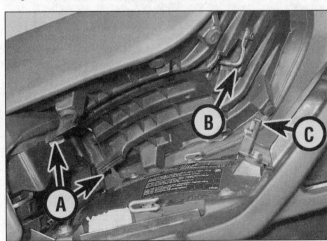

3.9 Locate the tabs (A) at the front, and the bar (B) in the lock (C)

7

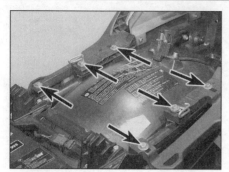

3.11a Unscrew the bolts (arrowed) . . .

3.11b . . . and remove the grab-rail

3.16 The top cover is secured by a screw and a bolt (arrowed)

3.17a Remove the four screws (arrowed) . . .

3.17b . . . to free the panel

Passenger grab-rail

10 Remove the seat (see above).
11 Unscrew the bolts securing each side of the grab-rail to the frame and remove the grab-rail **(see illustrations)**.
12 Installation is the reverse of removal.

Rear side panels

13 If fitted, remove the panniers.
14 Remove the passenger grab-rail (see above).
15 Remove the tail light (see Chapter 8).
16 Remove the screw and bolt securing each side of the top cover and remove the cover **(see illustration)**.

17 Remove the screws securing the side panel and remove the panel, noting how it fits **(see illustrations)**.
18 Installation is the reverse of removal. Make sure the lip on each panel locates over the cushioned support on the frame.

Fairing side panels

19 Remove the seat (see above).
20 On models with full fairing side panels, each panel is secured by eight fasteners **(see illustration)**. On models with half fairing side panels, each panel is secured by six fasteners. Working around the panel,

remove the various fasteners securing it to the fuel tank, sub-frame, engine brackets and fairing, noting carefully which type of fastener fits where. Carefully ease the panel off the bike.
21 Installation is the reverse of removal.

Fairing

22 Remove the fairing side panels (see above). Disconnect the battery negative (-ve) lead.
23 Remove the fasteners securing the windshield and remove the windshield **(see illustration)**.
24 Remove the screws securing the cockpit panels, and carefully release the panels, noting how they fit together **(see illustrations)**. As the wiring connectors for the switches on the left panel and the digital rider information display on the right panel become accessible, disconnect them **(see illustration)**.
25 On models with an adjustable windshield, unscrew and remove the adjuster **(see illustration)**.
26 Disconnect the headlight wiring connector, and remove the turn signal and sidelight bulbholders **(see illustrations)**.
27 Remove any remaining screws securing the fairing, then remove the nuts securing the headlight to the headlight frame and draw the fairing forward, noting how it fits **(see**

3.20 Fairing side panel is secured at eight points – models with full fairing

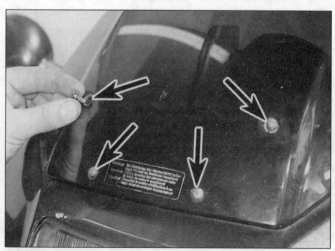

3.23 The windshield is secured by four screws (arrowed)

3.24a Remove the three front screws (arrowed) . . .

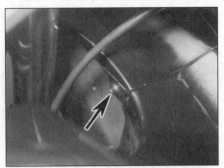

3.24b . . . and the side screw (arrowed) . . .

3.24c . . . and remove the panel

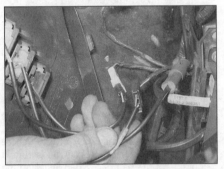

3.24d Disconnect the switch wiring

3.25 Unscrew and withdraw the windshield adjuster mechanism

3.26a Disconnect the headlight wiring connector . . .

illustrations). If required, remove the headlight from the fairing (see Chapter 8).

28 Installation is the reverse of removal.

Check that all the collars are fitted in the headlight mounts (see illustration). Make sure the various wiring connectors are

securely and correctly connected. Check the operation of the lights and turn signals before taking the bike on the road.

3.26b . . . and remove the turn signal bulb holder (arrowed) . . .

3.26c . . . and the sidelight bulb holder

3.27a Remove the screws securing the fairing to the tank . . .

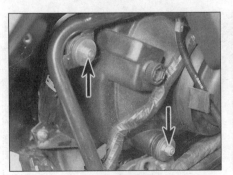

3.27b . . . and the headlight mounting nuts . . .

3.27c . . . and remove the fairing

3.28 Check that all the mounting grommets and collars are fitted

4.1 Remove the two bolts (arrowed) securing the front section

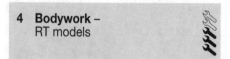

4 Bodywork – RT models

Front mudguard

1 Unscrew the two bolts securing the front section to the forks, noting that each top bolt also secure the sides of the rear section and that a collar fits between them, then draw the front section off the forks **(see illustration)**.
2 Unscrew the bolt securing the rear section to the underside of the fork bridge **(see illustration 3.3)** and remove the rear section.
3 Installation is the reverse of removal.

Seat

4 Insert the ignition key into the seat lock located below the tail light, and turn it clockwise to unlock the seat **(see illustration 3.5)**.

4.17 Detach the mirror and remove the turn signal bulb holder

5 Keeping the key turned, remove the passenger's section of the seat, followed by the rider's section, noting how they locate.
6 The rider's seat height can be adjusted to suit individual needs. When installing the seat, set it in the slots required in the mountings. There are three positions **(see illustration 3.7)**.
7 Install the rider's seat first, locating the mounting bar in the required slots in the mountings.
8 Locate the front of the passenger's seat, then locate the rear bar in the lock and press down on the seat to engage the latch **(see illustration 3.9)**.

Passenger grab-rail/ luggage rack

9 Detach the top box (where fitted) from the luggage rack. Remove the seat (see above).
10 Unscrew the three bolts securing each side of the grab-rail to the sub-frame **(see**

illustration 3.11a) and the two bolts which retain the luggage rack to the top cover, then remove the grab-rail and luggage rack as a unit.
11 Installation is the reverse of removal.

Rear side panels

12 Remove the panniers and top box, where fitted. Remove both seat sections and the passenger grab-rail/luggage rack (see above). Remove the tail light unit (see Chapter 8).
13 Release the fastener securing the side panel insert. Disengage the insert tabs from the footrest bracket and remove the insert.
14 To remove the side panels and top cover as an assembly, remove the two screws from underneath the panel on each side which retain it to the rear mudguard, plus the two screws which retain the top cover to the rear sub-frame. Ease the panel carefully away from the rear of the bike. If necessary, separate each side panel and from the top cover by removing the retaining screws.
15 Installation is the reverse of removal. Check that the top cover rubber-mountings on the sub-frame are correctly installed; the rubber dampers are marked R or L to denote right or left, and a headed collar is inserted in each side of the damper. When installing the side panel insert, make sure the tabs on its lower edge engage the slots in the footrest bracket.

Fairing side panels

16 Remove the rear side panels (see above).
17 Carefully pull the mirrors/turn signal assemblies out of their clips in the fairing, then turn the bulbholder anti-clockwise and withdraw it **(see illustration)**.
18 Each side panel is secured by seventeen fasteners, two of which secure the belly-pan section **(see illustrations)**. Working around the panel, remove the various fasteners securing it to the fuel tank cover, sub-frame, engine brackets and fairing, noting carefully which type of fastener fits where. Carefully release the panel fasteners and remove it. If both panels are being removed, remove the two fasteners securing the belly-pan to each side panel first and remove the belly-pan. Otherwise it can be left attached to one side panel while the other is removed.

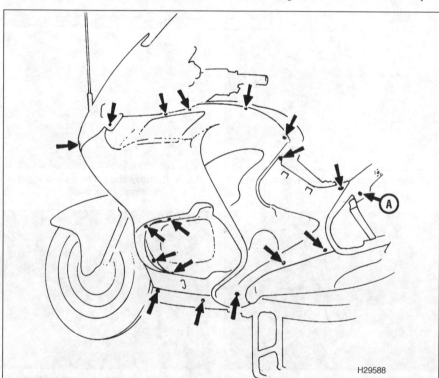

4.18a Fairing side panel fastener locations (arrowed). Note side panel insert fastener (A)

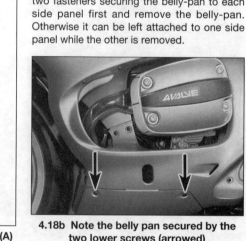

4.18b Note the belly pan secured by the two lower screws (arrowed)

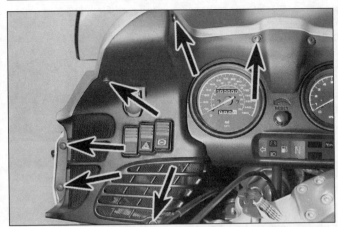

4.22 Each side of the cockpit panel is secured by six screws (arrowed), plus a screw accessed from the front of the fairing

5.2 Upper front mudguard lower mountings bolts (arrowed)

19 Installation is the reverse of removal. Check the operation of the turn signals before taking the bike on the road.

Fairing

20 Remove the fairing side panels (see above). Disconnect the battery negative (-ve) lead.
21 Remove the four screws securing the windshield to the height adjuster bracket pads and remove the windshield.
22 Remove the screws securing the cockpit panel to the fairing (see illustration), not forgetting the two screws which are accessed from the front of the fairing, under the windshield. Carefully release the panel, noting how its tabs located under the lip of the fairing. As the wiring connectors for the switches and auxiliary power socket on the left panel and the digital rider information display on the right panel become accessible, disconnect them.
23 Remove the instrument cluster (see Chapter 8).
24 Disconnect the headlight and sidelight wiring connectors (see Chapter 8). Where fitted, unscrew the nut securing the radio aerial and remove the aerial, noting how it fits.

25 Remove any remaining screws securing the fairing, then remove the nuts securing the headlight to the headlight frame and draw the fairing forward, guiding it over the windshield mounting bracket pads. If required, remove the headlight from the fairing (see Chapter 8).
26 Installation is the reverse of removal. Where a threaded pin fits into a metal collar, remove the collar and fit it onto the pin, then press the pin and collar into the mounting. Make sure the various wiring connectors are securely and correctly connected and that the aerial lead connection is sound. Check the operation of the lights and turn signals before taking the bike on the road.

5 Bodywork – GS models

Front mudguard

1 To remove the lower (main) mudguard, first remove the front wheel (see Chapter 6) and pass the speedometer cable through its guide on the mudguard. Remove the bolt and nut from each side which secure the mudguard to

the fork sliders, then remove the bolt from the underside of the mudguard which secures it to the fork bridge.
2 The upper (coloured) mudguard is retained to the headlight bracket by two bolts from the underside (see illustration) and a bolt on each side. Remove the side sections for access to the side bolts (see Step 10).
3 Installation is the reverse of removal.

Seat

4 Insert the ignition key into the seat lock located on the left-hand side, and turn it clockwise to unlock the seat.
5 Keeping the key turned, remove the rear section of the seat, followed by the front section, noting how they locate.
6 The rider's seat height can be adjusted to suit individual needs. When installing the seat, set it in the mounting slots required (see illustration). There are two positions.
7 Install the rider's seat first, locating the mounting bar in the required slots in the mountings (see illustration 5.6).
8 Locate the front of the passenger's seat, then locate the rear bar in the lock and press down on the seat to engage the latch (see illustration).

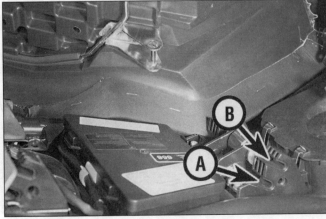

5.6 The seat has two positions, low (A) and high (B)

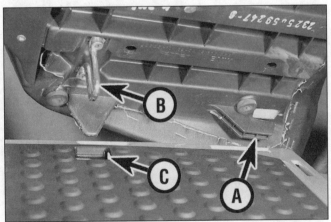

5.8 Locate the tabs (A) at the front, then fit the bar (B) into the latch (C)

5.9a Remove the two lower screws to remove the windshield – the two upper screws secure the inner trim

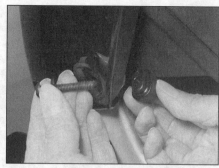

5.9b Note the threaded knob on the rear lower screw . . .

5.9c . . . and the collar and washers on the lower front screw

Windshield, front side panels and fuel tank trim

9 To remove the windshield, remove the two lower screws on each side, and remove the windshield complete with its inner pieces (see illustrations).

10 To remove the front side sections, release the two fasteners from their underside and withdraw the side sections, noting how they locate against the fuel tank. Access is now available to the upper mudguard side bolts – see Step 2.

11 To remove the fuel tank trim, carefully pull its lugs out of the rubber mounts on the tank (see illustration).

12 Installation is the reverse of removal. To adjust the height of the windshield slacken the lower front screw on each side and tilt the windshield as required. Retighten the screw to lock the setting.

Passenger grab-rail and luggage carrier

13 Remove the seat (see above).

14 Unscrew the two bolts securing the grab-rail and remove it (see illustration).

15 The luggage carrier is retained to the sub-frame by four bolts, the rear bolts are accessed after removing the toolbox cover.

16 Installation is the reverse of removal.

Sump guard

17 The sump guard is secured by four nuts and washers to the underside of the engine. Clean the threads of the studs before attempting removal.

18 With the sump guard removed, access can now be gained to the engine protector plate. Remove the four studs/rubber dampers and withdraw the plate from the engine.

20 Installation is the reverse of removal. If the stud/rubber dampers or nuts have corroded due to their exposed position, replace them with new ones.

5.11 Pull the tank trim out of its mounts (arrowed)

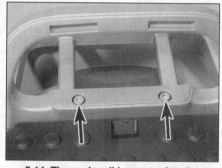

5.14 The grab-rail is secured to the luggage carrier by two bolts (arrowed)

Chapter 8
Electrical system

Contents

Degrees of difficulty

Easy, suitable for novice with little experience	**Fairly easy,** suitable for beginner with some experience	**Fairly difficult,** suitable for competent DIY mechanic 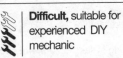	**Difficult,** suitable for experienced DIY mechanic 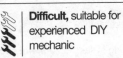	**Very difficult,** suitable for expert DIY or professional 

Specifications

Battery

Capacity . . . . 12 V, 19 Ah

Alternator

Type . . . . Three-phase AC with integral regulator/rectifier
Maximum output . . . . 700 W, 14 V
Maximum current
    at 1000 rpm . . . . 18 A
    at 4000 rpm . . . . 50 A
Resistance between slip rings . . . . 2.8 to 3.2 ohms
Resistance between slip rings and ground (earth) . . . . 0 ohms (no resistance)

Starter motor

Type . . . . Permanent magnet with planetary gear drive
Power rating . . . . 1.1 kW

Fuel pump

Fuel pressure . . . . 43.5 psi (3.0 Bar)
Delivery volume . . . . 110 litres/hr

8

Fuses

F1 – Instrument cluster and brake light .... 15 A
F2 – Sidelight and tail light .... 15 A
F3 – Turn signals, clock .... 15 A
F4 – Auxiliary power socket .... 15 A
F5 – Motronic control unit .... 15 A
F6 – Fuel pump .... 15 A
F7 – Heated handlebar grips .... 4 A
F8 – Radio (RT models) .... 15 A

Bulbs

Headlight .... 60/55 W H4 halogen
Sidelight .... 4 W
Brake/tail light .... 21/10 W
Turn signal lights .... 21 W
Instrument lighting and warning lights
   Turn signal and high beam warning light bulbs .... 3 W
   All other bulbs .... 1.7 W

Torque settings

Oil pressure switch .... 30 Nm
Fuel pump cover bolts .... 6 Nm
Starter motor mounting bolts .... 20 Nm
Starter motor lead nut .... 10 Nm
Starter motor cover bolt .... 7 Nm
Alternator drive belt pre-load (on adjusting bolt) .... 8 Nm
Alternator mounting bolt and nuts .... 20 Nm
Side rail to strut
   8.8 bolt (tensile marking on bolt head) .... 47 Nm
   10.9 bolt (tensile marking on bolt head) .... 58 Nm
Side rail to engine
   RS and RT models .... 47 Nm
   R and GS models .... 58 Nm

1 General information

All models have a 12-volt electrical system charged by a three-phase alternator. The alternator has integral regulator and rectifier units.

The regulator maintains the charging system output within the specified range to prevent overcharging, and the rectifier converts the ac (alternating current) output of the alternator to dc (direct current) to power the lights and other components and to charge the battery. The alternator is mounted centrally on top of the engine unit and is driven by belt off a pulley on the front end of the crankshaft.

The starter motor is mounted on the left-hand side of the engine. The starting system includes the starter motor, solenoid, starter relay and switches. If the engine kill switch is in the RUN position and the ignition (main) switch is ON, the starter relay allows the starter motor to operate only if the transmission is in neutral (neutral switch on) or, if the transmission is in gear with the clutch lever pulled into the handlebar and the side stand up.

Note: *Keep in mind that electrical parts, once purchased, cannot be returned. To avoid unnecessary expense, make very sure the faulty component has been positively identified before buying a replacement part.*

2 Electrical system – fault finding

 Warning: To prevent the risk of short circuits, the ignition (main) switch must always be OFF and the battery negative (-ve) terminal should be disconnected before any of the bike's other electrical components are disturbed. Don't forget to reconnect the terminal securely once work is finished or if battery power is needed for circuit testing.

1 A typical electrical circuit consists of an electrical component, the switches, relays, etc. related to that component and the wiring and connectors that hook the component to both the battery and the frame. To aid in locating a problem in any electrical circuit, refer to the wiring diagrams at the end of this Chapter.

2 Before tackling any troublesome electrical circuit, first study the wiring diagram (see end of Chapter) thoroughly to get a complete picture of what makes up that individual circuit. Trouble spots, for instance, can often be narrowed down by noting if other components related to that circuit are operating properly or not. If several components or circuits fail at one time, chances are the fault lies in the fuse or earth (ground) connection, as several circuits often are routed through the same fuse and earth (ground) connections.

3 Electrical problems often stem from simple causes, such as loose or corroded connections or a blown fuse. Prior to any electrical fault finding, always visually check the condition of the fuse, wires and connections in the problem circuit. Intermittent failures can be especially frustrating, since you can't always duplicate the failure when it's convenient to test. In such situations, a good practice is to clean all connections in the affected circuit, whether or not they appear to be good. All of the connections and wires should also be wiggled to check for looseness which can cause intermittent failure.

4 If testing instruments are going to be utilised, use the wiring diagram to plan where you will make the necessary connections in order to accurately pinpoint the trouble spot.

5 The basic tools needed for electrical fault finding include a battery and bulb test circuit, a continuity tester, a test light, and a jumper wire. A multimeter capable of reading volts, ohms and amps is also very useful as an alternative to the above, and is necessary for performing more extensive tests and checks.

 Refer to Fault Finding Equipment in the Reference section for details of how to use electrical test equipment.

3.2 Remove the tank bolt (arrowed) and raise the tank at the rear

3.3a Unscrew the bolt (arrowed) . . .

3.3b . . . and remove the air pipe

3 Battery – removal and installation

Caution: Be extremely careful when handling or working around the battery. The electrolyte is very caustic and an explosive gas (hydrogen) is given off when the battery is charging.

1 Remove the seat (see Chapter 7). On RS and RT models, remove the left-hand fairing side panel (see Chapter 7). On R and GS models, remove the fuel tank trim (see Chapter 7).
2 On R and GS models, remove the bolt securing the rear of the fuel tank to the right-hand side of the frame, then raise the tank and carefully support it in that position, making sure there is no strain on the fuel pipes **(see illustration)**.

3 Remove the air filter (see Chapter 1), then unscrew the bolt securing the air intake pipe and lift the pipe out of the air filter housing **(see illustrations)**.
4 Release the rubber strap retaining the battery, then detach the breather hose from its union **(see illustrations)**.
5 Draw the battery out slightly, then unscrew the negative (-ve) terminal nut and disconnect the lead(s) from the battery **(see illustration)**. Now draw the battery fully out, then unscrew the positive (+ve) terminal nut and disconnect the lead **(see illustration)**. Remove the battery.
6 On installation, clean the battery terminals and lead ends with a wire brush or knife and emery paper. Reconnect the leads, connecting the positive (+ve) terminal first, then connect the breather hose and secure the battery with its strap.

HAYNES HINT *Battery corrosion can be kept to a minimum by applying a layer of petroleum jelly to the terminals after the cables have been connected.*

7 Install all other components in a reverse of the removal procedure.

4 Battery – charging

Caution: Be extremely careful when handling or working around the battery. The electrolyte is very caustic and an explosive gas (hydrogen) is given off when the battery is charging.

1 Remove the battery (see Section 3).

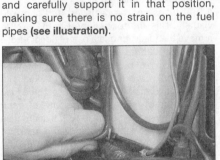

3.4a Unhook the battery strap . . .

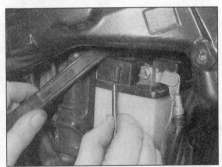

3.4b . . . and detach the breather hose

3.5a Unscrew the negative terminal nut . . .

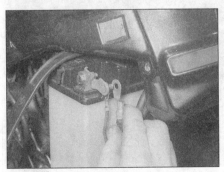

3.5b . . . and detach the leads

3.5c Draw the battery out . . .

3.5d . . . and detach the positive lead

8

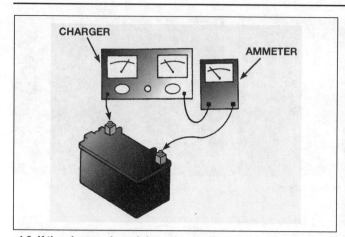

4.2 If the charger doesn't have an ammeter built in, connect one in series as shown. DO NOT connect the ammeter between the battery terminals or it will be ruined

5.2 Release the clips at the front and remove the lid

Connect the charger to the battery, making sure that the positive (+ve) lead on the charger is connected to the positive (+ve) terminal on the battery, and the negative (-ve) lead is connected to the negative (-ve) terminal.

2 BMW recommend that the battery is charged at a maximum rate of 1.9 amps for 10 hours. Exceeding this figure can cause the battery to overheat, buckling the plates and rendering it useless. Few owners will have access to an expensive current controlled charger, so if a normal domestic charger is used check that after a possible initial peak, the charge rate falls to a safe level **(see illustration)**. If the battery becomes hot during charging **stop**. Further charging will cause damage. **Note:** *In emergencies the battery can be charged at a higher rate of around 6.0 amps for a period of 1 hour. However, this is not recommended and the low amp charge is by far the safer method of charging the battery.*

3 If the recharged battery discharges rapidly if left disconnected it is likely that an internal

short caused by physical damage or sulphation has occurred. A new battery will be required. A sound item will tend to lose its charge at about 1% per day.

4 Install the battery (see Section 3).

5 If the motorcycle sits unused for long periods of time, charge the battery once every month to six weeks and leave it disconnected.

5 Fuses and relays – check and replacement

Fuses

1 The electrical system is protected by fuses of different ratings. All fuses are housed in the fuse/relay box, which is located under the seat.

2 To access the fuses, remove the seat (see Chapter 7) and unclip the fusebox lid **(see illustration)**.

3 From left to right when looking forward, the fuses are identified as follows **(see illustration):**

1 Instrument cluster and brake light
2 Sidelight and tail light
3 Turn signals, clock
4 Auxiliary power socket
5 Motronic control unit
6 Fuel pump
7 Heated handlebar grips
8 Radio (RT models), spare (all other models)
9 Spare
10 Spare
11 Fuse/relay removing tool

4 The fuses can be removed and checked visually. Use the tool clipped in the box to remove the fuses, or use a pair of needle-nose pliers **(see illustration)**. A blown fuse is easily identified by a break in the element **(see illustration)**. Each fuse is clearly marked with its rating and must only be replaced by a fuse of the correct rating. A spare fuse of each rating is housed in the fusebox. If a spare fuse is used, always replace it so that a spare of each rating is carried on the bike at all times.

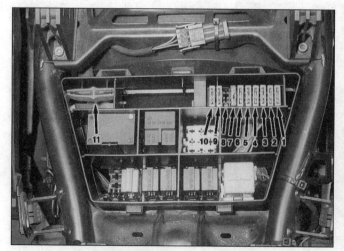

5.3 Fuse identification (see text)

5.4a Use the tool provided to remove a fuse

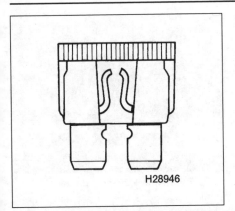

5.4b A blown fuse can be identified by a break in its element

⚠️ *Warning: Never put in a fuse of a higher rating or bridge the terminals with any other substitute, however temporary it may be. Serious damage may be done to the circuit, or a fire may start.*

5 If a fuse blows, be sure to check the wiring circuit very carefully for evidence of a short-circuit. Look for bare wires and chafed, melted or burned insulation. If the fuse is replaced before the cause is located, the new fuse will blow immediately.

6 Occasionally a fuse will blow or cause an open-circuit for no obvious reason. Corrosion of the fuse ends and fusebox terminals may occur and cause poor fuse contact. If this happens, remove the corrosion with a wire brush or emery paper, then spray the fuse end and terminals with electrical contact cleaner.

Relays

7 All relays are housed in the fuse/relay box, which is located under the seat.

8 To access the relays, remove the seat (see Chapter 7) and unclip the fusebox lid **(see illustration 5.2)**.

9 The relays are identified as follows **(see illustration)**:

1 *Encoding plug for Motronic and catalytic converter*
2 *Fuel level damping unit*
3 *Starter motor relay*
4 *Load-relief relay*
5 *Horn relay*
6 *Fuel pump relay*
7 *Motronic relay*
8 *ABS warning relay*
9 *Not in use*
10 *Turn signal relay*
11 *Relay removing tool*

10 The relays can be removed using the tool clipped in the box or a pair of needle-nose pliers **(see illustration)**.

11 Unfortunately BMW provide no test details for any of the relays except the starter motor relay (see Section 26). If there is a fault in a particular circuit and all other tests for that circuit indicate that a relay is at fault, the only way to be sure is to replace the suspect relay with one that is known to be good and seeing whether the fault is cured. It is not possible to repair faulty relays. Make sure the battery is fully charged when checking faulty circuits, as this could affect the operation of a relay.

> **HAYNES HiNT** *If you suspect a relay fault, before buying a replacement, check first whether any of the other relays are identical, ie have the same identification number on them. If so, swap the relays over and retest the circuit. If the fault is still apparent, the relay is proved good, whereas if it is cured, the relay is confirmed faulty. Never put in a relay other than that specified for the purpose. Serious damage may be done to the circuit, or a fire may start.*

6 Lighting system – check

1 The battery provides power for operation of the headlight, tail light, brake light and instrument cluster lights. If none of the lights operate, always check battery voltage before proceeding. Low battery voltage indicates either a faulty battery or a defective charging system. Refer to Chapter 1 for battery checks and Section 30 for charging system tests. Also, check the condition of the fuses.

Headlight

2 If the headlight fails to work, first check the bulb (see Section 7). If the bulb is sound, use jumper wires to connect the bulb directly to the battery terminals, paying attention to the bulb terminals. If the bulb illuminates, the problem lies in the wiring, lighting switch (UK models only) or load relief relay. Refer to Section 18 for the switch testing procedures, and also the wiring diagrams at the end of this Chapter.

Tail light

3 If the tail light fails to work, check the bulb and the bulb terminals first, then fuse F2, then check for battery voltage on the supply side of the tail light wiring connector. If voltage is present, check the earth (ground) circuit for an open or poor connection.

4 If no voltage is indicated, check the wiring between the tail light and the ignition switch, then check the switch. Also check the lighting switch on UK models.

Brake light

5 See Section 13 for brake switch check and Section 9 for bulb replacement.

Instrument and warning lights

6 See Section 15 for bulb replacement.

Turn signal lights

7 See Section 11 for the turn signal circuit check.

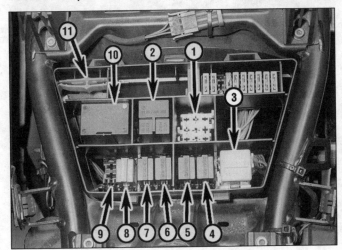

5.9 Relay identification (see text)

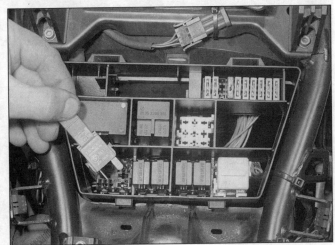

5.10 Use the tool provided to remove a relay

8

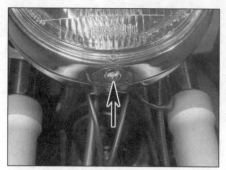

7.1a Remove the screw (arrowed) . . .

7.1b . . . and ease the rim off the shell

7.2 Remove the cockpit panel

7 Headlight bulb and sidelight bulb – replacement

Caution: The headlight bulb is of the quartz-halogen type. Do not touch the bulb glass as skin acids will shorten the bulb's service life. If the bulb is accidentally touched, it should be wiped carefully when cold with a rag soaked in methylated spirit (stoddard solvent) and dried before fitting. Use a paper towel or cloth when handling bulbs.

⚠ *Warning: Allow the bulb time to cool before removing it if the headlight has just been on.*

Headlight

1 On R models, remove the screw securing the headlight rim to the headlight shell, and ease the rim off the shell, noting how it fits **(see illustrations)**.

2 On RS models, turn the handlebars to full lock on either side, then remove the screws securing the right-hand cockpit panel, and carefully release the panel, noting how it fits **(see illustration)**. As the wiring connector for the digital rider information display becomes accessible, disconnect it.

3 On RT models, turn the handlebars to full right-hand lock, then reach under the cockpit panel to access the wiring connector.

4 On GS models, remove the three screws securing the headlight rim and remove the rim **(see illustration)**. Carefully draw the headlight out of the shell **(see illustration)**.

5 On all models, disconnect the wiring connector from the back of the headlight assembly and remove the rubber dust cover (where fitted), noting how it fits **(see illustrations)**.

6 Release the bulb retaining clip, noting how it fits, then remove the bulb **(see illustrations)**.

7 Fit the new bulb, bearing in mind the information in the **Caution** above. Make sure the tabs on the bulb fit correctly in the slots in the bulb housing, and secure it in position with the retaining clip.

8 Where fitted, install the rubber dust cover with the TOP mark at the top, making sure it is correctly seated, and connect the wiring connector.

9 Check the operation of the headlight.

Sidelight

10 On R models, remove the screw securing the headlight rim to the headlight shell, and ease the rim out of the shell, noting how it fits **(see illustrations 7.1a and b)**. Pull the bulbholder out of the back of the headlight **(see illustration)**.

7.4a Remove the screws . . .

7.4b . . . and the rim . . .

7.4c . . . and draw the headlight out

7.5a Disconnect the wiring connector (R models)

7.5b Disconnect the wiring connector . . .

7.5c . . . and remove the dust cover (RS and RT models)

7.6a Release the clip . . .

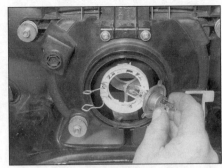

7.6b . . . and remove the bulb

7.10 Draw the bulb holder from the shell

11 On RS and RT models, pull the bulbholder out of the base of the headlight **(see illustration)**.
12 On GS models, remove the three screws securing the headlight rim and remove the rim **(see illustration 7.4a and b)**. Carefully draw the headlight out of the shell **(see illustration 7.4c)**. Pull the sidelight bulbholder out of the back of the headlight.
13 If the bulb is of a bayonet fitting in the bulbholder **(see illustration 7.10)**, carefully push the bulb into the holder and twist it anti-clockwise to release it. If a capless type bulb is fitted, gently pull it from the bulbholder **(see illustration)**.
14 Install the new bulb in the bulbholder, then install the bulbholder by pressing it in.
15 Check the operation of the sidelight.

7.11 Draw the bulb holder from the base of the headlight

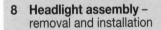

8 Headlight assembly – removal and installation

R models

Removal

1 Remove the screw securing the headlight rim to the headlight shell, and ease the rim out of the shell, noting how it fits **(see illustrations 7.1a and b)**. Disconnect the headlight wiring connector and withdraw the sidelight bulbholder from the back of the headlight assembly **(see illustration 7.5a and 7.10)**.
2 On models with a headlight cowl, disconnect the vertical beam height adjuster from the back of the shell. Unscrew the two bolts securing the headlight shell in the

7.13 Removing the sidelight bulb from its holder – capless type bulb

brackets and remove the shell, easing the wiring out of the back as you do **(see illustration)**.

Installation

3 Installation is the reverse of removal. Make sure all the wiring is correctly connected and secured. Check the operation of the headlight and sidelight.
4 Check the headlight aim (see Chapter 1).

RS and RT models

Removal

5 Remove the fairing (see Chapter 7).
6 Where fitted on RS models, remove the screws securing the windshield adjuster plate and remove the plate **(see illustrations)**.
7 Remove the screws securing the headlight unit to the fairing and remove the headlight, noting how it fits **(see illustration)**.

8.2 Remove the bolts securing the headlight shell

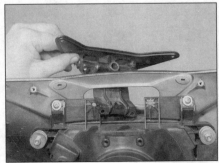

8.6a Remove the screws (arrowed) . . .

8.6b . . . and withdraw the windshield adjuster plate

8.7 Remove the screws securing the headlight to the fairing (RS model shown)

8

8.10 Detach the spring clip (arrowed) to disconnect the vertical beam adjuster

8.12 A single screw (arrowed) retains the instruments to the headlight shell

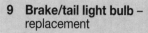

9 Brake/tail light bulb – replacement

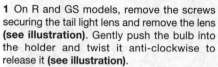

Installation

8 Installation is the reverse of removal. Make sure all the wiring is correctly connected and secured. Check the operation of the headlight and sidelight. Check the headlight aim (see Chapter 1).

GS models

Removal

9 Remove the windshield (see Chapter 7).
10 Detach the vertical beam adjuster mechanism from the back of the headlight shell (see illustration).
11 Remove the three screws securing the headlight rim and remove the rim (see illustration 7.4a and b). Carefully draw the headlight out of the shell (see illustration

7.4c). Disconnect the headlight wiring connector and withdraw the sidelight bulbholder from the back of the headlight assembly.
12 Unscrew the two bolts securing the headlight shell in the brackets, the screw with washer which retain the instrument cluster to the inside of the headlight shell (see illustration) and the two bolts which retain the instrument cluster to the top of the headlight shell. Ease the headlight shell out of position.

Installation

13 Installation is the reverse of removal. Make sure all the wiring is correctly connected and secured. Check the operation of the headlight and sidelight. Check the headlight aim (see Chapter 1).

1 On R and GS models, remove the screws securing the tail light lens and remove the lens (see illustration). Gently push the bulb into the holder and twist it anti-clockwise to release it (see illustration).
2 On RS and RT models, remove the tail light assembly (see Section 10). Turn the bulbholder anti-clockwise and withdraw it from the tail light, then gently push the bulb into the holder and twist it anti-clockwise to remove it (see illustrations).
3 Check the socket terminals for corrosion and clean them if necessary. Line up the pins of the new bulb with the slots in the socket, then push the bulb in and turn it clockwise until it locks into place. **Note:** *The pins on the bulb are offset so it can only be installed one way. It is a good idea to use a paper towel or dry cloth when handling the new bulb to prevent injury if the bulb should break and to increase bulb life.*
4 On R and GS models, check that the rubber seal is in place, then fit the tail light lens and tighten the screws, taking care not to overtighten them as the lens is easily cracked.
5 On RS and RT models, install the bulbholder into the tail light and turn it clockwise to secure it, then install the tail light assembly (see Section 10).

10 Tail light assembly (RS and RT models) – removal and installation

Removal

1 Remove the seat (see Chapter 7).
2 Unscrew the two knurled knobs securing the tail light assembly and draw the assembly off the back of the bike (see illustrations). Remove the bulbholder.

Installation

3 Installation is the reverse of removal. Check the operation of the tail light, the brake light and the rear turn signals.

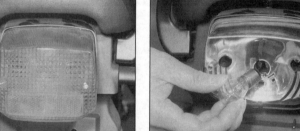

9.1a Remove the screw securing the lens

9.1b Push the bulb in and twist it anti-clockwise to remove it

9.2a Remove the bulb holder from the tail light . . .

9.2b . . . and the bulb from the holder

10.2a Unscrew the two knurled knobs (arrowed) . . .

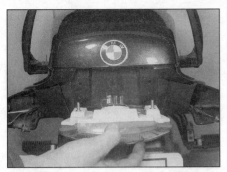

10.2b . . . and remove the tail light

11 Turn signal circuit – check

1 The battery provides power for operation of the turn signal lights, so if they do not operate, always check the battery voltage first. Low battery voltage indicates either a faulty battery or a defective charging system. Refer to Section 30 for charging system tests. Also, check fuse F3 (see Section 5) and the switch (see Section 18).

2 Most turn signal problems are the result of a burned out bulb or corroded socket. This is especially true when the turn signals function properly in one direction, but fail to flash in the other direction. Check the bulbs and the sockets (see Section 12). If it is found that the

turn signals remain on after the ignition main switch has been switched off, this is most likely due to a faulty turn signal relay - the relay was modified during 1996.

3 If the bulbs and sockets are good, using the appropriate wiring diagram at the end of this Chapter, check the wiring between the relay, turn signal switch and turn signal lights for continuity. If the wiring and switch are sound, replace the relay with a new one (see Section 5).

4 Note that the turn signal relay also incorporates a hazard circuit, where all turn signal bulbs flash simultaneously. Apart from testing the hazard switch as described in Section 18, and its wiring, if a fault is indicated the turn signal relay must be replaced – no test details are available.

12 Turn signal bulbs – replacement

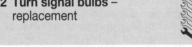

Front turn signals

1 On RT models, remove the rear view mirror/front turn signal housing by easing it out of its retaining clips in the fairing. It is a press fit. Turn the bulbholder anti-clockwise and withdraw it from the housing (see illustration).

2 On RS models, turn the handlebars to full lock on either side, then remove the screws

securing the relevant cockpit panel, and carefully release the panel, noting how it fits (see illustration). As the wiring connectors for either the digital rider information display or the switches become accessible, disconnect them. Turn the bulbholder anti-clockwise and withdraw it from the housing (see illustration).

3 On R and GS models, remove the screw securing the turn signal lens and remove the lens (see illustration).

4 Push the bulb into the holder and twist it anti-clockwise to remove it (see illustration). Check the socket terminals for corrosion and clean them if necessary. Line up the pins of the new bulb with the slots in the socket, then push the bulb in and turn it clockwise until it locks into place.

Rear turn signals

5 On RT models the rear turn signals are housed within the tail light assembly. Remove the assembly (see Section 10). Turn the bulbholder anti-clockwise and withdraw it from the tail light.

6 On R, RS and GS models, remove the screw securing the turn signal lens and remove the lens (see illustration).

7 Push the bulb into the holder and twist it anti-clockwise to remove it (see illustration). Check the socket terminals for corrosion and clean them if necessary. Line up the pins of the new bulb with the slots in the socket, then push the bulb in and turn it clockwise until it locks into place.

8

12.1 Pull the mirror off the fairing to access the bulb holder

12.2a Remove the relevant cockpit panel . . .

12.2b . . . and remove the bulb holder

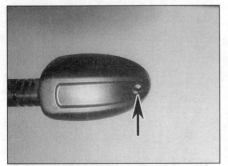

12.3 Remove the screw (arrowed) securing the lens

12.4 Push the bulb in and twist it anti-clockwise to remove it

12.6 Remove the screw securing the lens . . .

12.7 ... then push the bulb in and twist it anti-clockwise to remove it

13.6a Remove the screw (arrowed) ...

13.6b ... to release the master cylinder cover

> **HAYNES HiNT**
>
> *If the socket contacts are dirty or corroded, scrape them clean and spray with electrical contact cleaner before a new bulb is installed.*

13 Brake light switches – check and replacement

Circuit check

1 Before checking any electrical circuit, check the bulb (see Section 9) and fuse F1 (see Section 5).
2 Using a multimeter or test light connected to a good earth (ground), check for voltage at the brake light switch wiring connector. If there's no voltage present, check the wire between the switch and the ignition switch (see the *wiring diagrams* at the end of this Chapter).
3 If voltage is available, touch the probe of the test light to the other terminal of the switch, then pull the brake lever in or depress the brake pedal. If no reading is obtained or the test light doesn't light up, replace the switch.
4 If a reading is obtained or the test light does light up, check the wiring between the switch and the brake light bulb (see the *wiring diagrams* at the end of this Chapter).
5 Check that the contact plate on the rear brake switch is not dirty or damaged and spray it with contact cleaner if necessary.

Switch replacement

Front brake lever switch

6 The switch is mounted on the underside of the brake master cylinder. Remove the front cover to access the switch **(see illustrations)**. Trace the switch wiring and disconnect it at the connector, releasing it from any ties.
7 Remove the screws securing the switch to the bottom of the master cylinder and remove the switch **(see illustration)**.
8 Installation is the reverse of removal. The switch isn't adjustable.

Rear brake pedal switch

9 The switch is mounted on the inside of the right-hand footrest bracket **(see illustration)**.

Trace the switch wiring and disconnect it at the connector, then unscrew the bolt securing the switch and remove it.
10 Installation is the reverse of removal. Check that the switch is heard to click and the brake light comes on when the brake pedal is operated. Spray the switch contact with a water dispersant spray, such as WD40.

14 Speedometer cable – replacement

1 On RS models, turn the handlebars to full lock on either side, then remove the screws securing the left-hand cockpit panel, and carefully release the panel, noting how it fits **(see illustration 12.2a)**. As the wiring connector for the switches become accessible, disconnect them.
2 On RT models, remove the screws securing the cockpit panel to the fairing and carefully release the panel, noting how it fits **(see illustration)**. As the wiring connectors for the switches and auxiliary power socket on the left panel and the digital readout on the right panel become accessible, disconnect them.
3 On GS models, remove the windshield (see Chapter 7).
4 Unscrew the knurled ring securing the speedometer cable to the rear of the instrument cluster and detach the cable **(see illustration)**.

13.7 The front brake switch is secured by two screws (arrowed)

13.9 The rear brake switch is secured by a bolt (arrowed)

14.2 The panel is secured by six screws (arrowed) on each side

14.4 Unscrew the knurled ring and detach the cable

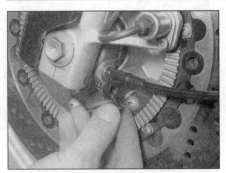

14.5a Remove the screw . . .

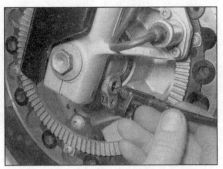

14.5b . . . and detach the cable

15.2a Pull the bulb holder out of its socket . . .

5 Remove the screw securing the lower end of the cable to the drive housing on the left-hand side of the front wheel and detach the cable **(see illustrations)**.

6 Withdraw the cable and remove it from the bike, noting its correct routing.

7 Route the cable to the back of the instrument cluster.

8 Connect the cable upper end to the speedometer and tighten the retaining ring securely **(see illustration 14.4)**.

9 Connect the cable lower end to the drive housing, aligning the slot in the cable end with the drive tab, and tighten the retaining screw securely **(see illustrations 14.5b and a)**.

10 Check that the cable doesn't restrict steering movement or interfere with any other components. On RS and RT models install the cockpit panel. Make sure the wiring connectors are secure.

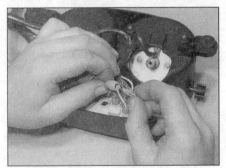

15.2b . . . then pull the bulb out of the holder

15.4 Instrument mounting bolts – R models without headlight cowl

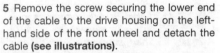

15 Instruments and warning lights – removal and installation

Warning light bulbs

1 Remove the following components to access the instrument and warning light bulbs, according to model:

R models without a headlight cowl – remove the chrome cover from the base of the instruments.

R models with a headlight cowl – remove the cowl side sections.

RS and RT models – remove the fairing cockpit panels for access to warning light bulbs.

GS models – remove the headlight assembly.

2 Gently pull the bulb holder out of position, then pull the bulb out of the bulbholder **(see illustrations)**. If the socket contacts are dirty or corroded, scrape them clean and spray with electrical contact cleaner before a new bulb is installed. Carefully push the new bulb into the holder, then push the holder into place.

3 Install any removed components and check the operation of the warning light.

Instrument cluster – R models (without headlight cowl)

4 Remove the headlight assembly for full access to the instruments (see Section 8). Unscrew the chrome caps from the base of

15.5a Lift each instrument out of the instrument cluster

each instrument and the clock. Remove the four screws which retain the chrome instrument lower cover and lower it off the base of the instruments. Unscrew the speedometer cable ring and pull the inner cable clear. Unplug the bulb holders and disconnect the wiring. Remove the two bolts and lift the instrument cluster off its mounting bracket **(see illustration)**.

5 Each instrument head can be lifted out of the instrument cluster, having first disconnected its bulb holders and wiring **(see illustration)**. On installation, ensure that the instrument is positioned so that the two pegs align with the holes in the chrome cover **(see illustration)**.

Instrument cluster – R models (with headlight cowl)

6 Remove the headlight assembly (see Section 8). Disconnect the front turn signal wiring and unscrew the mounting nut to release each turn signal stalk from the headlight bracket. Remove the four screws to release each cowl half.

7 Remove the trip reset knob, then slip the speedometer off its mounting pegs and disconnect its wiring. Note that the speedometer can be removed from its housing by removing the two nuts and washers from the underside of the housing. On installation, remember to fit the sealing ring between the speedometer and its mounting bracket.

8 The warning lights are contained in a separate unit mounted to the underside of the instrument mounting bracket by three screws.

15.5b Pegs must engage holes in chrome cover (arrowed)

8

15.9 Instrument cluster is retained by a bolt on each side on RS models

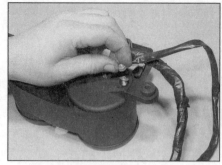

15.10a Remove the two nuts with wave washers . . .

15.10b . . . to release the instrument heads

Instrument cluster – RS models

9 Remove the fairing cockpit panels and windshield height adjuster (see Chapter 7). Remove the two bolts which retain the instruments, noting the headed spacers which engage the grommet on the mounting bracket **(see illustration)**. Ease the instruments upwards so that the speedometer cable can be unscrewed and pulled clear. Unplug the bulb holders and disconnect the wiring. On installation, make sure that the grommets are in place in the mounting bracket and that the headed spacers locate in each side of the grommet.

10 The speedometer and tachometer heads are retained by two nuts (under plastic caps) with wave washers on the base of the instrument cluster **(see illustration)**. Remove the nuts and ease the instrument out of its housing **(see illustration)**. The warning light panel is retained to the instrument cluster by two screws with washers **(see illustration)**. On installation, check that the speedometer trip reset knob engages the grommet in the instrument cluster housing.

Instrument cluster – RT models

11 Remove the fairing cockpit panel (see Chapter 7). Unscrew the speedometer cable ring and pull the inner cable clear. Remove the four bolts which retain the instrument cluster to its mounting bracket, ease it upwards to allow the bulb holders to be unplugged and all wiring disconnected **(see illustration)**. On installation, make sure that the grommets are

in place in the mounting bracket and that the headed spacers locate in each side of the grommet.

12 The speedometer and tachometer heads are retained by two nuts (under plastic caps) with wave washers on the base of the instrument cluster. Remove the nuts and ease the instrument out of its housing, noting that the trip link cable must first be disconnected from the side of the speedometer. The warning light panel is retained to the instrument cluster by two screws with washers.

Instrument cluster – GS models

13 Remove the headlight assembly (see Section 8). Unscrew the speedometer cable ring and pull the inner cable clear. Disconnect the instrument wiring and disconnect the bulb holders. Trace and disconnect the wiring to the digital rider information display and switches. Remove the two mounting nuts from the underside of the instrument cluster and lift it free. On installation ensure that the grommets and spacer are in place on the mounting bracket.

14 The speedometer and tachometer heads are retained by two nuts with wave washers on the base of the instrument cluster. Remove the nuts and ease the instrument out of its housing. The warning light panel is retained to the instrument cluster by two screws with washers. On installation, check that the speedometer trip reset knob engages the grommet in the instrument cluster housing.

HAYNES HiNT *Label all bulb holders as they are unplugged to ensure they are returned to their correct locations on installation*

Digital rider information display

15 Where fitted, the rider information display unit provides fuel volume, gear position, engine oil temperature and time detail. The unit is fitted in the right-hand fairing cockpit panel on RS models, main cockpit panel on RT models, and in the right-hand side of the instrument cluster on GS models.

16 No test details are available with which to test the unit. It complete failure is experienced first check that fuse F3 is intact. Using the wiring diagrams at the end of this manual, trace all wires from the RID to the supply senders or switches, checking for poor connections and damaged or pinched wiring. If the RID provides incomplete or intermittent information this could be due to an earth (ground) fault; check the security of the brown earth lead on the fuel lever sender. The RID unit is retained by three screws to the rear of the cockpit panel or instrument cluster, as applicable **(see illustration)**.

17 A modified rider information display was produced during 1996 and can be identified by the yellow sticker on its rear surface.

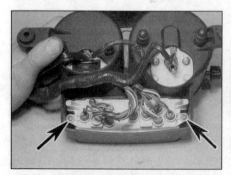

15.10c Warning light panel is retained by two screws (arrowed)

15.11 Instruments are retained by two bolts on each side (arrowed)

15.16 Rider information display unit is retained by three screws

16 Oil pressure switch – check, removal and installation

Check

1 The oil pressure warning light should come on when the ignition (main) switch is turned ON and extinguish a few seconds after the engine is started. If the oil pressure warning light comes on whilst the engine is running, stop the engine immediately and carry out an oil level check after leaving the oil level to stabilise for a few minutes. If the level is correct, and the warning light still stays on, carry out at oil pressure check (see Chapter 1).

Warning light doesn't come on

2 If the oil pressure warning light does not come on when the ignition is turned on, check the bulb (see Section 15) and fuse F1 (see Section 5).

3 The oil pressure switch is screwed into the crankcase below the left-hand cylinder (see illustration). On RS models with a full fairing and RT models, remove the left-hand fairing side panel for access (see Chapter 7). On GS models remove the sump guard for access (see Chapter 7). Detach the wiring connector (see illustration). With the ignition switched ON, earth (ground) the wire on the crankcase and check that the warning light comes on. If the light comes on, the switch is defective and must be replaced.

4 If the light still does not come on, check for voltage at the wire terminal. If there is no voltage present, check the wire between the switch, the instrument cluster and fusebox for continuity (see the wiring diagrams at the end of this Chapter).

Warning light comes on

5 If the warning light comes on whilst the engine is running, yet the oil pressure is satisfactory, remove the wire from the oil pressure switch. With the wire detached and the ignition switched ON the light should be out. If it is illuminated, the wire between the switch and instrument cluster must be earthed (grounded) at some point. If the wiring is good, the switch must be assumed faulty and replaced.

Removal

6 On RS models with a full fairing and RT models, remove the left-hand fairing side panel (see Chapter 7). On GS models remove the sump guard for access (see Chapter 7).

7 The oil pressure switch is screwed into the crankcase below the left-hand cylinder (see illustration 16.3a). Detach the wiring connector (see illustration 16.3b).

8 Unscrew the oil pressure switch and withdraw it from the crankcase.

Installation

9 Apply a suitable sealant to the upper portion of the switch threads near the switch

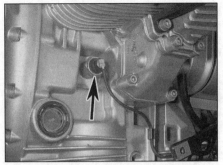

16.3a Oil pressure switch (arrowed)

body, leaving the bottom 3 to 4 mm of thread clean. Install the switch in the crankcase and tighten it to the torque setting specified at the beginning of the Chapter. Attach the wiring connector (see illustration 16.3b).

10 Run the engine and check that the switch operates correctly.

11 On RS models with a full fairing and RT models, install the fairing panel (see Chapter 7). On GS models install the sump guard.

17 Ignition (main) switch – check, removal and installation

⚠️ Warning: To prevent the risk of short circuits, disconnect the battery negative (-ve) lead before making any ignition (main) switch checks.

Check

1 Disconnect the battery negative (-ve) lead. Trace the ignition (main) switch wiring back from the base of the switch and disconnect it at the connector. Remove the cockpit panel(s), fairing or fuel tank as required according to your model to access the connector.

2 Using an ohmmeter or a continuity tester, check the continuity of the connector terminal pairs according to the switch table in the relevant wiring diagram at the end of this Chapter. Continuity should exist between the terminals connected by a solid line on the diagram when the switch is in the indicated position.

3 If the switch fails any of the tests, replace it.

Removal

Note: Support the bike on its centre stand or an auxiliary stand and tie the back end down so that all weight is off the front end of the bike.

4 Trace the ignition (main) switch wiring back from the base of the switch and disconnect it at the connector. Remove the cockpit panel(s), fairing or fuel tank as required according to your model to access the connector (see Chapter 7 or 3).

5 Refer to Chapter 5, Section 7 and remove the upper fork bridge.

16.3b Pull the connector off the switch

6 Special security bolts are used to mount the ignition switch (see illustration). Remove the two special bolts by drilling their heads off. New bolts must be used on installation.

Installation

7 Install the switch onto the fork bridge. Using new special Torx bolts, tighten them until either the tool slips round on the bolt head, or until the bolt head sheers off.

8 Installation is the reverse of removal. Make sure wiring is securely connected and correctly routed.

18 Handlebar switches – check

1 Generally speaking, the switches are reliable and trouble-free. Most troubles, when they do occur, are caused by dirty or corroded contacts, but wear and breakage of internal parts is a possibility that should not be overlooked. If breakage does occur, the entire switch and related wiring harness will have to be replaced with a new one, since individual parts are not available.

2 The switches can be checked for continuity using an ohmmeter or a continuity test light.

⚠️ Warning: Always disconnect the battery negative (-ve) lead, which will prevent the possibility of a short circuit, before making the checks.

3 Trace the wiring harness of the switch in question back to its connector and

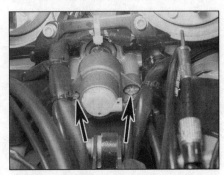

17.6 The ignition switch is secured by two special bolts

8

19.2a Remove the screw (arrowed) . . .

19.2b . . . and detach the panel

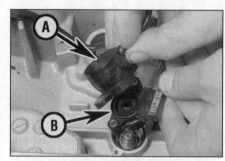

20.8 Unscrew the bolts and remove the gear position indicator (A) and the neutral switch (B)

disconnect it. Remove either the cockpit panel(s) or fuel tank as required according to your model to access the connector.

4 Check for continuity between the terminals of the switch harness with the switch in the various positions (ie switch off – no continuity, switch on – continuity) – see the *wiring diagrams* at the end of this Chapter.

5 If the continuity check indicates a problem exists, refer to Section 19, remove the switch and spray the switch contacts with electrical contact cleaner. If they are accessible, the contacts can be scraped clean with a knife or polished with crocus cloth. If switch components are damaged or broken, it will be obvious when the switch is disassembled.

19 Handlebar switches – removal and installation

Removal

1 If the switch panel is to be removed from the bike, rather than just displaced from the handlebar, trace the wiring harness back from the switch to the wiring connector and disconnect it. Remove the cockpit panel(s), fairing or fuel tank as required according to your model to access the connector. Work back along the harness, freeing it from all the relevant clips and ties, whilst noting its correct routing.

2 Remove the single screw securing the switch panel to either the brake master cylinder or the clutch lever bracket and detach the panel **(see illustrations)**.

Installation

3 Installation is the reverse of removal. Make sure the wiring is correctly routed and secured by any clips and ties, and the connector is securely connected.

20 Neutral and gear position indicator switch – check, removal and installation

Neutral switch check

1 Before checking the electrical circuit, check the neutral indicator bulb (see Section 15) and fuse F1 (see Section 5).

2 The switch is located in the back of the gearbox casing, in front of the swingarm. Where fitted, the gear position indicator switch is mounted onto the neutral switch. Trace the wiring from the top of the gearbox and disconnect it at the connector(s). Make sure the transmission is in neutral.

3 To test the neutral switch, check for continuity between the wiring terminals on the switch side of the wiring connector. In neutral, there should be continuity. In gear, there should be no continuity. If the tests prove otherwise, replace the switch, although check that the wiring between the connector and switch is sound.

Gear position switch check (where fitted)

4 No test details are available with which to check the gear position switch. First check that the fuse for the rider information display

unit is sound. Trace the four wires from the switch on the back of the gearbox to the digital rider information display unit. Check for signs of pinched or broken wires or corroded connectors. Make continuity checks across each of the four wires, from one end of the wiring to the other to confirm whether a wire breakage exists.

5 If the wiring is sound, remove the gear position switch (see below) and check its operation. Spray the switch contacts with electrical contact cleaner if necessary. If the wiring and switch are sound the fault must lie in the digital rider information display; note that this is likely to be an expensive unit to replace and confirmation of the unit's condition should be sought from a BMW dealer before buying a replacement.

Removal

6 Remove the swingarm (see Chapter 5).

7 Trace the wiring from the top of the gearbox and disconnect it at the connector(s). Free the wiring from its clips and ties.

8 Unscrew the two bolts securing the switch and remove it from the transmission casing **(see illustration)**.

Installation

9 Check the condition of the O-ring and oil seals and replace them if they are worn, damaged or deteriorated **(see illustrations)**.

10 Install the switch(es) and tighten the bolts securely. Connect the wiring at the connector and secure the wiring in its clip and ties **(see illustration)**. Note that the gear position switch connector must be fitted the correct

20.9a Check the condition of the O-ring . . .

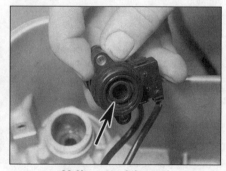

20.9b . . . and the seal

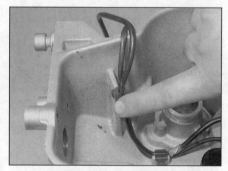

20.10 Secure the wiring in its clips

21.7 The side stand switch fits on the back of the side stand

22.1 Clutch switch location in clutch lever bracket

round otherwise inaccurate gear positions may be indicated on the rider information display.

11 Check the operation of the neutral light and gear position indicator (where fitted).

12 Install the swingarm (see Chapter 5).

21 Side stand switch – check and replacement

Check

1 The side stand switch is mounted on the back of the side stand pivot. The switch is part of the safety circuit which prevents or stops the engine running if the transmission is in gear whilst the side stand is down, and prevents the engine from starting if the transmission is in gear unless the side stand is up and unless the clutch lever is pulled in.

2 Place the machine on the centre stand. Trace the wiring back from the switch to its connector and disconnect it.

3 Check the operation of the switch using an ohmmeter or continuity test light. Connect the meter to the green/red and green/yellow wires on the switch side of the connector. With the side stand up there should be continuity (zero resistance) between the terminals, and with the stand down there should be no continuity (infinite resistance).

4 If the switch does not perform as expected, it is defective and must be replaced. Note that due to its exposed location, the switch may become waterlogged and corrode – spray it with a water dispersant contact cleaner to prevent future problems.

5 If the switch is good, check the wiring between the various components in the starter safety circuit (see the *wiring diagrams* at the end of this book).

Replacement

6 The side stand switch is mounted on the side stand pivot bracket. Place the machine on the centre stand. Trace the wiring back from the switch to its connector and disconnect it. Work back along the switch wiring, freeing it from any relevant retaining clips and ties, noting its correct routing.

7 Remove the side stand (see Chapter 5), and remove the switch, noting how it locates **(see illustration)**.

8 Fit the new switch, then install the side stand (see Chapter 5).

9 Make sure the wiring is correctly routed up to the connector and retained by all the necessary clips and ties.

10 Reconnect the wiring connector and check the operation of the side stand switch.

22 Clutch switch – check and replacement

Check

1 The clutch switch is mounted in the clutch lever bracket **(see illustration)**. The switch is part of the safety circuit which prevents or stops the engine running if the transmission is in gear whilst the side stand is down, and prevents the engine from starting if the transmission is in gear unless the side stand is up and the clutch lever is pulled in.

2 To check the switch, trace the wiring and disconnect it at the connector. Connect the probes of an ohmmeter or a continuity test light to the two connector terminals on the switch side. With the clutch lever pulled in, continuity should be indicated. With the clutch lever out, no continuity (infinite resistance) should be indicated.

3 If the switch is good, check the other components in the starter circuit as described in the relevant sections of this Chapter. If all

23.1 Disconnect the fuel tank wiring connector

components are good, check the wiring between the various components (see the *wiring diagrams* at the end of this book).

Replacement

4 Trace the wiring back from the switch to its connector and disconnect it. Work back along the switch wiring, freeing it from any relevant retaining clips and ties, noting its correct routing. Unscrew the retaining nut to release the switch from the clutch lever bracket, noting that removal of the bracket from the handlebar is advised for improved access.

5 Installation is the reverse of removal. The switch isn't adjustable.

23 Fuel level sender and warning light – check and replacement

⚠ *Warning: Petrol (gasoline) is extremely flammable, so take extra precautions when you work on any part of the fuel system. Don't smoke or allow open flames or bare light bulbs near the work area, and don't work in a garage where a natural gas-type appliance is present. If you spill any fuel on your skin, rinse it off immediately with soap and water. When you perform any kind of work on the fuel system, wear safety glasses and have a fire extinguisher suitable for a class B type fire (flammable liquids) on hand.*

Fuel level sender

Check

1 If the fuel level display fails to operate, trace the wiring back from the fuel level sender in the right-hand side of the fuel tank and disconnect it at the connector **(see illustration)**. Remove the seat, and if necessary the right-hand fairing side panel (RS and RT models) or fuel tank trim (R and GS models), to access the connector (see Chapter 7).

2 Unfortunately BMW provide no resistance specifications for the sender, making it difficult to accurately assess its condition. However it is possible to make a rough check of its condition. Using the wiring diagrams at the end of this chapter, identify the two sender unit wires in the connector. Using an ohmmeter set to ohms x 100 scale, connect its probes to the terminals on the sender side of the connector. Check the resistance reading with the tank empty and full. Alternatively, remove the sender from the tank (see Steps 4 and 5 below) and, with the meter connected as above, manually move the float up and down to emulate the different positions **(see illustrations)**.

3 If the readings show no difference between the full and empty positions, replace the sender. If the readings show a progressive resistance from one position to the other, check the wiring between the fuel tank and the digital rider information display (see *Wiring*

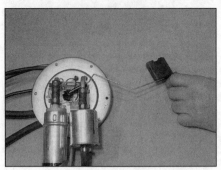

23.2a Fuel level sender in the full position . . .

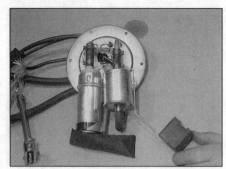

23.2b . . . and the empty position

Diagrams at the end of the Chapter). If the wiring is good, the digital display could be faulty – confirm your findings with a BMW dealer before buying a replacement display unit.

Replacement

4 The fuel level sender is mounted on the back of the cover in the fuel tank, and is integral with it. Remove the fuel pump and the fuel filter (see Section 24) and replace the sender/cover assembly.
5 Install the sender by reversing the fuel pump removal process (see Section 24).

Fuel level warning light and damping unit

6 The fuel level warning light should come on when there is only 4 litres of fuel remaining in the tank. If the light fails to come on, first check the bulb (see Section 15) and fuse F1 (see Section 5). If the bulb has not blown, the fault could lie in the fuel level damping unit (see Section 5 for location). The damping unit prevents the warning light coming on when cornering.
7 Referring to the engine system wiring diagrams at the end of this Chapter, check the wiring from the bulb to the damping unit, and from the damping unit to the fuel level sender.
8 On certain early RT models, the fuel warning light was found to come on when there was still sufficient fuel in the tank. As a result a modified fuel level sender unit and fuel level damping unit were introduced – refer to a BMW dealer for details.

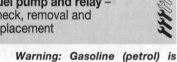

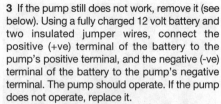

24 Fuel pump and relay – check, removal and replacement

⚠️ *Warning: Gasoline (petrol) is extremely flammable, so take extra precautions when you work on any part of the fuel system. Don't smoke or allow open flames or bare light bulbs near the work area, and don't work in a garage where a natural gas-type appliance (such as a water heater or clothes dryer) is present. If you spill any fuel on your skin, rinse it off immediately with soap and water. When you perform any kind of work on the fuel system, wear safety glasses and have a fire extinguisher suitable for a class B type fire (flammable liquids) on hand.*

Check

1 The fuel pump is located inside the fuel tank.
2 It should be possible to hear the fuel pump running whenever the engine is turning over – remove the seat (see Chapter 7) and place your ear close to the cover on the rear of the tank. If you can't hear anything, check fuse F6 and the fuel pump relay (see Section 5). If the fuse and relay are good, check the wiring and terminals for physical damage or loose or corroded connections and rectify as necessary.

3 If the pump still does not work, remove it (see below). Using a fully charged 12 volt battery and two insulated jumper wires, connect the positive (+ve) terminal of the battery to the pump's positive terminal, and the negative (-ve) terminal of the battery to the pump's negative terminal. The pump should operate. If the pump does not operate, replace it.
4 If the pump operates but is thought to be delivering an insufficient amount of fuel, first check that the fuel tank breather hose is unobstructed, that all fuel hoses are in good condition and not pinched or trapped. Check that the fuel filter is not blocked.
5 The fuel pump's output can be checked as follows: make sure the ignition switch is OFF. Remove the seat and the right-hand fairing side panel (RS and RT models) or fuel tank trim (R and GS models).
6 Disconnect the fuel tank outlet hose at the fuel pipe and place the end into a graduated beaker. The outlet hose comes from the central pipe union on the cover on the inside of the tank.
7 Turn the ignition switch ON and let fuel flow from the pump into the beaker for 5 seconds, then switch the ignition OFF.
8 Measure the amount of fuel that has flowed into the beaker, then multiply that amount by 12 to determine the fuel pump flow rate per minute. The minimum flow rate required is 1830 cc per minute. If the flow rate recorded is below the minimum required, then the fuel pump must be replaced.

Removal

9 Make sure the ignition is switched OFF. Drain and remove the fuel tank (see Chapter 3).
10 Unscrew the bolts securing the cover to the tank, noting the earth (ground) wire, then withdraw the pump assembly part-way and detach the vent hoses from their unions **(see illustrations)**. Carefully withdraw the pump assembly, taking care not to bend the float arm on the fuel level sender. Discard the cover O-ring as a new one must be used.
11 Unscrew the terminal nuts and detach the wires, noting which fits where, then release the clamp securing the pump to its hose and remove the pump **(see illustration)**.

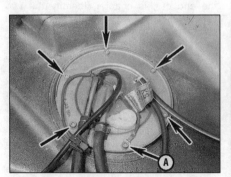

24.10a Unscrew the bolts (arrowed), noting the earth wire secured by the bolt (A) . . .

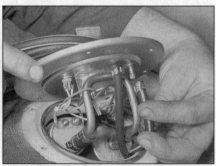

24.10b . . . then withdraw the pump assembly and detach the breather hoses

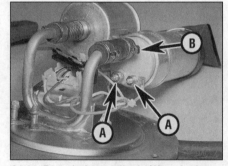

24.11 Detach the terminals (A) and release the hose clamp (B)

24.12a Check the gauze filter (arrowed) for damage

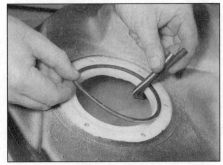

24.12b Fit a new O-ring . . .

24.12c . . . then install the assembly

Installation

Note: *The vent hoses on bikes up to frame No. 0 058 383 have been known to kink, causing a breakdown. If not already done, the hoses should be shortened by 50 mm.*

12 Installation is the reverse of removal. Check the condition of the gauze filter on the fuel pump and replace it if it is damaged **(see illustration)**. Also check the fuel filter (see Chapter 1). Make sure the fuel hose and wires are correctly and securely fitted to the pump. Install the pump assembly using a new cover O-ring and tighten the cover bolts to the torque setting specified at the beginning of the Chapter, not forgetting the earth wire **(see illustrations)**.

13 Start the engine and look carefully for any signs of leaks at the pipe connections; in particular check that there is no leakage from the external fuel hoses, particularly around the clamps. It is best to check for leakage with the ignition switched on, as the fuel pump will have pressurised the system. Make sure the wiring is correctly routed, making sure it cannot be trapped between the seat and the frame.

25 Horn(s) –
check and replacement

Check

1 The horn is mounted in between the forks below the front strut.

2 Unplug the wiring connectors from the horn. Using two jumper wires, apply battery voltage directly to the terminals on the horn. If the horn sounds, check the switch (see Section 19) and the wiring between the switch and the horn (see the *wiring diagrams* at the end of this Chapter).

3 If the horn doesn't sound, replace it.

Replacement

4 The horn is mounted in between the forks below the front strut.

5 Unplug the wiring connectors from the horn, then unscrew the nut securing the horn and remove it from the bike **(see illustrations)**.

6 Install the horn and securely tighten the bolt. Connect the wiring connectors to the horn.

26 Starter relay –
check and replacement

Check

1 If the starter circuit is faulty, first check that the battery is fully charged. Check that the engine kill switch is in the run, centre, position, that the side stand is up, and that the transmission is in neutral. If the starter motor still fails to operate, turn off the ignition and disconnect the battery negative (-ve) lead (see Section 3).

2 The starter relay is located in the fuse/relay box under the seat (see Section 5 for identification). Pull the relay out of its socket using the tool provided or a pair of pliers, to provide access to the rear terminals.

3 Set a multimeter to the ohms x 1 scale or use a continuity tester and connect it across the relay's No. 30 and No. 87 terminals. The multimeter should read 0 ohms (continuity). Using a fully-charged 12 volt battery and two insulated jumper wires, connect the positive (+ve) terminal of the battery to the No. 85 terminal of the relay, and the negative (-ve) terminal to the No. 86 terminal of the relay. At this point the multimeter should read 0 ohms (continuity). If this is the case the relay is proved good. If the relay indicates no

24.12d Do not forget to secure the earth lead with the bolt

continuity (infinite resistance) across its terminals with battery voltage applied, it is faulty and must be replaced.

4 If the relay is good, check for battery voltage on the relay's red wire (terminal 30) when the starter button is pressed. Check the other components in the starter circuit (ie the clutch switch, side stand switch, starter switch and kill switch) as described in the relevant sections of this Chapter. If all components are good, check the wiring between the various components (see the *wiring diagrams* at the end of this book). Note that if all components in the starter circuit are proved good, the fault could be due to a sticking or damaged starter motor solenoid (see Section 27).

Replacement

5 See Section 5.

8

25.5a Unplug the wiring connectors (arrowed) . . .

25.5b . . . then unscrew the nut and remove the horn

27.2a Unscrew the bolt (arrowed) and remove the cover . . .

27.2b . . . and detach the power socket wiring connector (arrowed)

27.3a Detach the solenoid wire . . .

27 Starter motor –
removal and installation

Removal

1 Remove the seat (see Chapter 7). Disconnect the battery negative (-ve) lead. The starter motor is mounted on the left-hand side of the engine. On RS and RT models, remove the left-hand fairing side panel for access (see Chapter 7).
2 On R, RS and GS models, remove the bolt securing the starter motor cover and draw the cover away (see illustration). Disconnect the auxiliary power socket wiring connector as it becomes accessible (see illustration).
3 Pull the solenoid wiring connector off its terminal, then unscrew the nut securing the starter lead to the motor and detach the lead (see illustrations).

4 Unscrew the starter motor mounting bolts, then draw the starter motor out of the casing and remove it (see illustration).

Installation

5 Manoeuvre the motor into position and slide it into the casing (see illustration). Install the mounting bolts and tighten them to the torque setting specified at the beginning of the Chapter (see illustrations).
6 Connect the starter lead to the motor and tighten the nut to the specified torque (see illustration 27.3b). Fit the solenoid wiring connector onto its terminal (see illustration 27.3a).
7 On R, RS and GS models, connect the auxiliary power socket wiring connector to the starter motor cover. Fit the cover, noting that the pegs on its forward end must engage the two grommets set in the casing, then tighten its bolt to the specified torque (see illustrations).

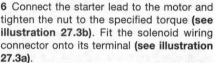

27.3b . . . then unscrew the nut and detach the starter motor lead

8 Connect the battery negative (-ve) lead. On RS and RT models install the left-hand fairing side panel, and on all models install the seat (see Chapter 7).

27.4 Unscrew the bolts (arrowed) and remove the starter motor

27.5a Fit the starter motor into the casing . . .

27.5b . . . then install the bolts . . .

27.5c . . . and tighten them to the specified torque

27.7a Connect the wiring connector . . .

27.7b . . . and fit the cover. Note peg (arrowed)

28 Starter motor –
disassembly, inspection and reassembly

Disassembly

Note: *Before disassembly, check whether replacement parts are available from a BMW dealer.*

1 Remove the starter motor (see Section 27).

2 Unscrew the nuts securing the rear cover and remove the cover, noting the rubber plug and how the brush terminal fits onto one of the studs **(see illustration)**.

3 Lift the brush spring ends, noting the small insulating pieces, and withdraw the brushes.

4 Remove the retaining clip from the rear end of the armature and remove the shims, noting how many. Carefully withdraw the brush holder and seal.

5 Unscrew the bolts securing the front cover and remove the cover.

6 Remove the screws securing the solenoid to the starter motor body and remove the solenoid.

7 Drift the solenoid lever arm pivot pin out

and remove the arm, noting how it engages with the pinion.

8 Remove the retaining clip from the front end of the armature shaft and slide the bush and the pinion off the shaft.

Inspection

9 The parts of the starter motor that are most likely to require attention are the brushes. Measure the length of the brushes. Unfortunately, BMW provide no specifications for the standard and minimum brush lengths, however some nearly new brushes were measured at 14 mm, indicating a minimum length of 7 mm. If the brushes are worn below this, replace them with new ones. If the brushes are not worn excessively, nor cracked, chipped, or otherwise damaged, they may be re-used.

10 Inspect the commutator bars on the armature for scoring, scratches and discoloration. The commutator can be cleaned and polished with crocus cloth, but do not use sandpaper or emery paper. After cleaning, wipe away any residue with a cloth soaked in electrical system cleaner or denatured alcohol.

11 Using an ohmmeter or a continuity test light, check for continuity between the commutator bars. Continuity should exist between each bar and all of the others. Also, check for continuity between the commutator bars and the armature shaft. There should be no continuity (infinite resistance) between the commutator and the shaft. If the checks indicate otherwise, the armature is defective.

12 Check the starter pinion gear for worn, cracked, chipped and broken teeth. If the gear is damaged or worn, replace the starter motor. Clean any old grease off the pinion and the shaft.

13 Check the condition of the shaft bushes and the lever arm pivot pin and replace them if they are worn or damaged.

Reassembly

14 Reassemble the motor in a reverse of the disassembly procedure. Lubricate the bushes, pinion, shaft and pivot pin with a smear of silicone grease. Check that the brushes slide freely in their holders. Make sure all the shims that were removed from between the retaining clip and the brush holder are installed.

15 Install the starter motor (see Section 27).

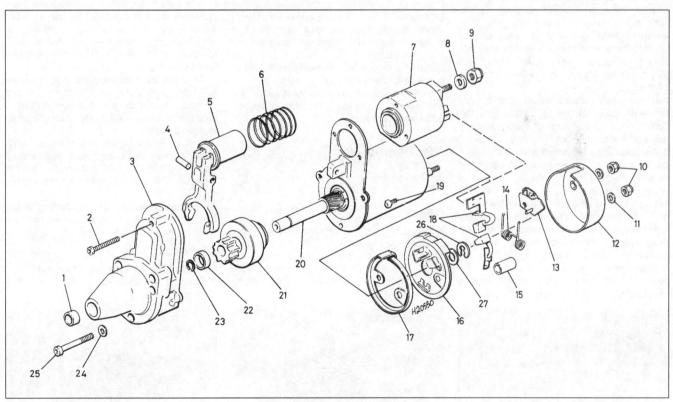

28.2 Starter motor components

1 Bush	6 Spring	12 Rear cover	18 Brushes	23 Retaining clip
2 Front cover bolts	7 Solenoid	13 Rubber plug	19 Screw	24 Washer
3 Front cover	8 Washer	14 Brush spring	20 Starter motor	25 Bolt
4 Pivot pin	9 Nut	15 Spacer	assembly/armature	26 Retaining clip
5 Solenoid lever arm	10 Rear cover nuts	16 Brush holder	21 Starter pinion	27 Shims(s)
	11 Washers	17 Seal	22 Bush	

8

30.1a Remove the screws (arrowed) . . .

30.1b . . . then ease out the clips and remove the cover

30.2a Unscrew the screws . . .

29 Charging system – testing

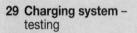

1 The charge warning lamp should light when the ignition is switched on and should remain lit as the engine is started, but should go out as soon as the engine speed increases significantly above idle. If this is not the case, first check the bulb itself and the connections to the instrument panel. Note that the lamp is connected directly to the alternator via the smaller, blue, wire which appears at the alternator connector plug, from the D + terminal. Note also that a faulty charge warning lamp operation is usually (but not always) caused by faulty brushes; a lot of time may be saved if these are checked first. See Section 30.

2 If the fault persists check the battery terminals (see Section 3). Check that the battery and alternator connections are securely fastened and that the battery is fully charged.

3 Accurate assessment of alternator output requires special equipment and a degree of skill. A rough idea of whether output is adequate can be gained by using a voltmeter (range 0 to 15 or 0 to 20 volts) as follows.

4 Connect the voltmeter across the battery terminals. Switch on the lights (UK models only) and note the voltage reading: it should be between 12 and 13 volts.

5 Start the engine and run it at a fast idle (approx. 1500 rpm). Read the voltmeter it should indicate 13 to 14 volts.

6 With the engine still running at a fast idle, switch on as many electrical consumers as possible (lights, stop lamp, turn signals and any accessories). The voltage at the battery should be maintained at 13 to 14 volts. Increase the engine speed slightly if necessary to keep the voltage up.

7 If alternator output is low or zero, check the brushes, as described in Section 30. If the brushes are in good condition the alternator requires attention.

8 Occasionally the condition may arise where the alternator output is excessive. Clues to this condition are constantly blowing bulbs; brightness of lights varying considerably with engine speed; overheating of alternator and battery, possibly with steam or fumes coming from the battery. This condition is almost certainly due to a defective voltage regulator, but expert advice should be sought.

9 Note that the voltage regulator is part of the alternator brush holder unit; refer to Section 30 for access details.

30 Alternator – check, removal and installation

Check

Note: *Apart from the combined brush holder and voltage regulator unit, no replacement parts are listed by BMW for the alternator. If a fault is indicated seek the advice of a Bosch specialist or BMW dealer on the best course of action to take.*

1 Remove the alternator (see below). Remove the three screws securing the alternator end cover, then release the clips and remove the cover **(see illustrations)**.

2 Remove the screws securing the brush holder/regulator and remove the holder, noting how it fits **(see illustrations)**. Inspect the holder for any signs of damage. Measure the brush lengths. Unfortunately, BMW provide no specifications for the standard and minimum brush lengths, however some nearly new brushes were measured at 14 mm, indicating a minimum length of 7 mm. If the brushes are worn below this, replace them with new ones. If the brushes are not worn excessively, nor cracked, chipped, or otherwise damaged, they may be re-used. Clean the slip rings with a rag moistened with some solvent.

3 To check the rotor coil resistance, measure the resistance between the slip rings and compare the reading to the Specifications **(see illustration)**. If it is higher than specified, replace the rotor.

4 Check for continuity between the slip rings and the armature housing **(see illustration)**. There should be no continuity (infinite resistance). If there is continuity, replace the rotor.

5 Any further testing or dismantling of the alternator assembly must be carried out by a Bosch specialist or BMW dealer.

Removal

6 Remove the fuel tank (see Chapter 3).

7 Remove the Motronic control unit (see Chapter 4).

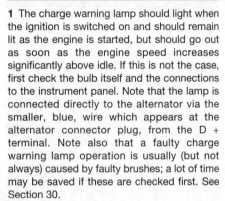

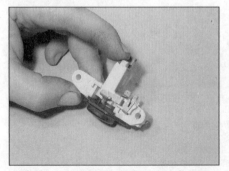

30.2b . . . and remove the brush holder

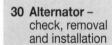

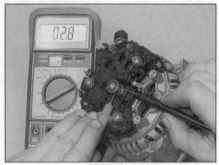

30.3 Checking rotor coil resistance

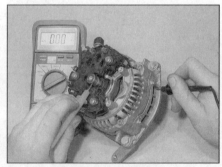

30.4 Checking for a short between the slip rings and the housing

30.8 Remove the screw and the bolt on each side (arrowed) to free the ABS unit

30.9a Remove the cap . . .

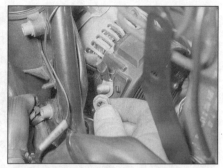

30.9b . . . then unscrew the nut and detach the lead . . .

8 On models equipped with ABS, unscrew the two bolts and the screw securing the ABS pressure modulator **(see illustration)**. This allows the unit to be displaced slightly to provide clearance for the alternator to be removed. Do not disconnect any of the hydraulic pipes or the modulator's wiring. Great care must be taken when displacing the unit to avoid any strain on the hydraulic pipes.

9 Remove the cap on the alternator terminal nut, then unscrew the nut and detach the lead from the B+ terminal. Pull the blue wire off the D+ terminal **(see illustrations)**.

10 Unscrew the nuts securing the left-hand side rail to the strut and to the engine, and remove the side rail **(see illustration)**. Depending on which way round the bolts have been installed, it may be necessary to

withdraw the right-hand side rail. If this is the case, refit the right-hand side rail and loosely install the bolts.

11 Unscrew the four bolts securing the engine front cover and remove the cover **(see illustration)**.

12 Slacken the two nuts and the bolt securing the alternator to the engine casing, then push the alternator down in its mountings to release the tension in the belt **(see illustration)**. Slip the belt off the pulley on the alternator. Remove the nuts and the bolt and manoeuvre the alternator out of the left-hand side of the bike, noting the adjuster nut that fits between the alternator and the casing on the left-hand mounting and the spacer that fits between the alternator and engine on the top mounting.

Installation

13 Manoeuvre the alternator into position and install the mounting bolts and nuts, but leave them loose **(see illustrations)**. Do not forget the adjuster nut that fits between the alternator and the casing and the left-hand mounting spacer between the alternator and the engine on the top mounting **(see illustration)**. Push the alternator down in its mountings to provide freeplay for the belt, then tighten the nut onto the adjusting bolt finger-tight, but still leave the others slack **(see illustration)**. Slip the alternator belt around the pulley, making sure its ribs sit correctly in the channels in the pulley.

14 Tighten the adjusting bolt using a torque wrench to pre-load the belt to the torque setting specified at the beginning of the

30.9c . . . and pull the wire off its terminal

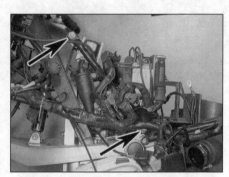

30.10 Left-hand side rail mountings (arrowed)

30.11 The front cover is secured by four bolts (arrowed)

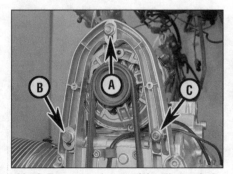

30.12 Top mounting nut (A), adjuster bolt nut (B), mounting bolt (C)

30.13a Fit the alternator . . .

30.13b . . . and its mounting nuts and bolt . . .

8

30.13c . . . not forgetting the adjuster nut

30.13d Tighten the adjuster bolt nut finger-tight

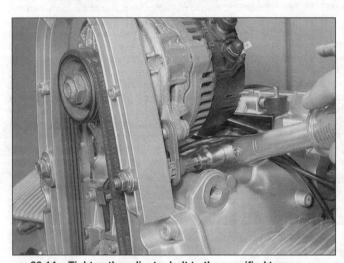

30.14a Tighten the adjuster bolt to the specified torque . . .

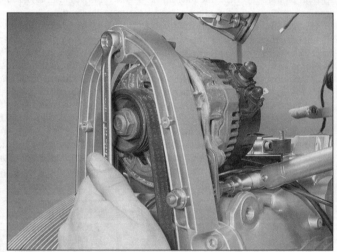

30.14b . . . then tighten the top nut while counter-holding the adjuster bolt

Chapter (see illustration). Keeping the torque wrench held on the adjuster to prevent it from slackening, tighten the top retaining nut (see illustration). Now tighten the bolt and the adjuster bolt nut to the specified torque settings.

15 Fit the alternator lead and wire onto their terminals, then tighten the lead nut and fit the terminal cap (see illustrations 31.9c, b and a).

16 Install the engine front cover, the sub-frame, the ABS modulator, Motronic unit and the fuel tank in a reverse of the removal procedure, referring to the relevant Chapters. Tighten the sub-frame mountings to the specified torque settings.

31 Wiring diagrams – notes

1 The wiring system is split into two main diagrams, engine and frame. Certain components are duplicated where they have a common application in both systems, eg the battery, fusebox and ignition switch, others are called up as a grid reference on other diagrams. Note that the grid references commence with the diagram number, followed by the grid reference on that diagram.

2 Additional diagrams are included for ABS, and on the RT model the electrically-controlled windshield and radio.

3 Where applicable, DIN terminal number details are given for electrical components, thus aiding testing and terminal identification. Likewise, connector block terminal identification is given and should correspond to that on the machine. Fuse numbering corresponds to the key given in Section 5 of this Chapter.

4 Wires can be identified by their colour and wire thickness details, express in millimetres.

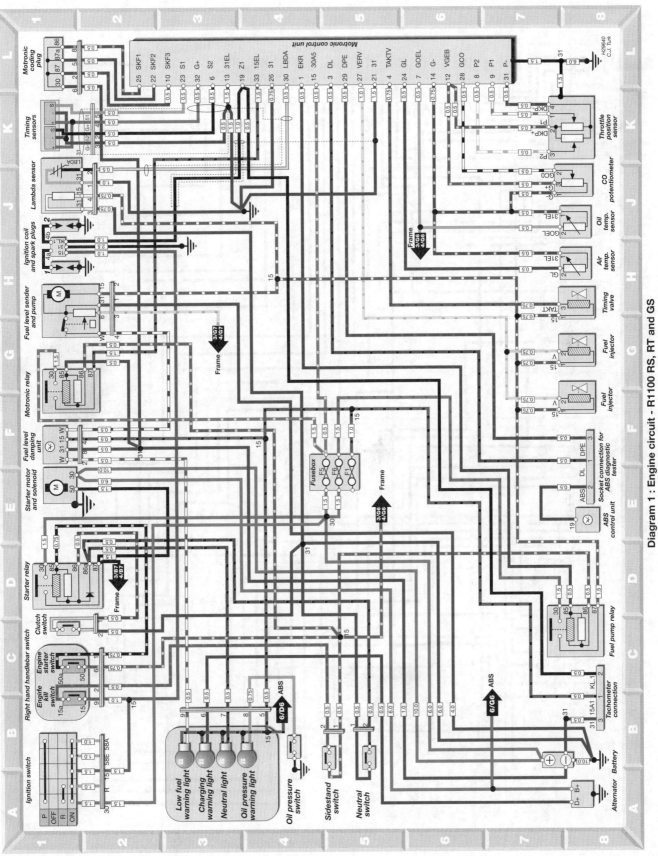

Diagram 1 : Engine circuit - R1100 RS, RT and GS

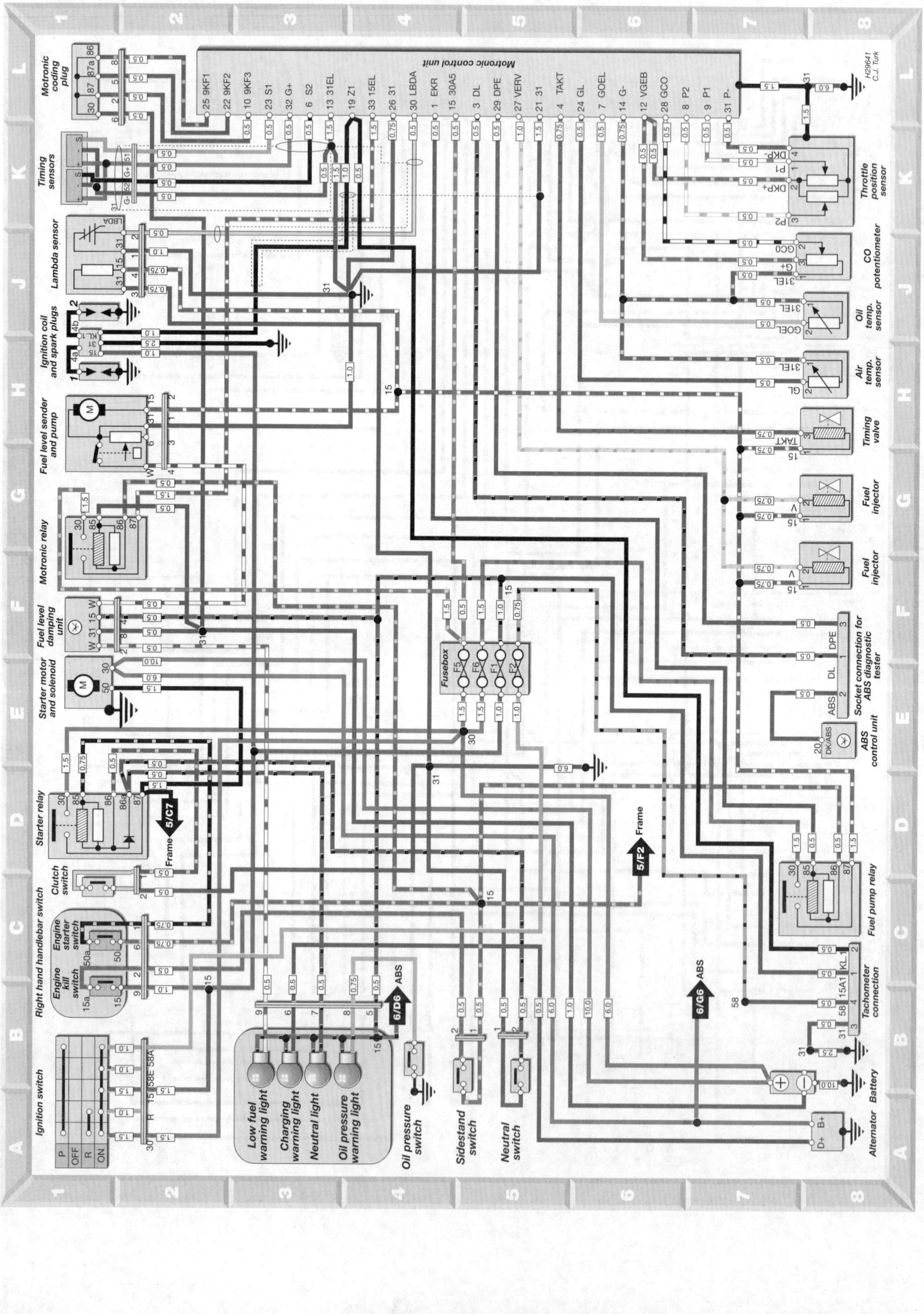

Diagram 2 : Engine circuit - R850/1100 R

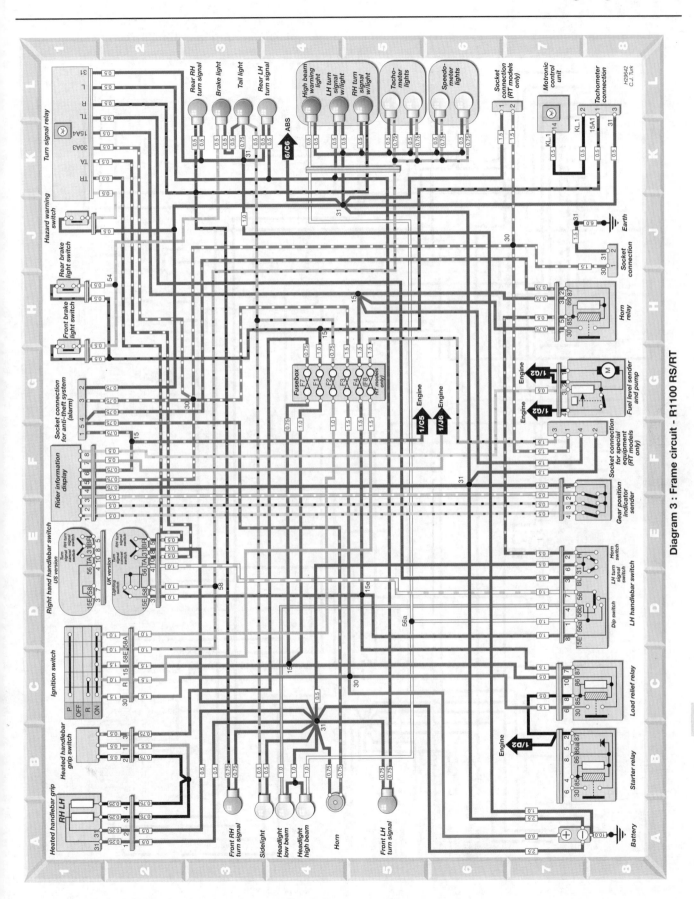

Diagram 3 : Frame circuit - R1100 RS/RT

8

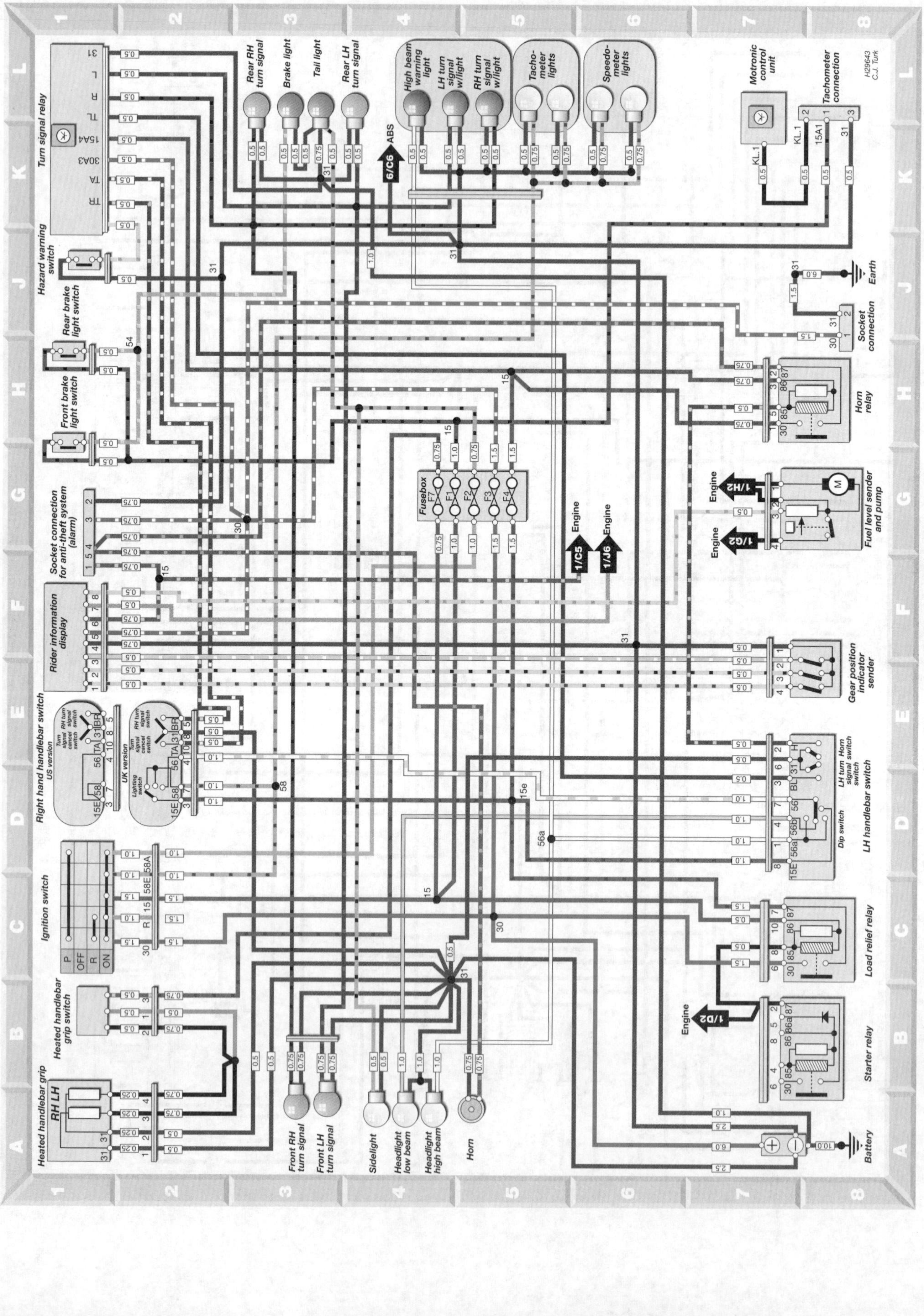

Diagram 4 : Frame circuit - R1100 GS

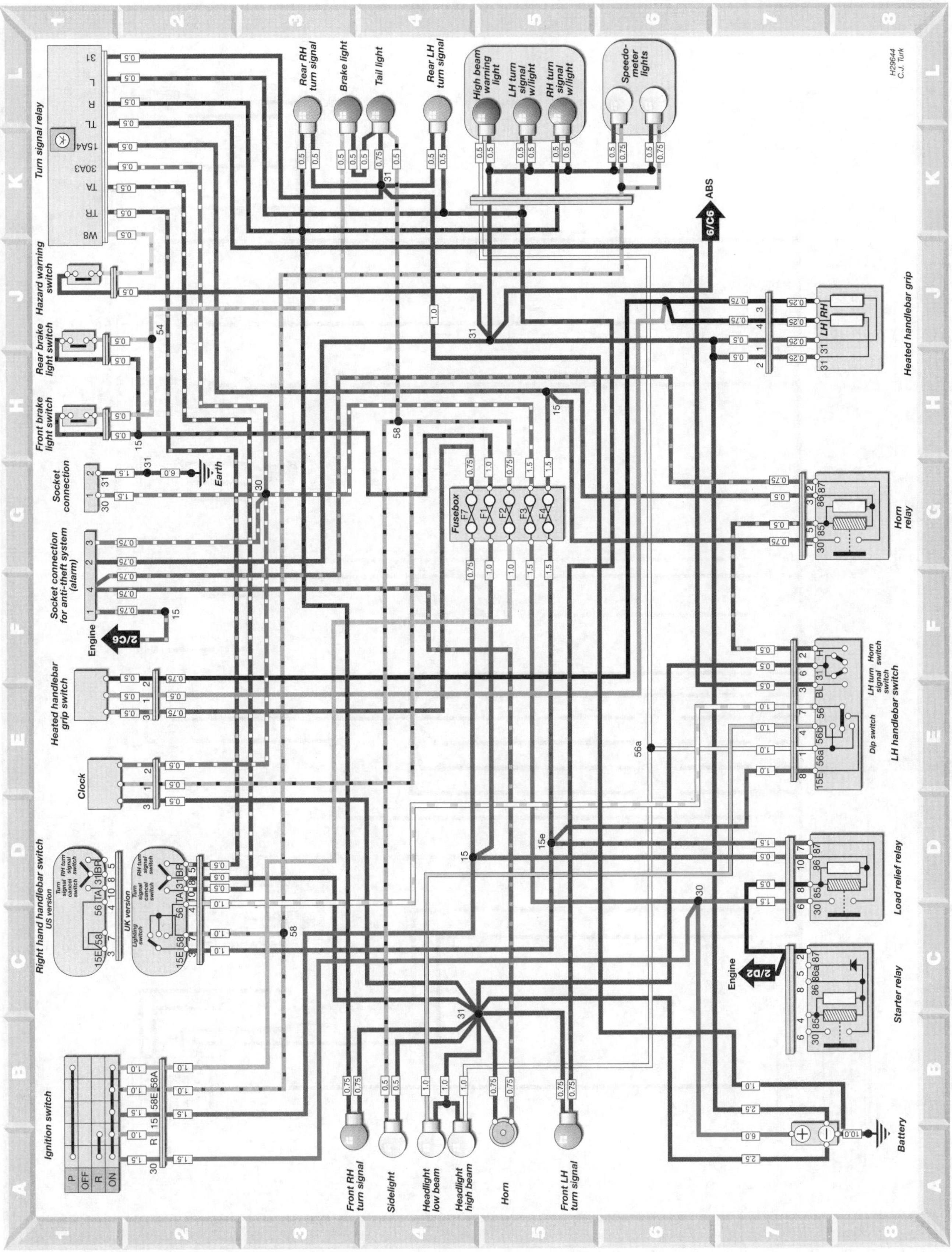

Diagram 5 : Frame circuit - R850/1100 R

H29644
C.J.Turk

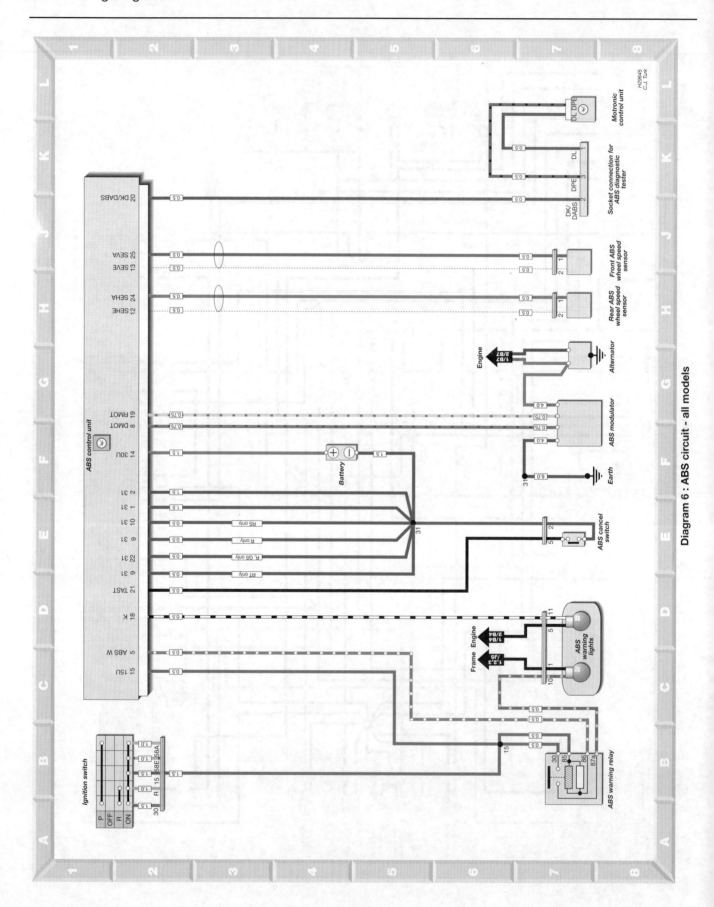

Diagram 6 : ABS circuit - all models

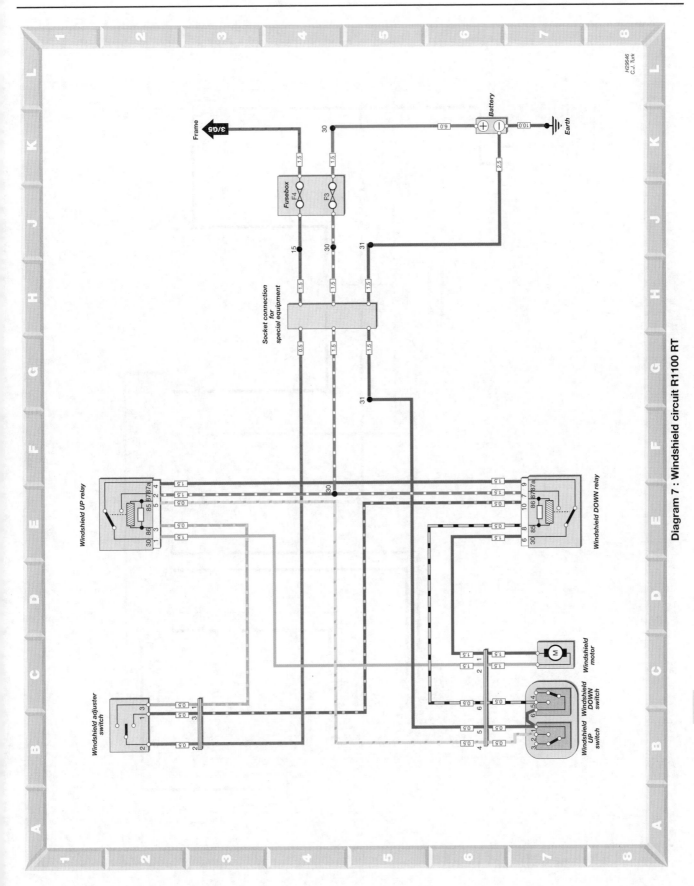

Diagram 7 : Windshield circuit R1100 RT

8

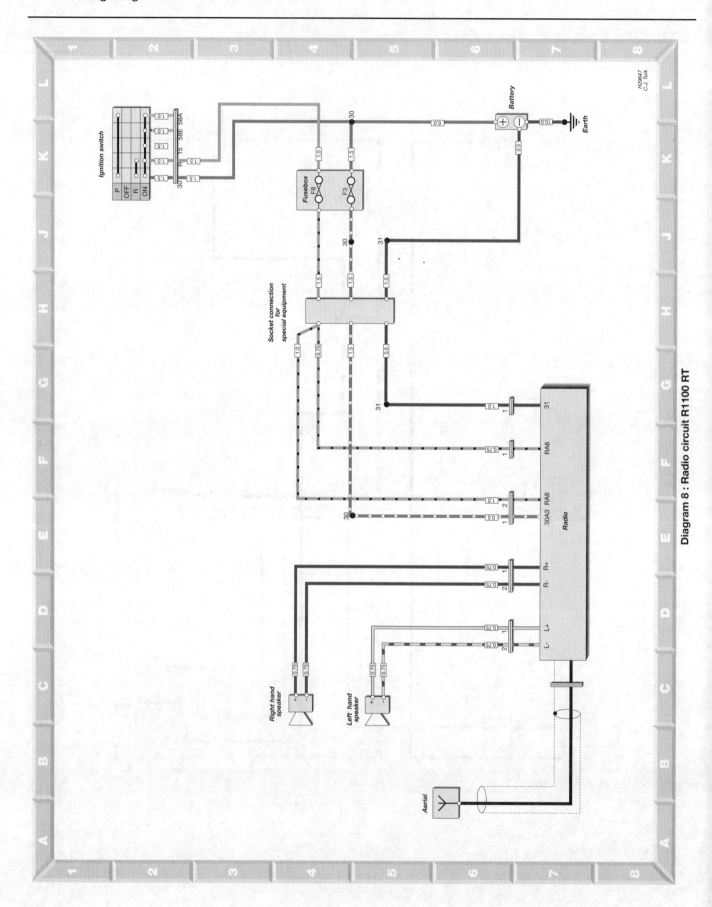

Diagram 8 : Radio circuit R1100 RT

Reference

Dimensions and Weights

R850R/R1100R

Wheelbase . . . . .1487 mm
Overall length . . . . .2197 mm
Overall width . . . . .898 mm
Overall height . . . . .1060 mm
Seat height (3-pos. adjustable) . . . . .760/780/800 mm
Ground clearance (normal load with rider) . . . . .138 mm
Weight
    Unladen, full fuel tank . . . . .235 kg
    Maximum (inc. rider, passenger, luggage, full fuel tank) . . . . .450 kg

R1100RS

Wheelbase . . . . .1473 mm
Overall length . . . . .2175 mm
Overall width . . . . .920 mm
Overall height . . . . .1286 mm
Seat height (3-pos. adjustable) . . . . .780/800/820 mm
Ground clearance (normal load with rider) . . . . .153 mm
Weight
    Unladen, full fuel tank . . . . .239 kg
    Maximum (inc. rider, passenger, luggage, full fuel tank) . . . . .450 kg

R1100RT

Wheelbase . . . . .1485 mm
Overall length . . . . .2205 mm
Overall width . . . . .900 mm
Overall height . . . . .1380 mm
Seat height (3-pos. adjustable) . . . . .780/800/820 mm
Ground clearance (normal load with rider) . . . . .153 mm
Weight
    Unladen, full fuel tank . . . . .282 kg
    Maximum (inc. rider, passenger, luggage, full fuel tank) . . . . .490 kg

R1100GS

Wheelbase . . . . .1509 mm
Overall length . . . . .2189 mm
Overall width . . . . .920 mm
Overall height . . . . .1366 mm
Seat height (2-pos. adjustable) . . . . .840/860 mm
Ground clearance (normal load with rider) . . . . .200 mm
Weight
    Unladen, full fuel tank . . . . .243 kg
    Maximum (inc. rider, passenger, luggage, full fuel tank) . . . . .450 kg

Buying tools

A toolkit is a fundamental requirement for servicing and repairing a motorcycle. Although there will be an initial expense in building up enough tools for servicing, this will soon be offset by the savings made by doing the job yourself. As experience and confidence grow, additional tools can be added to enable the repair and overhaul of the motorcycle. Many of the specialist tools are expensive and not often used so it may be preferable to hire them, or for a group of friends or motorcycle club to join in the purchase.

As a rule, it is better to buy more expensive, good quality tools. Cheaper tools are likely to wear out faster and need to be renewed more often, nullifying the original saving.

Warning: To avoid the risk of a poor quality tool breaking in use, causing injury or damage to the component being worked on, always aim to purchase tools which meet the relevant national safety standards.

The following lists of tools do not represent the manufacturer's service tools, but serve as a guide to help the owner decide which tools are needed for this level of work. In addition, items such as an electric drill, hacksaw, files, hammers, soldering iron and a workbench equipped with a vice, may be needed. Although not classed as tools, a selection of bolts, screws, nuts, washers and pieces of tubing always come in useful.

For more information about tools, refer to the Haynes *Motorcycle Workshop Practice Manual* (Bk. No. 1454).

Manufacturer's service tools

Inevitably certain tasks require the use of a service tool. Where possible an alternative tool or method of approach is recommended, but sometimes there is no option if personal injury or damage to the component is to be avoided. Where required, service tools are referred to in the relevant procedure.

Service tools can usually only be purchased from a motorcycle dealer and are identified by a part number. Some of the commonly-used tools, such as rotor pullers, are available in aftermarket form from mail-order motorcycle tool and accessory suppliers.

Maintenance and minor repair tools

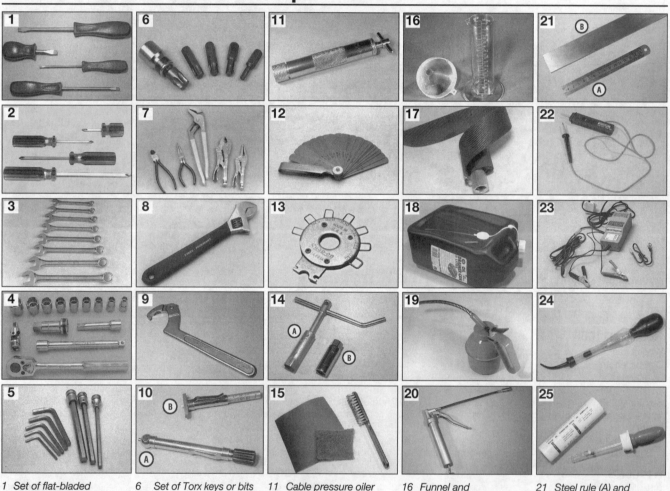

1 Set of flat-bladed screwdrivers
2 Set of Phillips head screwdrivers
3 Combination open-end & ring spanners
4 Socket set (3/8 inch or 1/2 inch drive)
5 Set of Allen keys or bits

6 Set of Torx keys or bits
7 Pliers and self-locking grips (Mole grips)
8 Adjustable spanner
9 C-spanner (ideally adjustable type)
10 Tyre pressure gauge (A) & tread depth gauge (B)

11 Cable pressure oiler
12 Feeler gauges
13 Spark plug gap measuring and adjusting tool
14 Spark plug spanner (A) or deep plug socket (B)
15 Wire brush and emery paper

16 Funnel and measuring vessel
17 Strap wrench, chain wrench or oil filter removal tool
18 Oil drainer can or tray
19 Pump type oil can
20 Grease gun

21 Steel rule (A) and straight-edge (B)
22 Continuity tester
23 Battery charger
24 Hydrometer (for battery specific gravity check)
25 Anti-freeze tester (for liquid-cooled engines)

Repair and overhaul tools

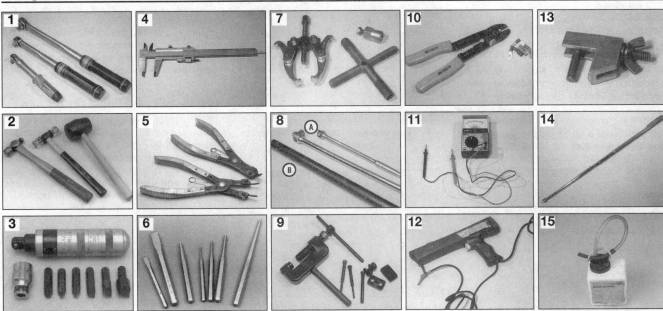

1 Torque wrench
   (small and mid-ranges)
2 Conventional, plastic or
   soft-faced hammers
3 Impact driver set
4 Vernier gauge
5 Circlip pliers (internal and
   external, or combination)
6 Set of punches
   and cold chisels
7 Selection of pullers
8 Breaker bars (A)
   and length of tubing (B)
9 Chain breaking/
   riveting tool
10 Wire crimper tool
11 Multimeter (measures
    amps, volts and ohms)
12 Stroboscope (for
    dynamic timing checks)
13 Hose clamp
    (wingnut type shown)
14 Magnetic arm
    (telescopic type shown)
15 One-man brake/clutch
    bleeder kit

Specialist tools

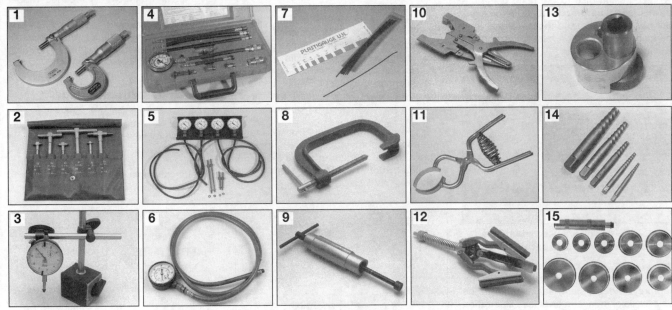

1 Micrometer
   (external type)
2 Telescoping gauges or
   small-hole gauges
3 Dial gauge
4 Cylinder
   compression gauge
5 Vacuum gauges (shown)
   or manometer
6 Oil pressure gauge
7 Plastigauge kit
8 Valve spring compressor
   (4-stroke engines)
9 Piston pin drawbolt tool
10 Piston ring removal and
    installation tool
11 Piston ring clamp
12 Cylinder bore hone
    (stone type shown)
13 Stud extractor
14 Screw extractor set
15 Bearing driver set

1 Workshop equipment and facilities

The workbench

● Work is made much easier by raising the bike up on a ramp - components are much more accessible if raised to waist level. The hydraulic or pneumatic types seen in the dealer's workshop are a sound investment if you undertake a lot of repairs or overhauls **(see illustration 1.1)**.

1.1 Hydraulic motorcycle ramp

● If raised off ground level, the bike must be supported on the ramp to avoid it falling. Most ramps incorporate a front wheel locating clamp which can be adjusted to suit different diameter wheels. When tightening the clamp, take care not to mark the wheel rim or damage the tyre - use wood blocks on each side to prevent this.
● Secure the bike to the ramp using tie-downs **(see illustration 1.2)**. If the bike has only a sidestand, and hence leans at a dangerous angle when raised, support the bike on an auxiliary stand.

1.2 Tie-downs are used around the passenger footrests to secure the bike

● Auxiliary (paddock) stands are widely available from mail order companies or motorcycle dealers and attach either to the wheel axle or swingarm pivot **(see illustration 1.3)**. If the motorcycle has a centrestand, you can support it under the crankcase to prevent it toppling whilst either wheel is removed **(see illustration 1.4)**.

1.3 This auxiliary stand attaches to the swingarm pivot

1.4 Always use a block of wood between the engine and jack head when supporting the engine in this way

Fumes and fire

● Refer to the Safety first! page at the beginning of the manual for full details. Make sure your workshop is equipped with a fire extinguisher suitable for fuel-related fires (Class B fire - flammable liquids) - it is not sufficient to have a water-filled extinguisher.
● Always ensure adequate ventilation is available. Unless an exhaust gas extraction system is available for use, ensure that the engine is run outside of the workshop.
● If working on the fuel system, make sure the workshop is ventilated to avoid a build-up of fumes. This applies equally to fume build-up when charging a battery. Do not smoke or allow anyone else to smoke in the workshop.

Fluids

● If you need to drain fuel from the tank, store it in an approved container marked as suitable for the storage of petrol (gasoline) **(see illustration 1.5)**. Do not store fuel in glass jars or bottles.

1.5 Use an approved can only for storing petrol (gasoline)

● Use proprietary engine degreasers or solvents which have a high flash-point, such as paraffin (kerosene), for cleaning off oil, grease and dirt - never use petrol (gasoline) for cleaning. Wear rubber gloves when handling solvent and engine degreaser. The fumes from certain solvents can be dangerous - always work in a well-ventilated area.

Dust, eye and hand protection

● Protect your lungs from inhalation of dust particles by wearing a filtering mask over the nose and mouth. Many frictional materials still contain asbestos which is dangerous to your health. Protect your eyes from spouts of liquid and sprung components by wearing a pair of protective goggles **(see illustration 1.6)**.

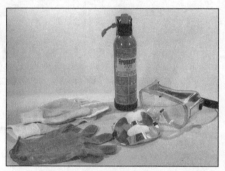

1.6 A fire extinguisher, goggles, mask and protective gloves should be at hand in the workshop

● Protect your hands from contact with solvents, fuel and oils by wearing rubber gloves. Alternatively apply a barrier cream to your hands before starting work. If handling hot components or fluids, wear suitable gloves to protect your hands from scalding and burns.

What to do with old fluids

● Old cleaning solvent, fuel, coolant and oils should not be poured down domestic drains or onto the ground. Package the fluid up in old oil containers, label it accordingly, and take it to a garage or disposal facility. Contact your local authority for location of such sites or ring the oil care hotline.

OIL CARE
FOLLOW THE CODE
OIL BANK LINE
0800 66 33 66

Note: It is antisocial and illegal to dump oil down the drain. To find the location of your local oil recycling bank, call this number free.

In the USA, note that any oil supplier must accept used oil for recycling.

2 Fasteners -
screws, bolts and nuts

Fastener types and applications

Bolts and screws

● Fastener head types are either of hexagonal, Torx or splined design, with internal and external versions of each type **(see illustrations 2.1 and 2.2)**; splined head fasteners are not in common use on motorcycles. The conventional slotted or Phillips head design is used for certain screws. Bolt or screw length is always measured from the underside of the head to the end of the item **(see illustration 2.11)**.

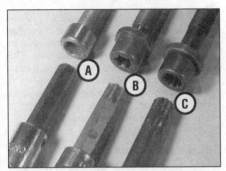

2.1 Internal hexagon/Allen (A), Torx (B) and splined (C) fasteners, with corresponding bits

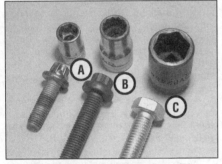

2.2 External Torx (A), splined (B) and hexagon (C) fasteners, with corresponding sockets

● Certain fasteners on the motorcycle have a tensile marking on their heads, the higher the marking the stronger the fastener. High tensile fasteners generally carry a 10 or higher marking. Never replace a high tensile fastener with one of a lower tensile strength.

Washers (see illustration 2.3)

● Plain washers are used between a fastener head and a component to prevent damage to the component or to spread the load when torque is applied. Plain washers can also be used as spacers or shims in certain assemblies. Copper or aluminium plain washers are often used as sealing washers on drain plugs.

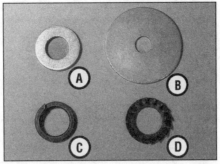

2.3 Plain washer (A), penny washer (B), spring washer (C) and serrated washer (D)

● The split-ring spring washer works by applying axial tension between the fastener head and component. If flattened, it is fatigued and must be renewed. If a plain (flat) washer is used on the fastener, position the spring washer between the fastener and the plain washer.

● Serrated star type washers dig into the fastener and component faces, preventing loosening. They are often used on electrical earth (ground) connections to the frame.

● Cone type washers (sometimes called Belleville) are conical and when tightened apply axial tension between the fastener head and component. They must be installed with the dished side against the component and often carry an OUTSIDE marking on their outer face. If flattened, they are fatigued and must be renewed.

● Tab washers are used to lock plain nuts or bolts on a shaft. A portion of the tab washer is bent up hard against one flat of the nut or bolt to prevent it loosening. Due to the tab washer being deformed in use, a new tab washer should be used every time it is disturbed.

● Wave washers are used to take up endfloat on a shaft. They provide light springing and prevent excessive side-to-side play of a component. Can be found on rocker arm shafts.

Nuts and split pins

● Conventional plain nuts are usually six-sided **(see illustration 2.4)**. They are sized by thread diameter and pitch. High tensile nuts carry a number on one end to denote their tensile strength.

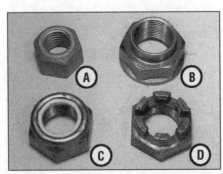

2.4 Plain nut (A), shouldered locknut (B), nylon insert nut (C) and castellated nut (D)

● Self-locking nuts either have a nylon insert, or two spring metal tabs, or a shoulder which is staked into a groove in the shaft - their advantage over conventional plain nuts is a resistance to loosening due to vibration. The nylon insert type can be used a number of times, but must be renewed when the friction of the nylon insert is reduced, ie when the nut spins freely on the shaft. The spring tab type can be reused unless the tabs are damaged. The shouldered type must be renewed every time it is disturbed.

● Split pins (cotter pins) are used to lock a castellated nut to a shaft or to prevent slackening of a plain nut. Common applications are wheel axles and brake torque arms. Because the split pin arms are deformed to lock around the nut a new split pin must always be used on installation - always fit the correct size split pin which will fit snugly in the shaft hole. Make sure the split pin arms are correctly located around the nut **(see illustrations 2.5 and 2.6)**.

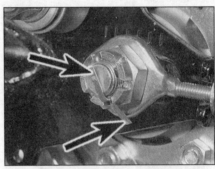

2.5 Bend split pin (cotter pin) arms as shown (arrows) to secure a castellated nut

2.6 Bend split pin (cotter pin) arms as shown to secure a plain nut

Caution: If the castellated nut slots do not align with the shaft hole after tightening to the torque setting, tighten the nut until the next slot aligns with the hole - never slacken the nut to align its slot.

● R-pins (shaped like the letter R), or slip pins as they are sometimes called, are sprung and can be reused if they are otherwise in good condition. Always install R-pins with their closed end facing forwards **(see illustration 2.7)**.

2.7 Correct fitting of R-pin. Arrow indicates forward direction

Circlips (see illustration 2.8)

● Circlips (sometimes called snap-rings) are used to retain components on a shaft or in a housing and have corresponding external or internal ears to permit removal. Parallel-sided (machined) circlips can be installed either way round in their groove, whereas stamped circlips (which have a chamfered edge on one face) must be installed with the chamfer facing away from the direction of thrust load **(see illustration 2.9)**.

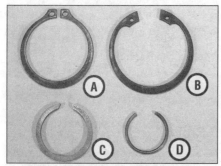

2.8 External stamped circlip (A), internal stamped circlip (B), machined circlip (C) and wire circlip (D)

● Always use circlip pliers to remove and install circlips; expand or compress them just enough to remove them. After installation, rotate the circlip in its groove to ensure it is securely seated. If installing a circlip on a splined shaft, always align its opening with a shaft channel to ensure the circlip ends are well supported and unlikely to catch **(see illustration 2.10)**.

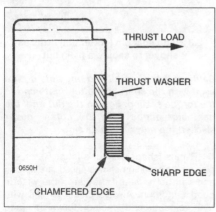

2.9 Correct fitting of a stamped circlip

THRUST LOAD
THRUST WASHER
SHARP EDGE
CHAMFERED EDGE
0650H

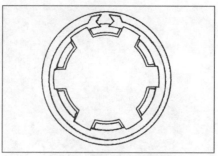

2.10 Align circlip opening with shaft channel

● Circlips can wear due to the thrust of components and become loose in their grooves, with the subsequent danger of becoming dislodged in operation. For this reason, renewal is advised every time a circlip is disturbed.
● Wire circlips are commonly used as piston pin retaining clips. If a removal tang is provided, long-nosed pliers can be used to dislodge them, otherwise careful use of a small flat-bladed screwdriver is necessary. Wire circlips should be renewed every time they are disturbed.

Thread diameter and pitch

● Diameter of a male thread (screw, bolt or stud) is the outside diameter of the threaded portion **(see illustration 2.11)**. Most motorcycle manufacturers use the ISO (International Standards Organisation) metric system expressed in millimetres, eg M6 refers to a 6 mm diameter thread. Sizing is the same for nuts, except that the thread diameter is measured across the valleys of the nut.
● Pitch is the distance between the peaks of the thread **(see illustration 2.11)**. It is expressed in millimetres, thus a common bolt size may be expressed as 6.0 x 1.0 mm (6 mm thread diameter and 1 mm pitch). Generally pitch increases in proportion to thread diameter, although there are always exceptions.
● Thread diameter and pitch are related for conventional fastener applications and the following table can be used as a guide. Additionally, the AF (Across Flats), spanner or socket size dimension of the bolt or nut **(see illustration 2.11)** is linked to thread and pitch specification. Thread pitch can be measured with a thread gauge **(see illustration 2.12)**.

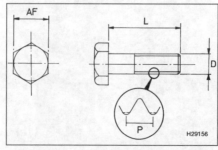

2.11 Fastener length (L), thread diameter (D), thread pitch (P) and head size (AF)

AF
L
D
P
H29156

2.12 Using a thread gauge to measure pitch

AF size	Thread diameter x pitch (mm)
8 mm	M5 x 0.8
8 mm	M6 x 1.0
10 mm	M6 x 1.0
12 mm	M8 x 1.25
14 mm	M10 x 1.25
17 mm	M12 x 1.25

● The threads of most fasteners are of the right-hand type, ie they are turned clockwise to tighten and anti-clockwise to loosen. The reverse situation applies to left-hand thread fasteners, which are turned anti-clockwise to tighten and clockwise to loosen. Left-hand threads are used where rotation of a component might loosen a conventional right-hand thread fastener.

Seized fasteners

● Corrosion of external fasteners due to water or reaction between two dissimilar metals can occur over a period of time. It will build up sooner in wet conditions or in countries where salt is used on the roads during the winter. If a fastener is severely corroded it is likely that normal methods of removal will fail and result in its head being ruined. When you attempt removal, the fastener thread should be heard to crack free and unscrew easily - if it doesn't, stop there before damaging something.
● A smart tap on the head of the fastener will often succeed in breaking free corrosion which has occurred in the threads **(see illustration 2.13)**.
● An aerosol penetrating fluid (such as WD-40) applied the night beforehand may work its way down into the thread and ease removal. Depending on the location, you may be able to make up a Plasticine well around the fastener head and fill it with penetrating fluid.

2.13 A sharp tap on the head of a fastener will often break free a corroded thread

● If you are working on an engine internal component, corrosion will most likely not be a problem due to the well lubricated environment. However, components can be very tight and an impact driver is a useful tool in freeing them **(see illustration 2.14)**.

2.14 Using an impact driver to free a fastener

● Where corrosion has occurred between dissimilar metals (eg steel and aluminium alloy), the application of heat to the fastener head will create a disproportionate expansion rate between the two metals and break the seizure caused by the corrosion. Whether heat can be applied depends on the location of the fastener - any surrounding components likely to be damaged must first be removed **(see illustration 2.15)**. Heat can be applied using a paint stripper heat gun or clothes iron, or by immersing the component in boiling water - wear protective gloves to prevent scalding or burns to the hands.

2.15 Using heat to free a seized fastener

● As a last resort, it is possible to use a hammer and cold chisel to work the fastener head unscrewed **(see illustration 2.16)**. This will damage the fastener, but more importantly extreme care must be taken not to damage the surrounding component.

Caution: Remember that the component being secured is generally of more value than the bolt, nut or screw - when the fastener is freed, do not unscrew it with force, instead work the fastener back and forth when resistance is felt to prevent thread damage.

2.16 Using a hammer and chisel to free a seized fastener

Broken fasteners and damaged heads

● If the shank of a broken bolt or screw is accessible you can grip it with self-locking grips. The knurled wheel type stud extractor tool or self-gripping stud puller tool is particularly useful for removing the long studs which screw into the cylinder mouth surface of the crankcase or bolts and screws from which the head has broken off **(see illustration 2.17)**. Studs can also be removed by locking two nuts together on the threaded end of the stud and using a spanner on the lower nut **(see illustration 2.18)**.

2.17 Using a stud extractor tool to remove a broken crankcase stud

2.18 Two nuts can be locked together to unscrew a stud from a component

● A bolt or screw which has broken off below or level with the casing must be extracted using a screw extractor set. Centre punch the fastener to centralise the drill bit, then drill a hole in the fastener **(see illustration 2.19)**. Select a drill bit which is

2.19 When using a screw extractor, first drill a hole in the fastener . . .

approximately half to three-quarters the diameter of the fastener and drill to a depth which will accommodate the extractor. Use the largest size extractor possible, but avoid leaving too small a wall thickness otherwise the extractor will merely force the fastener walls outwards wedging it in the casing thread.

● If a spiral type extractor is used, thread it anti-clockwise into the fastener. As it is screwed in, it will grip the fastener and unscrew it from the casing **(see illustration 2.20)**.

2.20 . . . then thread the extractor anti-clockwise into the fastener

● If a taper type extractor is used, tap it into the fastener so that it is firmly wedged in place. Unscrew the extractor (anti-clockwise) to draw the fastener out.

⚠ *Warning: Stud extractors are very hard and may break off in the fastener if care is not taken - ask an engineer about spark erosion if this happens.*

● Alternatively, the broken bolt/screw can be drilled out and the hole retapped for an oversize bolt/screw or a diamond-section thread insert. It is essential that the drilling is carried out squarely and to the correct depth, otherwise the casing may be ruined - if in doubt, entrust the work to an engineer.

● Bolts and nuts with rounded corners cause the correct size spanner or socket to slip when force is applied. Of the types of spanner/socket available always use a six-point type rather than an eight or twelve-point type - better grip

2.21 Comparison of surface drive ring spanner (left) with 12-point type (right)

is obtained. Surface drive spanners grip the middle of the hex flats, rather than the corners, and are thus good in cases of damaged heads **(see illustration 2.21)**.

● Slotted-head or Phillips-head screws are often damaged by the use of the wrong size screwdriver. Allen-head and Torx-head screws are much less likely to sustain damage. If enough of the screw head is exposed you can use a hacksaw to cut a slot in its head and then use a conventional flat-bladed screwdriver to remove it. Alternatively use a hammer and cold chisel to tap the head of the fastener round to slacken it. Always replace damaged fasteners with new ones, preferably Torx or Allen-head type.

HAYNES HiNT

A dab of valve grinding compound between the screw head and screw-driver tip will often give a good grip.

Thread repair

● Threads (particularly those in aluminium alloy components) can be damaged by overtightening, being assembled with dirt in the threads, or from a component working loose and vibrating. Eventually the thread will fail completely, and it will be impossible to tighten the fastener.

● If a thread is damaged or clogged with old locking compound it can be renovated with a thread repair tool (thread chaser) **(see illustrations 2.22 and 2.23)**; special thread

2.22 A thread repair tool being used to correct an internal thread

2.23 A thread repair tool being used to correct an external thread

chasers are available for spark plug hole threads. The tool will not cut a new thread, but clean and true the original thread. Make sure that you use the correct diameter and pitch tool. Similarly, external threads can be cleaned up with a die or a thread restorer file **(see illustration 2.24)**.

2.24 Using a thread restorer file

● It is possible to drill out the old thread and retap the component to the next thread size. This will work where there is enough surrounding material and a new bolt or screw can be obtained. Sometimes, however, this is not possible - such as where the bolt/screw passes through another component which must also be suitably modified, also in cases where a spark plug or oil drain plug cannot be obtained in a larger diameter thread size.

● The diamond-section thread insert (often known by its popular trade name of Heli-Coil) is a simple and effective method of renewing the thread and retaining the original size. A kit can be purchased which contains the tap, insert and installing tool **(see illustration 2.25)**. Drill out the damaged thread with the size drill specified **(see illustration 2.26)**. Carefully retap the thread **(see illustration 2.27)**. Install the

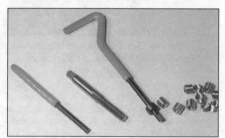

2.25 Obtain a thread insert kit to suit the thread diameter and pitch required

2.26 To install a thread insert, first drill out the original thread . . .

2.27 . . . tap a new thread . . .

2.28 . . . fit insert on the installing tool . . .

2.29 . . . and thread into the component . . .

2.30 . . . break off the tang when complete

insert on the installing tool and thread it slowly into place using a light downward pressure **(see illustrations 2.28 and 2.29)**. When positioned between a 1/4 and 1/2 turn below the surface withdraw the installing tool and use the break-off tool to press down on the tang, breaking it off **(see illustration 2.30)**.

● There are epoxy thread repair kits on the market which can rebuild stripped internal threads, although this repair should not be used on high load-bearing components.

Thread locking and sealing compounds

● Locking compounds are used in locations where the fastener is prone to loosening due to vibration or on important safety-related items which might cause loss of control of the motorcycle if they fail. It is also used where important fasteners cannot be secured by other means such as lockwashers or split pins.

● Before applying locking compound, make sure that the threads (internal and external) are clean and dry with all old compound removed. Select a compound to suit the component being secured - a non-permanent general locking and sealing type is suitable for most applications, but a high strength type is needed for permanent fixing of studs in castings. Apply a drop or two of the compound to the first few threads of the fastener, then thread it into place and tighten to the specified torque. Do not apply excessive thread locking compound otherwise the thread may be damaged on subsequent removal.

● Certain fasteners are impregnated with a dry film type coating of locking compound on their threads. Always renew this type of fastener if disturbed.

● Anti-seize compounds, such as copper-based greases, can be applied to protect threads from seizure due to extreme heat and corrosion. A common instance is spark plug threads and exhaust system fasteners.

3	Measuring tools and gauges

Feeler gauges

● Feeler gauges (or blades) are used for measuring small gaps and clearances (see illustration 3.1). They can also be used to measure endfloat (sideplay) of a component on a shaft where access is not possible with a dial gauge.

● Feeler gauge sets should be treated with care and not bent or damaged. They are etched with their size on one face. Keep them clean and very lightly oiled to prevent corrosion build-up.

3.1 Feeler gauges are used for measuring small gaps and clearances - thickness is marked on one face of gauge

● When measuring a clearance, select a gauge which is a light sliding fit between the two components. You may need to use two gauges together to measure the clearance accurately.

Micrometers

● A micrometer is a precision tool capable of measuring to 0.01 or 0.001 of a millimetre. It should always be stored in its case and not in the general toolbox. It must be kept clean and never dropped, otherwise its frame or measuring anvils could be distorted resulting in inaccurate readings.

● External micrometers are used for measuring outside diameters of components and have many more applications than internal micrometers. Micrometers are available in different size ranges, eg 0 to 25 mm, 25 to 50 mm, and upwards in 25 mm steps; some large micrometers have interchangeable anvils to allow a range of measurements to be taken. Generally the largest precision measurement you are likely to take on a motorcycle is the piston diameter.

● Internal micrometers (or bore micrometers) are used for measuring inside diameters, such as valve guides and cylinder bores. Telescoping gauges and small hole gauges are used in conjunction with an external micrometer, whereas the more expensive internal micrometers have their own measuring device.

External micrometer

Note: *The conventional analogue type instrument is described. Although much easier to read, digital micrometers are considerably more expensive.*

● Always check the calibration of the micrometer before use. With the anvils closed (0 to 25 mm type) or set over a test gauge (for

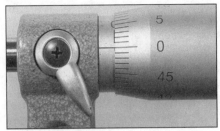

3.2 Check micrometer calibration before use

the larger types) the scale should read zero (see illustration 3.2); make sure that the anvils (and test piece) are clean first. Any discrepancy can be adjusted by referring to the instructions supplied with the tool. Remember that the micrometer is a precision measuring tool - don't force the anvils closed, use the ratchet (4) on the end of the micrometer to close it. In this way, a measured force is always applied.

● To use, first make sure that the item being measured is clean. Place the anvil of the micrometer (1) against the item and use the thimble (2) to bring the spindle (3) lightly into contact with the other side of the item (see illustration 3.3). Don't tighten the thimble down because this will damage the micrometer - instead use the ratchet (4) on the end of the micrometer. The ratchet mechanism applies a measured force preventing damage to the instrument.

● The micrometer is read by referring to the linear scale on the sleeve and the annular scale on the thimble. Read off the sleeve first to obtain the base measurement, then add the fine measurement from the thimble to obtain the overall reading. The linear scale on the sleeve represents the measuring range of the micrometer (eg 0 to 25 mm). The annular scale

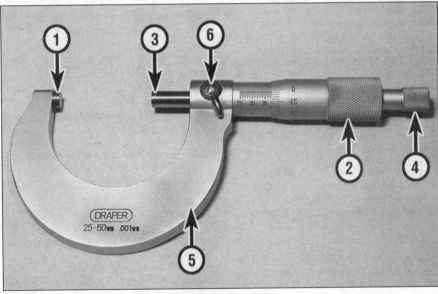

3.3 Micrometer component parts

1 *Anvil*	3 *Spindle*	5 *Frame*
2 *Thimble*	4 *Ratchet*	6 *Locking lever*

on the thimble will be in graduations of 0.01 mm (or as marked on the frame) - one full revolution of the thimble will move 0.5 mm on the linear scale. Take the reading where the datum line on the sleeve intersects the thimble's scale. Always position the eye directly above the scale otherwise an inaccurate reading will result.

In the example shown the item measures 2.95 mm **(see illustration 3.4)**:

Linear scale	2.00 mm
Linear scale	0.50 mm
Annular scale	0.45 mm
Total figure	**2.95 mm**

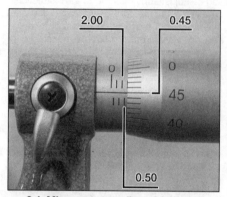

3.4 Micrometer reading of 2.95 mm

Most micrometers have a locking lever (6) on the frame to hold the setting in place, allowing the item to be removed from the micrometer.
● Some micrometers have a vernier scale on their sleeve, providing an even finer measurement to be taken, in 0.001 increments of a millimetre. Take the sleeve and thimble measurement as described above, then check which graduation on the vernier scale aligns with that of the annular scale on the thimble **Note**: *The eye must be perpendicular to the scale when taking the vernier reading - if necessary rotate the body of the micrometer to ensure this.* Multiply the vernier scale figure by 0.001 and add it to the base and fine measurement figures.

In the example shown the item measures 46.994 mm **(see illustrations 3.5 and 3.6)**:

Linear scale (base)	46.000 mm
Linear scale (base)	00.500 mm
Annular scale (fine)	00.490 mm
Vernier scale	00.004 mm
Total figure	**46.994 mm**

Internal micrometer

● Internal micrometers are available for measuring bore diameters, but are expensive and unlikely to be available for home use. It is suggested that a set of telescoping gauges and small hole gauges, both of which must be used with an external micrometer, will suffice for taking internal measurements on a motorcycle.

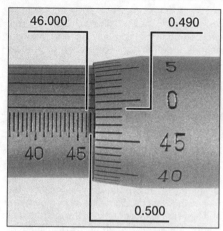

3.5 Micrometer reading of 46.99 mm on linear and annular scales . . .

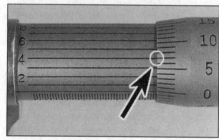

3.6 . . . and 0.004 mm on vernier scale

● Telescoping gauges can be used to measure internal diameters of components. Select a gauge with the correct size range, make sure its ends are clean and insert it into the bore. Expand the gauge, then lock its position and withdraw it from the bore **(see illustration 3.7)**. Measure across the gauge ends with a micrometer **(see illustration 3.8)**.

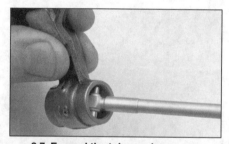

3.7 Expand the telescoping gauge in the bore, lock its position . . .

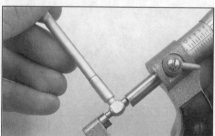

3.8 . . . then measure the gauge with a micrometer

3.9 Expand the small hole gauge in the bore, lock its position . . .

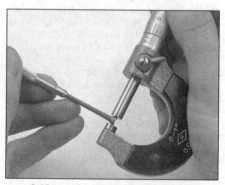

3.10 . . . then measure the gauge with a micrometer

● Very small diameter bores (such as valve guides) are measured with a small hole gauge. Once adjusted to a slip-fit inside the component, its position is locked and the gauge withdrawn for measurement with a micrometer **(see illustrations 3.9 and 3.10)**.

Vernier caliper

Note: *The conventional linear and dial gauge type instruments are described. Digital types are easier to read, but are far more expensive.*
● The vernier caliper does not provide the precision of a micrometer, but is versatile in being able to measure internal and external diameters. Some types also incorporate a depth gauge. It is ideal for measuring clutch plate friction material and spring free lengths.
● To use the conventional linear scale vernier, slacken off the vernier clamp screws (1) and set its jaws over (2), or inside (3), the item to be measured **(see illustration 3.11)**. Slide the jaw into contact, using the thumb-wheel (4) for fine movement of the sliding scale (5) then tighten the clamp screws (1). Read off the main scale (6) where the zero on the sliding scale (5) intersects it, taking the whole number to the left of the zero; this provides the base measurement. View along the sliding scale and select the division which lines up exactly with any of the divisions on the main scale, noting that the divisions usually represents 0.02 of a millimetre. Add this fine measurement to the base measurement to obtain the total reading.

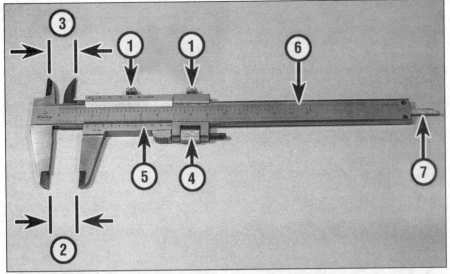

3.11 Vernier component parts (linear gauge)

1 Clamp screws   3 Internal jaws   5 Sliding scale   7 Depth gauge
2 External jaws   4 Thumbwheel   6 Main scale

Plastigauge

● Plastigauge is a plastic material which can be compressed between two surfaces to measure the oil clearance between them. The width of the compressed Plastigauge is measured against a calibrated scale to determine the clearance.

● Common uses of Plastigauge are for measuring the clearance between crankshaft journal and main bearing inserts, between crankshaft journal and big-end bearing inserts, and between camshaft and bearing surfaces. The following example describes big-end oil clearance measurement.

● Handle the Plastigauge material carefully to prevent distortion. Using a sharp knife, cut a length which corresponds with the width of the bearing being measured and place it carefully across the journal so that it is parallel with the shaft (see illustration 3.15). Carefully install both bearing shells and the connecting rod. Without rotating the rod on the journal tighten its bolts or nuts (as applicable) to the specified torque. The connecting rod and bearings are then disassembled and the crushed Plastigauge examined.

In the example shown the item measures 55.92 mm (see illustration 3.12):

Base measurement	55.00 mm
Fine measurement	00.92 mm
Total figure	**55.92 mm**

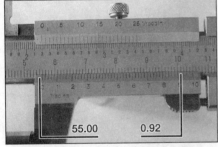

3.12 Vernier gauge reading of 55.92 mm

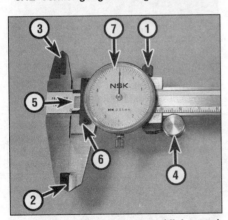

3.13 Vernier component parts (dial gauge)

1 Clamp screw   5 Main scale
2 External jaws   6 Sliding scale
3 Internal jaws   7 Dial gauge
4 Thumbwheel

● Some vernier calipers are equipped with a dial gauge for fine measurement. Before use, check that the jaws are clean, then close them fully and check that the dial gauge reads zero. If necessary adjust the gauge ring accordingly. Slacken the vernier clamp screw (1) and set its jaws over (2), or inside (3), the item to be measured (see illustration 3.13). Slide the jaws into contact, using the thumbwheel (4) for fine movement. Read off the main scale (5) where the edge of the sliding scale (6) intersects it, taking the whole number to the left of the zero; this provides the base measurement. Read off the needle position on the dial gauge (7) scale to provide the fine measurement; each division represents 0.05 of a millimetre. Add this fine measurement to the base measurement to obtain the total reading.

In the example shown the item measures 55.95 mm (see illustration 3.14):

Base measurement	55.00 mm
Fine measurement	00.95 mm
Total figure	**55.95 mm**

3.14 Vernier gauge reading of 55.95 mm

3.15 Plastigauge placed across shaft journal

● Using the scale provided in the Plastigauge kit, measure the width of the material to determine the oil clearance (see illustration 3.16). Always remove all traces of Plastigauge after use using your fingernails.

Caution: Arriving at the correct clearance demands that the assembly is torqued correctly, according to the settings and sequence (where applicable) provided by the motorcycle manufacturer.

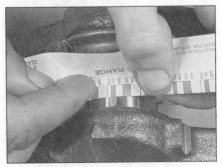

3.16 Measuring the width of the crushed Plastigauge

Dial gauge or DTI (Dial Test Indicator)

● A dial gauge can be used to accurately measure small amounts of movement. Typical uses are measuring shaft runout or shaft endfloat (sideplay) and setting piston position for ignition timing on two-strokes. A dial gauge set usually comes with a range of different probes and adapters and mounting equipment.

● The gauge needle must point to zero when at rest. Rotate the ring around its periphery to zero the gauge.

● Check that the gauge is capable of reading the extent of movement in the work. Most gauges have a small dial set in the face which records whole millimetres of movement as well as the fine scale around the face periphery which is calibrated in 0.01 mm divisions. Read off the small dial first to obtain the base measurement, then add the measurement from the fine scale to obtain the total reading.

In the example shown the gauge reads 1.48 mm **(see illustration 3.17)**:

Base measurement	1.00 mm
Fine measurement	0.48 mm
Total figure	**1.48 mm**

3.17 Dial gauge reading of 1.48 mm

● If measuring shaft runout, the shaft must be supported in vee-blocks and the gauge mounted on a stand perpendicular to the shaft. Rest the tip of the gauge against the centre of the shaft and rotate the shaft slowly whilst watching the gauge reading **(see illustration 3.18)**. Take several measurements along the length of the shaft and record the

3.18 Using a dial gauge to measure shaft runout

maximum gauge reading as the amount of runout in the shaft. **Note:** *The reading obtained will be total runout at that point - some manufacturers specify that the runout figure is halved to compare with their specified runout limit.*

● Endfloat (sideplay) measurement requires that the gauge is mounted securely to the surrounding component with its probe touching the end of the shaft. Using hand pressure, push and pull on the shaft noting the maximum endfloat recorded on the gauge **(see illustration 3.19)**.

3.19 Using a dial gauge to measure shaft endfloat

● A dial gauge with suitable adapters can be used to determine piston position BTDC on two-stroke engines for the purposes of ignition timing. The gauge, adapter and suitable length probe are installed in the place of the spark plug and the gauge zeroed at TDC. If the piston position is specified as 1.14 mm BTDC, rotate the engine back to 2.00 mm BTDC, then slowly forwards to 1.14 mm BTDC.

Cylinder compression gauges

● A compression gauge is used for measuring cylinder compression. Either the rubber-cone type or the threaded adapter type can be used. The latter is preferred to ensure a perfect seal against the cylinder head. A 0 to 300 psi (0 to 20 Bar) type gauge (for petrol/gasoline engines) will be suitable for motorcycles.

● The spark plug is removed and the gauge either held hard against the cylinder head (cone type) or the gauge adapter screwed into the cylinder head (threaded type) **(see illustration 3.20)**. Cylinder compression is measured with the engine turning over, but not running - carry out the compression test as described in

3.20 Using a rubber-cone type cylinder compression gauge

Fault Finding Equipment. The gauge will hold the reading until manually released.

Oil pressure gauge

● An oil pressure gauge is used for measuring engine oil pressure. Most gauges come with a set of adapters to fit the thread of the take-off point **(see illustration 3.21)**. If the take-off point specified by the motorcycle manufacturer is an external oil pipe union, make sure that the specified replacement union is used to prevent oil starvation.

3.21 Oil pressure gauge and take-off point adapter (arrow)

● Oil pressure is measured with the engine running (at a specific rpm) and often the manufacturer will specify pressure limits for a cold and hot engine.

Straight-edge and surface plate

● If checking the gasket face of a component for warpage, place a steel rule or precision straight-edge across the gasket face and measure any gap between the straight-edge and component with feeler gauges **(see illustration 3.22)**. Check diagonally across the component and between mounting holes **(see illustration 3.23)**.

3.22 Use a straight-edge and feeler gauges to check for warpage

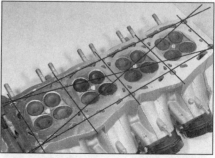

3.23 Check for warpage in these directions

● Checking individual components for warpage, such as clutch plain (metal) plates, requires a perfectly flat plate or piece of plate glass and feeler gauges.

4 Torque and leverage

What is torque?

● Torque describes the twisting force about a shaft. The amount of torque applied is determined by the distance from the centre of the shaft to the end of the lever and the amount of force being applied to the end of the lever; distance multiplied by force equals torque.

● The manufacturer applies a measured torque to a bolt or nut to ensure that it will not slacken in use and to hold two components securely together without movement in the joint. The actual torque setting depends on the thread size, bolt or nut material and the composition of the components being held.

● Too little torque may cause the fastener to loosen due to vibration, whereas too much torque will distort the joint faces of the component or cause the fastener to shear off. Always stick to the specified torque setting.

Using a torque wrench

● Check the calibration of the torque wrench and make sure it has a suitable range for the job. Torque wrenches are available in Nm (Newton-metres), kgf m (kilograms-force metre), lbf ft (pounds-feet), lbf in (inch-pounds). Do not confuse lbf ft with lbf in.

● Adjust the tool to the desired torque on the scale (see illustration 4.1). If your torque wrench is not calibrated in the units specified, carefully convert the figure (see Conversion Factors). A manufacturer sometimes gives a torque setting as a range (8 to 10 Nm) rather than a single figure - in this case set the tool midway between the two settings. The same torque may be expressed as 9 Nm ± 1 Nm. Some torque wrenches have a method of locking the setting so that it isn't inadvertently altered during use.

4.1 Set the torque wrench index mark to the setting required, in this case 12 Nm

● Install the bolts/nuts in their correct location and secure them lightly. Their threads must be clean and free of any old locking compound. Unless specified the threads and flange should be dry - oiled threads are necessary in certain circumstances and the manufacturer will take this into account in the specified torque figure. Similarly, the manufacturer may also specify the application of thread-locking compound.

● Tighten the fasteners in the specified sequence until the torque wrench clicks, indicating that the torque setting has been reached. Apply the torque again to double-check the setting. Where different thread diameter fasteners secure the component, as a rule tighten the larger diameter ones first.

● When the torque wrench has been finished with, release the lock (where applicable) and fully back off its setting to zero - do not leave the torque wrench tensioned. Also, do not use a torque wrench for slackening a fastener.

Angle-tightening

● Manufacturers often specify a figure in degrees for final tightening of a fastener. This usually follows tightening to a specific torque setting.

● A degree disc can be set and attached to the socket (see illustration 4.2) or a protractor can be used to mark the angle of movement on the bolt/nut head and the surrounding casting (see illustration 4.3).

4.2 Angle tightening can be accomplished with a torque-angle gauge . . .

4.3 . . . or by marking the angle on the surrounding component

Loosening sequences

● Where more than one bolt/nut secures a component, loosen each fastener evenly a little at a time. In this way, not all the stress of the joint is held by one fastener and the components are not likely to distort.

● If a tightening sequence is provided, work in the REVERSE of this, but if not, work from the outside in, in a criss-cross sequence (see illustration 4.4).

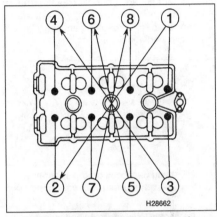

4.4 When slackening, work from the outside inwards

Tightening sequences

● If a component is held by more than one fastener it is important that the retaining bolts/nuts are tightened evenly to prevent uneven stress build-up and distortion of sealing faces. This is especially important on high-compression joints such as the cylinder head.

● A sequence is usually provided by the manufacturer, either in a diagram or actually marked in the casting. If not, always start in the centre and work outwards in a criss-cross pattern (see illustration 4.5). Start off by securing all bolts/nuts finger-tight, then set the torque wrench and tighten each fastener by a small amount in sequence until the final torque is reached. By following this practice,

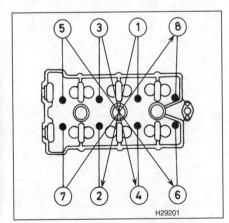

4.5 When tightening, work from the inside outwards

the joint will be held evenly and will not be distorted. Important joints, such as the cylinder head and big-end fasteners often have two- or three-stage torque settings.

Applying leverage

● Use tools at the correct angle. Position a socket wrench or spanner on the bolt/nut so that you pull it towards you when loosening. If this can't be done, push the spanner without curling your fingers around it **(see illustration 4.6)** - the spanner may slip or the fastener loosen suddenly, resulting in your fingers being crushed against a component.

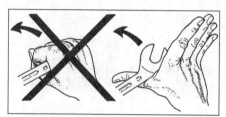

4.6 If you can't pull on the spanner to loosen a fastener, push with your hand open

● Additional leverage is gained by extending the length of the lever. The best way to do this is to use a breaker bar instead of the regular length tool, or to slip a length of tubing over the end of the spanner or socket wrench.
● If additional leverage will not work, the fastener head is either damaged or firmly corroded in place (see *Fasteners*).

5 Bearings

Bearing removal and installation

Drivers and sockets

● Before removing a bearing, always inspect the casing to see which way it must be driven out - some casings will have retaining plates or a cast step. Also check for any identifying markings on the bearing and if installed to a certain depth, measure this at this stage. Some roller bearings are sealed on one side - take note of the original fitted position.
● Bearings can be driven out of a casing using a bearing driver tool (with the correct size head) or a socket of the correct diameter. Select the driver head or socket so that it contacts the outer race of the bearing, not the balls/rollers or inner race. Always support the casing around the bearing housing with wood blocks, otherwise there is a risk of fracture. The bearing is driven out with a few blows on the driver or socket from a heavy mallet. Unless access is severely restricted (as with wheel bearings), a pin-punch is not recommended unless it is moved around the bearing to keep it square in its housing.

● The same equipment can be used to install bearings. Make sure the bearing housing is supported on wood blocks and line up the bearing in its housing. Fit the bearing as noted on removal - generally they are installed with their marked side facing outwards. Tap the bearing squarely into its housing using a driver or socket which bears only on the bearing's outer race - contact with the bearing balls/rollers or inner race will destroy it **(see illustrations 5.1 and 5.2)**.
● Check that the bearing inner race and balls/rollers rotate freely.

5.1 Using a bearing driver against the bearing's outer race

5.2 Using a large socket against the bearing's outer race

Pullers and slide-hammers

● Where a bearing is pressed on a shaft a puller will be required to extract it **(see illustration 5.3)**. Make sure that the puller clamp or legs fit securely behind the bearing and are unlikely to slip out. If pulling a bearing

5.3 This bearing puller clamps behind the bearing and pressure is applied to the shaft end to draw the bearing off

off a gear shaft for example, you may have to locate the puller behind a gear pinion if there is no access to the race and draw the gear pinion off the shaft as well **(see illustration 5.4)**. *Caution: Ensure that the puller's centre bolt locates securely against the end of the shaft and will not slip when pressure is applied. Also ensure that puller does not damage the shaft end.*

5.4 Where no access is available to the rear of the bearing, it is sometimes possible to draw off the adjacent component

● Operate the puller so that its centre bolt exerts pressure on the shaft end and draws the bearing off the shaft.
● When installing the bearing on the shaft, tap only on the bearing's inner race - contact with the balls/rollers or outer race with destroy the bearing. Use a socket or length of tubing as a drift which fits over the shaft end **(see illustration 5.5)**.

5.5 When installing a bearing on a shaft use a piece of tubing which bears only on the bearing's inner race

● Where a bearing locates in a blind hole in a casing, it cannot be driven or pulled out as described above. A slide-hammer with knife-edged bearing puller attachment will be required. The puller attachment passes through the bearing and when tightened expands to fit firmly behind the bearing **(see illustration 5.6)**. By operating the slide-hammer part of the tool the bearing is jarred out of its housing **(see illustration 5.7)**.
● It is possible, if the bearing is of reasonable weight, for it to drop out of its housing if the casing is heated as described below. If this

5.6 Expand the bearing puller so that it locks behind the bearing . . .

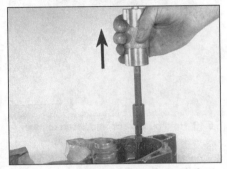

5.7 . . . attach the slide hammer to the bearing puller

method is attempted, first prepare a work surface which will enable the casing to be tapped face down to help dislodge the bearing - a wood surface is ideal since it will not damage the casing's gasket surface. Wearing protective gloves, tap the heated casing several times against the work surface to dislodge the bearing under its own weight (**see illustration 5.8**).

5.8 Tapping a casing face down on wood blocks can often dislodge a bearing

● Bearings can be installed in blind holes using the driver or socket method described above.

Drawbolts

● Where a bearing or bush is set in the eye of a component, such as a suspension linkage arm or connecting rod small-end, removal by drift may damage the component. Furthermore, a rubber bushing in a shock absorber eye cannot successfully be driven out of position. If access is available to a engineering press, the task is straightforward. If not, a drawbolt can be fabricated to extract the bearing or bush.

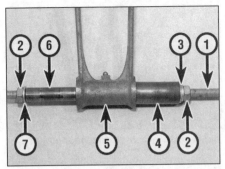

5.9 Drawbolt component parts assembled on a suspension arm

1 *Bolt or length of threaded bar*
2 *Nuts*
3 *Washer (external diameter greater than tubing internal diameter)*
4 *Tubing (internal diameter sufficient to accommodate bearing)*
5 *Suspension arm with bearing*
6 *Tubing (external diameter slightly smaller than bearing)*
7 *Washer (external diameter slightly smaller than bearing)*

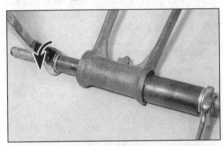

5.10 Drawing the bearing out of the suspension arm

● To extract the bearing/bush you will need a long bolt with nut (or piece of threaded bar with two nuts), a piece of tubing which has an internal diameter larger than the bearing/bush, another piece of tubing which has an external diameter slightly smaller than the bearing/ bush, and a selection of washers (**see illustrations 5.9 and 5.10**). Note that the pieces of tubing must be of the same length, or longer, than the bearing/bush.
● The same kit (without the pieces of tubing) can be used to draw the new bearing/bush back into place (**see illustration 5.11**).

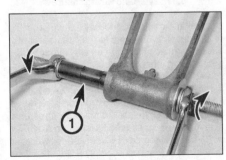

5.11 Installing a new bearing (1) in the suspension arm

Temperature change

● If the bearing's outer race is a tight fit in the casing, the aluminium casing can be heated to release its grip on the bearing. Aluminium will expand at a greater rate than the steel bearing outer race. There are several ways to do this, but avoid any localised extreme heat (such as a blow torch) - aluminium alloy has a low melting point.
● Approved methods of heating a casing are using a domestic oven (heated to 100°C) or immersing the casing in boiling water (**see illustration 5.12**). Low temperature range localised heat sources such as a paint stripper heat gun or clothes iron can also be used (**see illustration 5.13**). Alternatively, soak a rag in boiling water, wring it out and wrap it around the bearing housing.

⚠ *Warning: All of these methods require care in use to prevent scalding and burns to the hands. Wear protective gloves when handling hot components.*

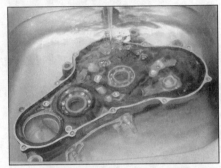

5.12 A casing can be immersed in a sink of boiling water to aid bearing removal

5.13 Using a localised heat source to aid bearing removal

● If heating the whole casing note that plastic components, such as the neutral switch, may suffer - remove them beforehand.
● After heating, remove the bearing as described above. You may find that the expansion is sufficient for the bearing to fall out of the casing under its own weight or with a light tap on the driver or socket.
● If necessary, the casing can be heated to aid bearing installation, and this is sometimes the recommended procedure if the motorcycle manufacturer has designed the housing and bearing fit with this intention.

● Installation of bearings can be eased by placing them in a freezer the night before installation. The steel bearing will contract slightly, allowing easy insertion in its housing. This is often useful when installing steering head outer races in the frame.

Bearing types and markings

● Plain shell bearings, ball bearings, needle roller bearings and tapered roller bearings will all be found on motorcycles **(see illustrations 5.14 and 5.15)**. The ball and roller types are usually caged between an inner and outer race, but uncaged variations may be found.

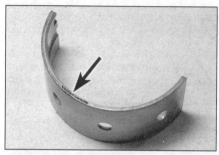

5.14 Shell bearings are either plain or grooved. They are usually identified by colour code (arrow)

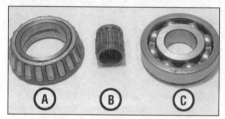

5.15 Tapered roller bearing (A), needle roller bearing (B) and ball journal bearing (C)

● Shell bearings (often called inserts) are usually found at the crankshaft main and connecting rod big-end where they are good at coping with high loads. They are made of a phosphor-bronze material and are impregnated with self-lubricating properties.
● Ball bearings and needle roller bearings consist of a steel inner and outer race with the balls or rollers between the races. They require constant lubrication by oil or grease and are good at coping with axial loads. Taper roller bearings consist of rollers set in a tapered cage set on the inner race; the outer race is separate. They are good at coping with axial loads and prevent movement along the shaft - a typical application is in the steering head.
● Bearing manufacturers produce bearings to ISO size standards and stamp one face of the bearing to indicate its internal and external diameter, load capacity and type **(see illustration 5.16)**.
● Metal bushes are usually of phosphor-bronze material. Rubber bushes are used in suspension mounting eyes. Fibre bushes have also been used in suspension pivots.

5.16 Typical bearing marking

Bearing fault finding

● If a bearing outer race has spun in its housing, the housing material will be damaged. You can use a bearing locking compound to bond the outer race in place if damage is not too severe.
● Shell bearings will fail due to damage of their working surface, as a result of lack of lubrication, corrosion or abrasive particles in the oil **(see illustration 5.17)**. Small particles of dirt in the oil may embed in the bearing material whereas larger particles will score the bearing and shaft journal. If a number of short journeys are made, insufficient heat will be generated to drive off condensation which has built up on the bearings.

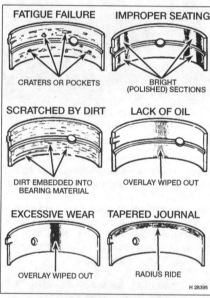

5.17 Typical bearing failures

● Ball and roller bearings will fail due to lack of lubrication or damage to the balls or rollers. Tapered-roller bearings can be damaged by overloading them. Unless the bearing is sealed on both sides, wash it in paraffin (kerosene) to remove all old grease then allow it to dry. Make a visual inspection looking to dented balls or rollers, damaged cages and worn or pitted races **(see illustration 5.18)**.
● A ball bearing can be checked for wear by listening to it when spun. Apply a film of light oil to the bearing and hold it close to the ear - hold the outer race with one hand and spin the inner

5.18 Example of ball journal bearing with damaged balls and cages

5.19 Hold outer race and listen to inner race when spun

race with the other hand **(see illustration 5.19)**. The bearing should be almost silent when spun; if it grates or rattles it is worn.

6 Oil seals

Oil seal removal and installation

● Oil seals should be renewed every time a component is dismantled. This is because the seal lips will become set to the sealing surface and will not necessarily reseal.
● Oil seals can be prised out of position using a large flat-bladed screwdriver **(see illustration 6.1)**. In the case of crankcase seals, check first that the seal is not lipped on the inside, preventing its removal with the crankcases joined.

6.1 Prise out oil seals with a large flat-bladed screwdriver

● New seals are usually installed with their marked face (containing the seal reference code) outwards and the spring side towards the fluid being retained. In certain cases, such as a two-stroke engine crankshaft seal, a double lipped seal may be used due to there being fluid or gas on each side of the joint.

• Use a bearing driver or socket which bears only on the outer hard edge of the seal to install it in the casing - tapping on the inner edge will damage the sealing lip.

Oil seal types and markings

• Oil seals are usually of the single-lipped type. Double-lipped seals are found where a liquid or gas is on both sides of the joint.
• Oil seals can harden and lose their sealing ability if the motorcycle has been in storage for a long period - renewal is the only solution.
• Oil seal manufacturers also conform to the ISO markings for seal size - these are moulded into the outer face of the seal (see illustration 6.2).

6.2 These oil seal markings indicate inside diameter, outside diameter and seal thickness

7 Gaskets and sealants

Types of gasket and sealant

• Gaskets are used to seal the mating surfaces between components and keep lubricants, fluids, vacuum or pressure contained within the assembly. Aluminium gaskets are sometimes found at the cylinder joints, but most gaskets are paper-based. If the mating surfaces of the components being joined are undamaged the gasket can be installed dry, although a dab of sealant or grease will be useful to hold it in place during assembly.
• RTV (Room Temperature Vulcanising) silicone rubber sealants cure when exposed to moisture in the atmosphere. These sealants are good at filling pits or irregular gasket faces, but will tend to be forced out of the joint under very high torque. They can be used to replace a paper gasket, but first make sure that the width of the paper gasket is not essential to the shimming of internal components. RTV sealants should not be used on components containing petrol (gasoline).
• Non-hardening, semi-hardening and hard setting liquid gasket compounds can be used with a gasket or between a metal-to-metal joint. Select the sealant to suit the application: universal non-hardening sealant can be used on virtually all joints; semi-hardening on joint faces which are rough or damaged; hard setting sealant on joints which require a permanent bond and are subjected to high temperature and pressure. **Note:** *Check first if*

the paper gasket has a bead of sealant impregnated in its surface before applying additional sealant.
• When choosing a sealant, make sure it is suitable for the application, particularly if being applied in a high-temperature area or in the vicinity of fuel. Certain manufacturers produce sealants in either clear, silver or black colours to match the finish of the engine. This has a particular application on motorcycles where much of the engine is exposed.
• Do not over-apply sealant. That which is squeezed out on the outside of the joint can be wiped off, whereas an excess of sealant on the inside can break off and clog oilways.

Breaking a sealed joint

• Age, heat, pressure and the use of hard setting sealant can cause two components to stick together so tightly that they are difficult to separate using finger pressure alone. Do not resort to using levers unless there is a pry point provided for this purpose (see illustration 7.1) or else the gasket surfaces will be damaged.
• Use a soft-faced hammer (see illustration 7.2) or a wood block and conventional hammer to strike the component near the mating surface. Avoid hammering against cast extremities since they may break off. If this method fails, try using a wood wedge between the two components.
Caution: If the joint will not separate, double-check that you have removed all the fasteners.

7.1 If a pry point is provided, apply gently pressure with a flat-bladed screwdriver

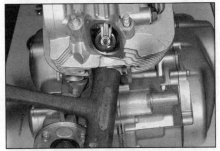

7.2 Tap around the joint with a soft-faced mallet if necessary - don't strike cooling fins

Removal of old gasket and sealant

• Paper gaskets will most likely come away complete, leaving only a few traces stuck on

Most components have one or two hollow locating dowels between the two gasket faces. If a dowel cannot be removed, do not resort to gripping it with pliers - it will almost certainly be distorted. Install a close-fitting socket or Phillips screwdriver into the dowel and then grip the outer edge of the dowel to free it.

the sealing faces of the components. It is imperative that all traces are removed to ensure correct sealing of the new gasket.
• Very carefully scrape all traces of gasket away making sure that the sealing surfaces are not gouged or scored by the scraper (see illustrations 7.3, 7.4 and 7.5). Stubborn deposits can be removed by spraying with an aerosol gasket remover. Final preparation of

7.3 Paper gaskets can be scraped off with a gasket scraper tool . . .

7.4 . . . a knife blade . . .

7.5 . . . or a household scraper

7.6 Fine abrasive paper is wrapped around a flat file to clean up the gasket face

7.7 A kitchen scourer can be used on stubborn deposits

the gasket surface can be made with very fine abrasive paper or a plastic kitchen scourer **(see illustrations 7.6 and 7.7).**

● Old sealant can be scraped or peeled off components, depending on the type originally used. Note that gasket removal compounds are available to avoid scraping the components clean; make sure the gasket remover suits the type of sealant used.

8 Chains

Breaking and joining final drive chains

● Drive chains for all but small bikes are continuous and do not have a clip-type connecting link. The chain must be broken using a chain breaker tool and the new chain securely riveted together using a new soft rivet-type link. Never use a clip-type connecting link instead of a rivet-type link, except in an emergency. Various chain breaking and riveting tools are available, either as separate tools or combined as illustrated in the accompanying photographs - read the instructions supplied with the tool carefully.

⚠ *Warning: The need to rivet the new link pins correctly cannot be overstressed - loss of control of the motorcycle is very likely to result if the chain breaks in use.*

● Rotate the chain and look for the soft link. The soft link pins look like they have been deeply centre-punched instead of peened over

8.1 Tighten the chain breaker to push the pin out of the link . . .

8.2 . . . withdraw the pin, remove the tool . . .

8.3 . . . and separate the chain link

like all the other pins **(see illustration 8.9)** and its sideplate may be a different colour. Position the soft link midway between the sprockets and assemble the chain breaker tool over one of the soft link pins **(see illustration 8.1).** Operate the tool to push the pin out through the chain **(see illustration 8.2).** On an O-ring chain, remove the O-rings **(see illustration 8.3).** Carry out the same procedure on the other soft link pin.

Caution: Certain soft link pins (particularly on the larger chains) may require their ends to be filed or ground off before they can be pressed out using the tool.

● Check that you have the correct size and strength (standard or heavy duty) new soft link - do not reuse the old link. Look for the size marking on the chain sideplates **(see illustration 8.10).**

● Position the chain ends so that they are engaged over the rear sprocket. On an O-ring chain, install a new O-ring over each pin of the link and insert the link through the two chain

8.4 Insert the new soft link, with O-rings, through the chain ends . . .

8.5 . . . install the O-rings over the pin ends . . .

8.6 . . . followed by the sideplate

ends **(see illustration 8.4).** Install a new O-ring over the end of each pin, followed by the sideplate (with the chain manufacturer's marking facing outwards) **(see illustrations 8.5 and 8.6).** On an unsealed chain, insert the link through the two chain ends, then install the sideplate with the chain manufacturer's marking facing outwards.

● Note that it may not be possible to install the sideplate using finger pressure alone. If using a joining tool, assemble it so that the plates of the tool clamp the link and press the sideplate over the pins **(see illustration 8.7).** Otherwise, use two small sockets placed over

8.7 Push the sideplate into position using a clamp

8.8 Assemble the chain riveting tool over one pin at a time and tighten it fully

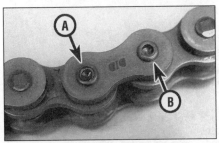

8.9 Pin end correctly riveted (A), pin end unriveted (B)

the rivet ends and two pieces of the wood between a G-clamp. Operate the clamp to press the sideplate over the pins.

● Assemble the joining tool over one pin (following the maker's instructions) and tighten the tool down to spread the pin end securely **(see illustrations 8.8 and 8.9)**. Do the same on the other pin.

 Warning: Check that the pin ends are secure and that there is no danger of the sideplate coming loose. If the pin ends are cracked the soft link must be renewed.

Final drive chain sizing

● Chains are sized using a three digit number, followed by a suffix to denote the chain type **(see illustration 8.10)**. Chain type is either standard or heavy duty (thicker sideplates), and also unsealed or O-ring/X-ring type.

● The first digit of the number relates to the pitch of the chain, ie the distance from the centre of one pin to the centre of the next pin **(see illustration 8.11)**. Pitch is expressed in eighths of an inch, as follows:

8.10 Typical chain size and type marking

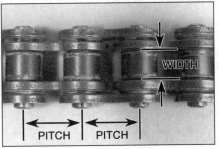

8.11 Chain dimensions

Sizes commencing with a 4 (eg 428) have a pitch of 1/2 inch (12.7 mm)

Sizes commencing with a 5 (eg 520) have a pitch of 5/8 inch (15.9 mm)

Sizes commencing with a 6 (eg 630) have a pitch of 3/4 inch (19.1 mm)

● The second and third digits of the chain size relate to the width of the rollers, again in imperial units, eg the 525 shown has 5/16 inch (7.94 mm) rollers **(see illustration 8.11)**.

9 Hoses

Clamping to prevent flow

● Small-bore flexible hoses can be clamped to prevent fluid flow whilst a component is worked on. Whichever method is used, ensure that the hose material is not permanently distorted or damaged by the clamp.

a) A brake hose clamp available from auto accessory shops **(see illustration 9.1)**.

b) A wingnut type hose clamp **(see illustration 9.2)**.

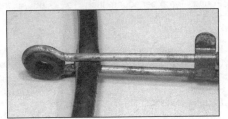

9.1 Hoses can be clamped with an automotive brake hose clamp . . .

9.2 . . . a wingnut type hose clamp . . .

c) Two sockets placed each side of the hose and held with straight-jawed self-locking grips **(see illustration 9.3)**.

d) Thick card each side of the hose held between straight-jawed self-locking grips **(see illustration 9.4)**.

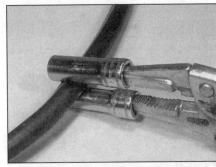

9.3 . . . two sockets and a pair of self-locking grips . . .

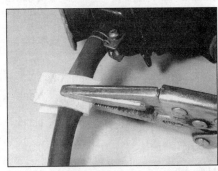

9.4 . . . or thick card and self-locking grips

Freeing and fitting hoses

● Always make sure the hose clamp is moved well clear of the hose end. Grip the hose with your hand and rotate it whilst pulling it off the union. If the hose has hardened due to age and will not move, slit it with a sharp knife and peel its ends off the union **(see illustration 9.5)**.

● Resist the temptation to use grease or soap on the unions to aid installation; although it helps the hose slip over the union it will equally aid the escape of fluid from the joint. It is preferable to soften the hose ends in hot water and wet the inside surface of the hose with water or a fluid which will evaporate.

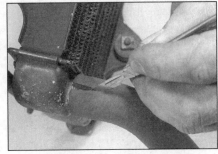

9.5 Cutting a coolant hose free with a sharp knife

Conversion Factors

Length (distance)

Inches (in)	x 25.4	= Millimetres (mm)	x 0.0394	= Inches (in)
Feet (ft)	x 0.305	= Metres (m)	x 3.281	= Feet (ft)
Miles	x 1.609	= Kilometres (km)	x 0.621	= Miles

Volume (capacity)

Cubic inches (cu in; in³)	x 16.387	= Cubic centimetres (cc; cm³)	x 0.061	= Cubic inches (cu in; in³)
Imperial pints (Imp pt)	x 0.568	= Litres (l)	x 1.76	= Imperial pints (Imp pt)
Imperial quarts (Imp qt)	x 1.137	= Litres (l)	x 0.88	= Imperial quarts (Imp qt)
Imperial quarts (Imp qt)	x 1.201	= US quarts (US qt)	x 0.833	= Imperial quarts (Imp qt)
US quarts (US qt)	x 0.946	= Litres (l)	x 1.057	= US quarts (US qt)
Imperial gallons (Imp gal)	x 4.546	= Litres (l)	x 0.22	= Imperial gallons (Imp gal)
Imperial gallons (Imp gal)	x 1.201	= US gallons (US gal)	x 0.833	= Imperial gallons (Imp gal)
US gallons (US gal)	x 3.785	= Litres (l)	x 0.264	= US gallons (US gal)

Mass (weight)

Ounces (oz)	x 28.35	= Grams (g)	x 0.035	= Ounces (oz)
Pounds (lb)	x 0.454	= Kilograms (kg)	x 2.205	= Pounds (lb)

Force

Ounces-force (ozf; oz)	x 0.278	= Newtons (N)	x 3.6	= Ounces-force (ozf; oz)
Pounds-force (lbf; lb)	x 4.448	= Newtons (N)	x 0.225	= Pounds-force (lbf; lb)
Newtons (N)	x 0.1	= Kilograms-force (kgf; kg)	x 9.81	= Newtons (N)

Pressure

Pounds-force per square inch (psi; lbf/in²; lb/in²)	x 0.070	= Kilograms-force per square centimetre (kgf/cm²; kg/cm²)	x 14.223	= Pounds-force per square inch (psi; lbf/in²; lb/in²)
Pounds-force per square inch (psi; lbf/in²; lb/in²)	x 0.068	= Atmospheres (atm)	x 14.696	= Pounds-force per square inch (psi; lbf/in²; lb/in²)
Pounds-force per square inch (psi; lbf/in²; lb/in²)	x 0.069	= Bars	x 14.5	= Pounds-force per square inch (psi; lbf/in²; lb/in²)
Pounds-force per square inch (psi; lbf/in²; lb/in²)	x 6.895	= Kilopascals (kPa)	x 0.145	= Pounds-force per square inch (psi; lbf/in²; lb/in²)
Kilopascals (kPa)	x 0.01	= Kilograms-force per square centimetre (kgf/cm²; kg/cm²)	x 98.1	= Kilopascals (kPa)
Millibar (mbar)	x 100	= Pascals (Pa)	x 0.01	= Millibar (mbar)
Millibar (mbar)	x 0.0145	= Pounds-force per square inch (psi; lbf/in²; lb/in²)	x 68.947	= Millibar (mbar)
Millibar (mbar)	x 0.75	= Millimetres of mercury (mmHg)	x 1.333	= Millibar (mbar)
Millibar (mbar)	x 0.401	= Inches of water (inH₂O)	x 2.491	= Millibar (mbar)
Millimetres of mercury (mmHg)	x 0.535	= Inches of water (inH₂O)	x 1.868	= Millimetres of mercury (mmHg)
Inches of water (inH₂O)	x 0.036	= Pounds-force per square inch (psi; lbf/in²; lb/in²)	x 27.68	= Inches of water (inH₂O)

Torque (moment of force)

Pounds-force inches (lbf in; lb in)	x 1.152	= Kilograms-force centimetre (kgf cm; kg cm)	x 0.868	= Pounds-force inches (lbf in; lb in)
Pounds-force inches (lbf in; lb in)	x 0.113	= Newton metres (Nm)	x 8.85	= Pounds-force inches (lbf in; lb in)
Pounds-force inches (lbf in; lb in)	x 0.083	= Pounds-force feet (lbf ft; lb ft)	x 12	= Pounds-force inches (lbf in; lb in)
Pounds-force feet (lbf ft; lb ft)	x 0.138	= Kilograms-force metres (kgf m; kg m)	x 7.233	= Pounds-force feet (lbf ft; lb ft)
Pounds-force feet (lbf ft; lb ft)	x 1.356	= Newton metres (Nm)	x 0.738	= Pounds-force feet (lbf ft; lb ft)
Newton metres (Nm)	x 0.102	= Kilograms-force metres (kgf m; kg m)	x 9.804	= Newton metres (Nm)

Power

Horsepower (hp)	x 745.7	= Watts (W)	x 0.0013	= Horsepower (hp)

Velocity (speed)

Miles per hour (miles/hr; mph)	x 1.609	= Kilometres per hour (km/hr; kph)	x 0.621	= Miles per hour (miles/hr; mph)

Fuel consumption*

Miles per gallon (mpg)	x 0.354	= Kilometres per litre (km/l)	x 2.825	= Miles per gallon (mpg)

Temperature

Degrees Fahrenheit = (°C x 1.8) + 32   Degrees Celsius (Degrees Centigrade; °C) = (°F - 32) x 0.56

It is common practice to convert from miles per gallon (mpg) to litres/100 kilometres (l/100km), where mpg x l/100 km = 282

Motorcycle chemicals and lubricants REF•21

A number of chemicals and lubricants are available for use in motorcycle maintenance and repair. They include a wide variety of products ranging from cleaning solvents and degreasers to lubricants and protective sprays for rubber, plastic and vinyl.

● **Contact point/spark plug cleaner** is a solvent used to clean oily film and dirt from points, grime from electrical connectors and oil deposits from spark plugs. It is oil free and leaves no residue. It can also be used to remove gum and varnish from carburettor jets and other orifices.

● **Carburettor cleaner** is similar to contact point/spark plug cleaner but it usually has a stronger solvent and may leave a slight oily reside. It is not recommended for cleaning electrical components or connections.

● **Brake system cleaner** is used to remove grease or brake fluid from brake system components (where clean surfaces are absolutely necessary and petroleum-based solvents cannot be used); it also leaves no residue.

● **Silicone-based lubricants** are used to protect rubber parts such as hoses and grommets, and are used as lubricants for hinges and locks.

● **Multi-purpose grease** is an all purpose lubricant used wherever grease is more practical than a liquid lubricant such as oil. Some multi-purpose grease is coloured white and specially formulated to be more resistant to water than ordinary grease.

● **Gear oil** (sometimes called gear lube) is a specially designed oil used in transmissions and final drive units, as well as other areas where high friction, high temperature lubrication is required. It is available in a number of viscosities (weights) for various applications.

● **Motor oil**, of course, is the lubricant specially formulated for use in the engine. It normally contains a wide variety of additives to prevent corrosion and reduce foaming and wear. Motor oil comes in various weights (viscosity ratings) of from 5 to 80. The recommended weight of the oil depends on the seasonal temperature and the demands on the engine. Light oil is used in cold climates and under light load conditions; heavy oil is used in hot climates and where high loads are encountered. Multi-viscosity oils are designed to have characteristics of both light and heavy oils and are available in a number of weights from 5W-20 to 20W-50.

● **Petrol additives** perform several functions, depending on their chemical makeup. They usually contain solvents that help dissolve gum and varnish that build up on carburettor and inlet parts. They also serve to break down carbon deposits that form on the inside surfaces of the combustion chambers. Some additives contain upper cylinder lubricants for valves and piston rings.

● **Brake and clutch fluid** is a specially formulated hydraulic fluid that can withstand the heat and pressure encountered in brake/clutch systems. Care must be taken that this fluid does not come in contact with painted surfaces or plastics. An opened container should always be resealed to prevent contamination by water or dirt.

● **Chain lubricants** are formulated especially for use on motorcycle final drive chains. A good chain lube should adhere well and have good penetrating qualities to be effective as a lubricant inside the chain and on the side plates, pins and rollers. Most chain lubes are either the foaming type or quick drying type and are usually marketed as sprays. Take care to use a lubricant marked as being suitable for O-ring chains.

● **Degreasers** are heavy duty solvents used to remove grease and grime that may accumulate on engine and frame components. They can be sprayed or brushed on and, depending on the type, are rinsed with either water or solvent.

● **Solvents** are used alone or in combination with degreasers to clean parts and assemblies during repair and overhaul. The home mechanic should use only solvents that are non-flammable and that do not produce irritating fumes.

● **Gasket sealing compounds** may be used in conjunction with gaskets, to improve their sealing capabilities, or alone, to seal metal-to-metal joints. Many gasket sealers can withstand extreme heat, some are impervious to petrol and lubricants, while others are capable of filling and sealing large cavities. Depending on the intended use, gasket sealers either dry hard or stay relatively soft and pliable. They are usually applied by hand, with a brush, or are sprayed on the gasket sealing surfaces.

● **Thread locking compound** is an adhesive locking compound that prevents threaded fasteners from loosening because of vibration. It is available in a variety of types for different applications.

● **Moisture dispersants** are usually sprays that can be used to dry out electrical components such as the fuse block and wiring connectors. Some types can also be used as treatment for rubber and as a lubricant for hinges, cables and locks.

● **Waxes and polishes** are used to help protect painted and plated surfaces from the weather. Different types of paint may require the use of different types of wax polish. Some polishes utilise a chemical or abrasive cleaner to help remove the top layer of oxidised (dull) paint on older vehicles. In recent years, many non-wax polishes (that contain a wide variety of chemicals such as polymers and silicones) have been introduced. These non-wax polishes are usually easier to apply and last longer than conventional waxes and polishes.

About the MOT Test

In the UK, all vehicles more than three years old are subject to an annual test to ensure that they meet minimum safety requirements. A current test certificate must be issued before a machine can be used on public roads, and is required before a road fund licence can be issued. Riding without a current test certificate will also invalidate your insurance.

For most owners, the MOT test is an annual cause for anxiety, and this is largely due to owners not being sure what needs to be checked prior to submitting the motorcycle for testing. The simple answer is that a fully roadworthy motorcycle will have no difficulty in passing the test.

This is a guide to getting your motorcycle through the MOT test. Obviously it will not be possible to examine the motorcycle to the same standard as the professional MOT tester, particularly in view of the equipment required for some of the checks. However, working through the following procedures will enable you to identify any problem areas before submitting the motorcycle for the test.

It has only been possible to summarise the test requirements here, based on the regulations in force at the time of printing. Test standards are becoming increasingly stringent, although there are some exemptions for older vehicles. More information about the MOT test can be obtained from the HMSO publications, *How Safe is your Motorcycle* and *The MOT Inspection Manual for Motorcycle Testing*.

Many of the checks require that one of the wheels is raised off the ground. If the motorcycle doesn't have a centre stand, note that an auxiliary stand will be required. Additionally, the help of an assistant may prove useful.

Certain exceptions apply to machines under 50 cc, machines without a lighting system, and Classic bikes - if in doubt about any of the requirements listed below seek confirmation from an MOT tester prior to submitting the motorcycle for the test.

Check that the frame number is clearly visible.

> **HAYNES HiNT**
> *If a component is in borderline condition, the tester has discretion in deciding whether to pass or fail it. If the motorcycle presented is clean and evidently well cared for, the tester may be more inclined to pass a borderline component than if the motorcycle is scruffy and apparently neglected.*

Electrical System

Lights, turn signals, horn and reflector

✔ With the ignition on, check the operation of the following electrical components. **Note:** *The electrical components on certain small-capacity machines are powered by the generator, requiring that the engine is run for this check.*

a) *Headlight and tail light. Check that both illuminate in the low and high beam switch positions.*

b) *Position lights. Check that the front position (or sidelight) and tail light illuminate in this switch position.*

c) *Turn signals. Check that all flash at the correct rate, and that the warning light(s) function correctly. Check that the turn signal switch works correctly.*

c) *Hazard warning system (where fitted). Check that all four turn signals flash in this switch position.*

d) *Brake stop light. Check that the light comes on when the front and rear brakes are independently applied. Models first used on or after 1st April 1986 must have a brake light switch on each brake.*

e) *Horn. Check that the sound is continuous and of reasonable volume.*

✔ Check that there is a red reflector on the rear of the machine, either mounted separately or as part of the tail light lens.

✔ Check the condition of the headlight, tail light and turn signal lenses.

Headlight beam height

✔ The MOT tester will perform a headlight beam height check using specialised beam setting equipment **(see illustration 1)**. This equipment will not be available to the home mechanic, but if you suspect that the headlight is incorrectly set or may have been maladjusted in the past, you can perform a rough test as follows.

✔ Position the bike in a straight line facing a brick wall. The bike must be off its stand, upright and with a rider seated. Measure the height from the ground to the centre of the headlight and mark a horizontal line on the wall at this height. Position the motorcycle 3.8 metres from the wall and draw a vertical

Headlight beam height checking equipment

line up the wall central to the centreline of the motorcycle. Switch to dipped beam and check that the beam pattern falls slightly lower than the horizontal line and to the left of the vertical line **(see illustration 2)**.

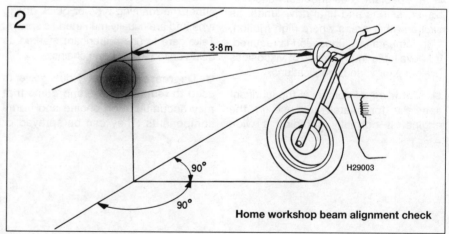

3·8 m

90°

90°

H29003

Home workshop beam alignment check

Exhaust System and Final Drive

Exhaust

✔ Check that the exhaust mountings are secure and that the system does not foul any of the rear suspension components.
✔ Start the motorcycle. When the revs are increased, check that the exhaust is neither holed nor leaking from any of its joints. On a linked system, check that the collector box is not leaking due to corrosion.

✔ Note that the exhaust decibel level ("loudness" of the exhaust) is assessed at the discretion of the tester. If the motorcycle was first used on or after 1st January 1985 the silencer must carry the BSAU 193 stamp, or a marking relating to its make and model, or be of OE (original equipment) manufacture. If the silencer is marked NOT FOR ROAD USE, RACING USE ONLY or similar, it will fail the MOT.

Final drive

✔ On chain or belt drive machines, check that the chain/belt is in good condition and does not have excessive slack. Also check that the sprocket is securely mounted on the rear wheel hub. Check that the chain/belt guard is in place.
✔ On shaft drive bikes, check for oil leaking from the drive unit and fouling the rear tyre.

Steering and Suspension

Steering

✔ With the front wheel raised off the ground, rotate the steering from lock to lock. The handlebar or switches must not contact the fuel tank or be close enough to trap the rider's hand. Problems can be caused by damaged lock stops on the lower yoke and frame, or by the fitting of non-standard handlebars.
✔ When performing the lock to lock check, also ensure that the steering moves freely without drag or notchiness. Steering movement can be impaired by poorly routed cables, or by overtight head bearings or worn bearings. The

tester will perform a check of the steering head bearing lower race by mounting the front wheel on a surface plate, then performing a lock to lock check with the weight of the machine on the lower bearing (**see illustration 3**).
✔ Grasp the fork sliders (lower legs) and attempt to push and pull on the forks (**see illustration 4**). Any play in the steering head bearings will be felt. Note that in extreme cases, wear of the front fork bushes can be misinterpreted for head bearing play.
✔ Check that the handlebars are securely mounted.
✔ Check that the handlebar grip rubbers are secure. They should by bonded to the bar left end and to the throttle cable pulley on the right end.

Front suspension

✔ With the motorcycle off the stand, hold the front brake on and pump the front forks up and down (**see illustration 5**). Check that they are adequately damped.
✔ Inspect the area above and around the front fork oil seals (**see illustration 6**). There should be no sign of oil on the fork tube (stanchion) nor leaking down the slider (lower leg). On models so equipped, check that there is no oil leaking from the anti-dive units.
✔ On models with swingarm front suspension, check that there is no freeplay in the linkage when moved from side to side.

Rear suspension

✔ With the motorcycle off the stand and an assistant supporting the motorcycle by its handlebars, bounce the rear suspension (**see illustration 7**). Check that the suspension components do not foul on any of the cycle parts and check that the shock absorber(s) provide adequate damping.

3 Front wheel mounted on a surface plate for steering head bearing lower race check

4 Checking the steering head bearings for freeplay

5 Hold the front brake on and pump the front forks up and down to check operation

6 Inspect the area around the fork dust seal for oil leakage (arrow)

7 Bounce the rear of the motorcycle to check rear suspension operation

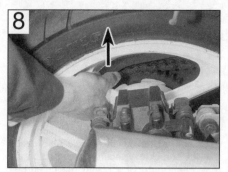

Checking for rear suspension linkage play

Worn suspension linkage pivots (arrows) are usually the cause of play in the rear suspension

Grasp the swingarm at the ends to check for play in its pivot bearings

✔ Visually inspect the shock absorber(s) and check that there is no sign of oil leakage from its damper. This is somewhat restricted on certain single shock models due to the location of the shock absorber.

✔ With the rear wheel raised off the ground, grasp the wheel at the highest point and attempt to pull it up (see illustration 8). Any play in the swingarm pivot or suspension linkage bearings will be felt as movement. Note: *Do not confuse play with actual suspension movement.* Failure to lubricate suspension linkage bearings can lead to bearing failure (see illustration 9).

✔ With the rear wheel raised off the ground, grasp the swingarm ends and attempt to move the swingarm from side to side and forwards and backwards - any play indicates wear of the swingarm pivot bearings (see illustration 10).

Brakes, Wheels and Tyres

Brakes

✔ With the wheel raised off the ground, apply the brake then free it off, and check that the wheel is about to revolve freely without brake drag.

✔ On disc brakes, examine the disc itself. Check that it is securely mounted and not cracked.

✔ On disc brakes, view the pad material through the caliper mouth and check that the pads are not worn down beyond the limit (see illustration 11).

✔ On drum brakes, check that when the brake is applied the angle between the operating lever and cable or rod is not too great (see illustration 12). Check also that the operating lever doesn't foul any other components.

✔ On disc brakes, examine the flexible hoses from top to bottom. Have an assistant hold the brake on so that the fluid in the hose is under pressure, and check that there is no sign of fluid leakage, bulges or cracking. If there are any metal brake pipes or unions, check that these are free from corrosion and damage. Where a brake-linked anti-dive system is fitted, check the hoses to the anti-dive in a similar manner.

✔ Check that the rear brake torque arm is secure and that its fasteners are secured by self-locking nuts or castellated nuts with split-pins or R-pins (see illustration 13).

✔ On models with ABS, check that the self-check warning light in the instrument panel works.

✔ The MOT tester will perform a test of the motorcycle's braking efficiency based on a calculation of rider and motorcycle weight. Although this cannot be carried out at home, you can at least ensure that the braking systems are properly maintained. For hydraulic disc brakes, check the fluid level, lever/pedal feel (bleed of air if its spongy) and pad material. For drum brakes, check adjustment, cable or rod operation and shoe lining thickness.

Wheels and tyres

✔ Check the wheel condition. Cast wheels should be free from cracks and if of the built-up design, all fasteners should be secure. Spoked wheels should be checked for broken, corroded, loose or bent spokes.

✔ With the wheel raised off the ground, spin the wheel and visually check that the tyre and wheel run true. Check that the tyre does not foul the suspension or mudguards.

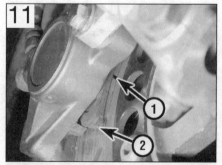

Brake pad wear can usually be viewed without removing the caliper. Most pads have wear indicator grooves (1) and some also have indicator tangs (2)

On drum brakes, check the angle of the operating lever with the brake fully applied. Most drum brakes have a wear indicator pointer and scale.

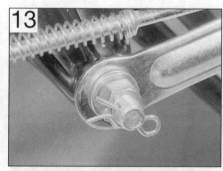

Brake torque arm must be properly secured at both ends

Check for wheel bearing play by trying to move the wheel about the axle (spindle)

Checking the tyre tread depth

Tyre direction of rotation arrow can be found on tyre sidewall

Castellated type wheel axle (spindle) nut must be secured by a split pin or R-pin

18

Two straightedges are used to check wheel alignment

✔ With the wheel raised off the ground, grasp the wheel and attempt to move it about the axle (spindle) **(see illustration 14)**. Any play felt here indicates wheel bearing failure.
✔ Check the tyre tread depth, tread condition and sidewall condition **(see illustration 15)**.
✔ Check the tyre type. Front and rear tyre types must be compatible and be suitable for road use. Tyres marked NOT FOR ROAD USE, COMPETITION USE ONLY or similar, will fail the MOT.
✔ If the tyre sidewall carries a direction of rotation arrow, this must be pointing in the direction of normal wheel rotation **(see illustration 16)**.
✔ Check that the wheel axle (spindle) nuts (where applicable) are properly secured. A self-locking nut or castellated nut with a split-pin or R-pin can be used **(see illustration 17)**.
✔ Wheel alignment is checked with the motorcycle off the stand and a rider seated. With the front wheel pointing straight ahead, two perfectly straight lengths of metal or wood and placed against the sidewalls of both tyres **(see illustration 18)**. The gap each side of the front tyre must be equidistant on both sides. Incorrect wheel alignment may be due to a cocked rear wheel (often as the result of poor chain adjustment) or in extreme cases, a bent frame.

General checks and condition

✔ Check the security of all major fasteners, bodypanels, seat, fairings (where fitted) and mudguards.

✔ Check that the rider and pillion footrests, handlebar levers and brake pedal are securely mounted.

✔ Check for corrosion on the frame or any load-bearing components. If severe, this may affect the structure, particularly under stress.

Sidecars

A motorcycle fitted with a sidecar requires additional checks relating to the stability of the machine and security of attachment and swivel joints, plus specific wheel alignment (toe-in) requirements. Additionally, tyre and lighting requirements differ from conventional motorcycle use. Owners are advised to check MOT test requirements with an official test centre.

Preparing for storage

Before you start

If repairs or an overhaul is needed, see that this is carried out now rather than left until you want to ride the bike again.

Give the bike a good wash and scrub all dirt from its underside. Make sure the bike dries completely before preparing for storage.

Engine

● Remove the spark plug(s) and lubricate the cylinder bores with approximately a teaspoon of motor oil using a spout-type oil can **(see illustration 1)**. Reinstall the spark plug(s). Crank the engine over a couple of times to coat the piston rings and bores with oil. If the bike has a kickstart, use this to turn the engine over. If not, flick the kill switch to the OFF position and crank the engine over on the starter **(see illustration 2)**. If the nature on the ignition system prevents the starter operating with the kill switch in the OFF position,

Squirt a drop of motor oil into each cylinder

Flick the kill switch to OFF . . .

. . . and ensure that the metal bodies of the plugs (arrows) are earthed against the cylinder head

remove the spark plugs and fit them back in their caps; ensure that the plugs are earthed (grounded) against the cylinder head when the starter is operated **(see illustration 3)**.

 Warning: It is important that the plugs are earthed (grounded) away from the spark plug holes otherwise there is a risk of atomised fuel from the cylinders igniting.

 On a single cylinder four-stroke engine, you can seal the combustion chamber completely by positioning the piston at TDC on the compression stroke.

Connect a hose to the carburettor float chamber drain stub (arrow) and unscrew the drain screw

● Drain the carburettor(s) otherwise there is a risk of jets becoming blocked by gum deposits from the fuel **(see illustration 4)**.

● If the bike is going into long-term storage, consider adding a fuel stabiliser to the fuel in the tank. If the tank is drained completely, corrosion of its internal surfaces may occur if left unprotected for a long period. The tank can be treated with a rust preventative especially for this purpose. Alternatively, remove the tank and pour half a litre of motor oil into it, install the filler cap and shake the tank to coat its internals with oil before draining off the excess. The same effect can also be achieved by spraying WD40 or a similar water-dispersant around the inside of the tank via its flexible nozzle.

● Make sure the cooling system contains the correct mix of antifreeze. Antifreeze also contains important corrosion inhibitors.

● The air intakes and exhaust can be sealed off by covering or plugging the openings. Ensure that you do not seal in any condensation; run the engine until it is hot, then switch off and allow to cool. Tape a piece of thick plastic over the silencer end(s) **(see illustration 5)**. Note that some advocate pouring a tablespoon of motor oil into the silencer(s) before sealing them off.

Exhausts can be sealed off with a plastic bag

Battery

● Remove it from the bike - in extreme cases of cold the battery may freeze and crack its case **(see illustration 6)**.

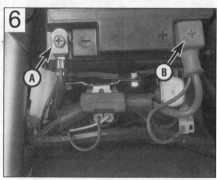
Disconnect the negative lead (A) first, followed by the positive lead (B)

● Check the electrolyte level and top up if necessary (conventional refillable batteries). Clean the terminals.
● Store the battery off the motorcycle and away from any sources of fire. Position a wooden block under the battery if it is to sit on the ground.
● Give the battery a trickle charge for a few hours every month **(see illustration 7)**.

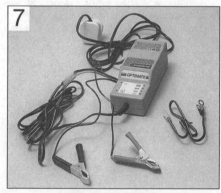
Use a suitable battery charger - this kit also assess battery condition

Tyres

● Place the bike on its centrestand or an auxiliary stand which will support the motorcycle in an upright position. Position wood blocks under the tyres to keep them off the ground and to provide insulation from damp. If the bike is being put into long-term storage, ideally both tyres should be off the ground; not only will this protect the tyres, but will also ensure that no load is placed on the steering head or wheel bearings.
● Deflate each tyre by 5 to 10 psi, no more or the beads may unseat from the rim, making subsequent inflation difficult on tubeless tyres.

Pivots and controls

● Lubricate all lever, pedal, stand and footrest pivot points. If grease nipples are fitted to the rear suspension components, apply lubricant to the pivots.
● Lubricate all control cables.

Cycle components

● Apply a wax protectant to all painted and plastic components. Wipe off any excess, but don't polish to a shine. Where fitted, clean the screen with soap and water.
● Coat metal parts with Vaseline (petroleum jelly). When applying this to the fork tubes, do not compress the forks otherwise the seals will rot from contact with the Vaseline.
● Apply a vinyl cleaner to the seat.

Storage conditions

● Aim to store the bike in a shed or garage which does not leak and is free from damp.
● Drape an old blanket or bedspread over the bike to protect it from dust and direct contact with sunlight (which will fade paint). This also hides the bike from prying eyes. Beware of tight-fitting plastic covers which may allow condensation to form and settle on the bike.

Getting back on the road

Engine and transmission

● Change the oil and replace the oil filter. If this was done prior to storage, check that the oil hasn't emulsified - a thick whitish substance which occurs through condensation.
● Remove the spark plugs. Using a spout-type oil can, squirt a few drops of oil into the cylinder(s). This will provide initial lubrication as the piston rings and bores comes back into contact. Service the spark plugs, or fit new ones, and install them in the engine.
● Check that the clutch isn't stuck on. The plates can stick together if left standing for some time, preventing clutch operation. Engage a gear and try rocking the bike back and forth with the clutch lever held against the handlebar. If this doesn't work on cable-operated clutches, hold the clutch lever back against the handlebar with a strong elastic band or cable tie for a couple of hours (see illustration 8).

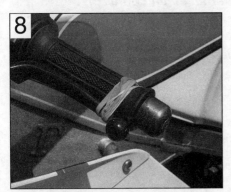

Hold clutch lever back against the handlebar with elastic bands or a cable tie

● If the air intakes or silencer end(s) were blocked off, remove the bung or cover used.
● If the fuel tank was coated with a rust preventative, oil or a stabiliser added to the fuel, drain and flush the tank and dispose of the fuel sensibly. If no action was taken with the fuel tank prior to storage, it is advised that the old fuel is disposed of since it will go off over a period of time. Refill the fuel tank with fresh fuel.

Frame and running gear

● Oil all pivot points and cables.
● Check the tyre pressures. They will definitely need inflating if pressures were reduced for storage.
● Lubricate the final drive chain (where applicable).
● Remove any protective coating applied to the fork tubes (stanchions) since this may well destroy the fork seals. If the fork tubes weren't protected and have picked up rust spots, remove them with very fine abrasive paper and refinish with metal polish.
● Check that both brakes operate correctly. Apply each brake hard and check that it's not possible to move the motorcycle forwards, then check that the brake frees off again once released. Brake caliper pistons can stick due to corrosion around the piston head, or on the sliding caliper types, due to corrosion of the slider pins. If the brake doesn't free after repeated operation, take the caliper off for examination. Similarly drum brakes can stick due to a seized operating cam, cable or rod linkage.
● If the motorcycle has been in long-term storage, renew the brake fluid and clutch fluid (where applicable).
● Depending on where the bike has been stored, the wiring, cables and hoses may have been nibbled by rodents. Make a visual check and investigate disturbed wiring loom tape.

Battery

● If the battery has been previously removal and given top up charges it can simply be reconnected. Remember to connect the positive cable first and the negative cable last.
● On conventional refillable batteries, if the battery has not received any attention, remove it from the motorcycle and check its electrolyte level. Top up if necessary then charge the battery. If the battery fails to hold a charge and a visual checks show heavy white sulphation of the plates, the battery is probably defective and must be renewed. This is particularly likely if the battery is old. Confirm battery condition with a specific gravity check.
● On sealed (MF) batteries, if the battery has not received any attention, remove it from the motorcycle and charge it according to the information on the battery case - if the battery fails to hold a charge it must be renewed.

Starting procedure

● If a kickstart is fitted, turn the engine over a couple of times with the ignition OFF to distribute oil around the engine. If no kickstart is fitted, flick the engine kill switch OFF and the ignition ON and crank the engine over a couple of times to work oil around the upper cylinder components. If the nature of the ignition system is such that the starter won't work with the kill switch OFF, remove the spark plugs, fit them back into their caps and earth (ground) their bodies on the cylinder head. Reinstall the spark plugs afterwards.
● Switch the kill switch to RUN, operate the choke and start the engine. If the engine won't start don't continue cranking the engine - not only will this flatten the battery, but the starter motor will overheat. Switch the ignition off and try again later. If the engine refuses to start, go through the fault finding procedures in this manual. **Note:** *If the bike has been in storage for a long time, old fuel or a carburettor blockage may be the problem. Gum deposits in carburettors can block jets - if a carburettor cleaner doesn't prove successful the carburettors must be dismantled for cleaning.*
● Once the engine has started, check that the lights, turn signals and horn work properly.
● Treat the bike gently for the first ride and check all fluid levels on completion. Settle the bike back into the maintenance schedule.

This Section provides an easy reference-guide to the more common faults that are likely to afflict your machine. Obviously, the opportunities are almost limitless for faults to occur as a result of obscure failures, and to try and cover all eventualities would require a book. Indeed, a number have been written on the subject.

Successful troubleshooting is not a mysterious 'black art' but the application of a bit of knowledge combined with a systematic and logical approach to the problem. Approach any troubleshooting by first accurately identifying the symptom and then checking through the list of possible causes, starting with the simplest or most obvious and progressing in stages to the most complex.

Take nothing for granted, but above all apply liberal quantities of common sense.

The main symptom of a fault is given in the text as a major heading below which are listed the various systems or areas which may contain the fault. Details of each possible cause for a fault and the remedial action to be taken are given, in brief, in the paragraphs below each heading. Further information should be sought in the relevant Chapter.

1 Engine doesn't start or is difficult to start

- [ ] Starter motor doesn't rotate
- [ ] Starter motor rotates but engine does not turn over
- [ ] Starter works but engine won't turn over (seized)
- [ ] No fuel flow
- [ ] Engine flooded
- [ ] No spark or weak spark
- [ ] Compression low
- [ ] Stalls after starting
- [ ] Rough idle

2 Poor running at low speed

- [ ] Spark weak
- [ ] Fuel/air mixture incorrect
- [ ] Compression low
- [ ] Poor acceleration

3 Poor running or no power at high speed

- [ ] Firing incorrect
- [ ] Fuel/air mixture incorrect
- [ ] Compression low
- [ ] Knocking or pinging
- [ ] Miscellaneous causes

4 Overheating

- [ ] Engine overheats
- [ ] Firing incorrect
- [ ] Fuel/air mixture incorrect
- [ ] Compression too high
- [ ] Engine load excessive
- [ ] Lubrication inadequate
- [ ] Miscellaneous causes

5 Clutch problems

- [ ] Clutch slipping
- [ ] Clutch not disengaging completely

6 Gearchanging problems

- [ ] Doesn't go into gear, or lever doesn't return
- [ ] Jumps out of gear
- [ ] Overshifts

7 Abnormal engine noise

- [ ] Knocking or pinking
- [ ] Piston slap or rattling
- [ ] Valve noise
- [ ] Other noise

8 Abnormal driveline noise

- [ ] Clutch noise
- [ ] Transmission noise
- [ ] Final drive noise

9 Abnormal frame and suspension noise

- [ ] Front end noise
- [ ] Shock absorber noise
- [ ] Brake noise

10 Oil pressure warning light comes on

- [ ] Engine lubrication system
- [ ] Electrical system

11 Excessive exhaust smoke

- [ ] White smoke
- [ ] Black smoke
- [ ] Brown smoke

12 Poor handling or stability

- [ ] Handlebar hard to turn
- [ ] Handlebar shakes or vibrates excessively
- [ ] Handlebar pulls to one side
- [ ] Poor shock absorbing qualities

13 Braking problems

- [ ] Brakes are spongy, don't hold
- [ ] Brake lever or pedal pulsates
- [ ] Brakes drag

14 Electrical problems

- [ ] Battery dead or weak
- [ ] Battery overcharged

1 Engine doesn't start or is difficult to start

Starter motor doesn't rotate

- [ ] Engine kill switch OFF.
- [ ] Fuse blown. Check fuse (Chapter 8).
- [ ] Battery voltage low. Check and recharge battery (Chapter 8).
- [ ] Starter motor defective. Make sure the wiring to the starter is secure. Make sure the starter relay clicks when the start button is pushed. If the relay clicks, then the fault is in the wiring or motor.
- [ ] Starter relay faulty. Check it according to the procedure in Chapter 8.
- [ ] Starter switch not contacting. The contacts could be wet, corroded or dirty. Disassemble and clean the switch (Chapter 8).
- [ ] Wiring open or shorted. Check all wiring connections and harnesses to make sure that they are dry, tight and not corroded. Also check for broken or frayed wires that can cause a short to ground (earth) (see wiring diagram, Chapter 8).
- [ ] Ignition (main) switch defective. Check the switch according to the procedure in Chapter 8. Replace the switch with a new one if it is defective.
- [ ] Engine kill switch defective. Check for wet, dirty or corroded contacts. Clean or replace the switch as necessary (Chapter 8).
- [ ] Faulty neutral, side stand or clutch switch. Check the wiring to each switch and the switch itself according to the procedures in Chapter 8.

Starter motor rotates but engine does not turn over

- [ ] Starter motor solenoid or lever arm defective. Inspect and repair or replace (Chapter 8).
- [ ] Damaged starter pinion or flywheel. Inspect and replace the damaged parts (Chapters 2 and 8).

Starter works but engine won't turn over (seized)

- [ ] Seized engine caused by one or more internally damaged components. Failure due to wear, abuse or lack of lubrication. Damage can include seized valves, camshafts, pistons, crankshaft, connecting rod bearings, or transmission gears or bearings. Refer to Chapter 2 for engine disassembly.

No fuel flow

- [ ] No fuel in tank.
- [ ] Fuel pump failure or filter blockage (see Chapters 8 and 1 respectively).
- [ ] Fuel tank breather hose obstructed.
- [ ] Fuel line clogged. Pull the fuel line loose and carefully blow through it.
- [ ] Injector clogged. For the injectors to be clogged, either a very bad batch of fuel with an unusual additive has been used, or some other foreign material has entered the system. Many times after a machine has been stored for many months without running, the fuel turns to a varnish-like liquid and forms deposits.

Engine flooded

- [ ] Starting technique incorrect. Under normal circumstances the machine should start with little or no throttle. When the engine is cold, the choke should be operated and the engine started without opening the throttle. When the engine is at operating temperature, only a very slight amount of throttle should be necessary. If the engine is flooded, hold the throttle open while cranking the engine. This will allow additional air to reach the cylinders.

No spark or weak spark

- [ ] Ignition switch OFF.
- [ ] Engine kill switch turned to the OFF position.
- [ ] Battery voltage low. Check and recharge the battery as necessary (Chapter 8).

- [ ] Spark plugs dirty, defective or worn out. Locate reason for fouled plugs using spark plug condition chart and follow the plug maintenance procedures (Chapter 1).
- [ ] Spark plug caps or secondary (HT) wiring faulty. Check condition. Replace either or both components if cracks or deterioration are evident (Chapter 4).
- [ ] Spark plug caps not making good contact. Make sure that the plug caps fit snugly over the plug ends.
- [ ] Motronic control unit defective. Refer to a BMW dealer equipped with the diagnostic tester.
- [ ] Timing sensor defective. Check the unit, referring to Chapter 4 for details.
- [ ] Ignition HT coils defective. Check the coils, referring to Chapter 4.
- [ ] Ignition or kill switch shorted. This is usually caused by water, corrosion, damage or excessive wear. The switches can be disassembled and cleaned with electrical contact cleaner. If cleaning does not help, replace the switches (Chapter 8).
- [ ] Wiring shorted or broken between:
  - a) Ignition (main) switch and engine kill switch (or blown fuse)
  - b) Motronic control unit and engine kill switch
  - c) Motronic control unit and ignition HT coils
  - d) Ignition HT coils and spark plugs
  - e) Motronic control unit and timing sensor
- [ ] Make sure that all wiring connections are clean, dry and tight. Look for chafed and broken wires (Chapters 4 and 8).

Compression low

- [ ] Spark plugs loose. Remove the plugs and inspect their threads. Reinstall and tighten to the specified torque (Chapter 1).
- [ ] Cylinder head not sufficiently tightened down. If the cylinder head is suspected of being loose, then there's a chance that the gasket or head is damaged if the problem has persisted for any length of time. The head nuts should be tightened to the proper torque in the correct sequence (Chapter 2).
- [ ] Improper valve clearance. This means that the valve is not closing completely and compression pressure is leaking past the valve. Check and adjust the valve clearances (Chapter 1).
- [ ] Cylinder and/or piston worn. Excessive wear will cause compression pressure to leak past the rings. This is usually accompanied by worn rings as well. A top-end overhaul is necessary (Chapter 2).
- [ ] Piston rings worn, weak, broken, or sticking. Broken or sticking piston rings usually indicate a lubrication or carburation problem that causes excess carbon deposits or seizures to form on the pistons and rings. Top-end overhaul is necessary (Chapter 2).
- [ ] Piston ring-to-groove clearance excessive. This is caused by excessive wear of the piston ring lands. Piston replacement is necessary (Chapter 2).
- [ ] Cylinder head gasket damaged. If the head is allowed to become loose, or if excessive carbon build-up on the piston crown and combustion chamber causes extremely high compression, the head gasket may leak. Retorquing the head is not always sufficient to restore the seal, so gasket replacement is necessary (Chapter 2).
- [ ] Cylinder head warped. This is caused by overheating or improperly tightened head bolts. Machine shop resurfacing or head replacement is necessary (Chapter 2).
- [ ] Valve spring broken or weak. Caused by component failure or wear; the springs must be replaced (Chapter 2).
- [ ] Valve not seating properly. This is caused by a bent valve (from over-revving or improper valve adjustment), burned valve or seat or an accumulation of carbon deposits on the seat. The valves must be cleaned and/or replaced and the seats serviced (Chapter 2).

1 Engine doesn't start or is difficult to start (continued)

Stalls after starting

☐ Improper choke action. Check the cable (Chapter 3).
☐ Ignition malfunction. See Chapter 4.
☐ Injector malfunction. See Chapter 3.
☐ Fuel contaminated. The fuel can be contaminated with either dirt or water, or can change chemically if the machine is allowed to sit for several months or more. Drain the tank and fuel pipes (Chapter 3).
☐ Intake air leak. Check for loose throttle body-to-intake manifold and air duct connections, loose or missing vacuum gauge caps, or loose injectors (Chapter 3).
☐ Engine idle speed incorrect. Turn idle adjusting screw until the engine idles at the specified rpm (Chapter 1).

Rough idle

☐ Ignition malfunction. See Chapter 4.
☐ Idle speed incorrect. See Chapter 1.
☐ Throttles not synchronised. Adjust throttle body synchronisation as described in Chapter 1.
☐ Injector malfunction. See Chapter 3.
☐ Fuel contaminated. The fuel can be contaminated with either dirt or water, or can change chemically if the machine is allowed to sit for several months or more. Drain the tank (Chapter 3).
☐ Intake air leak. Check for loose throttle body-to-intake manifold and air duct connections, loose or missing vacuum gauge caps, or loose injectors (Chapter 3).
☐ Air filter clogged. Replace the air filter element (Chapter 1).

2 Poor running at low speeds

Spark weak

☐ Battery voltage low. Check and recharge battery (Chapter 8).
☐ Spark plugs fouled, defective or worn out. Refer to Chapter 1 for spark plug maintenance.
☐ Spark plug cap or HT wiring defective. Refer to Chapters 1 and 4 for details on the ignition system.
☐ Spark plug caps not making contact.
☐ Incorrect spark plugs. Wrong type, heat range or cap configuration. Check and install correct plugs listed in Chapter 1.
☐ Motronic control unit defective. Refer to a BMW dealer equipped with the diagnostic tester.
☐ Timing sensor defective. See Chapter 4.
☐ Ignition HT coils defective. See Chapter 4.

Fuel/air mixture incorrect

☐ Injector clogged. Dirt, water or other contaminants can clog the injectors. Renew the fuel filter (Chapter 1).
☐ Air filter clogged, poorly sealed or missing (Chapter 1).
☐ Air filter housing poorly sealed. Look for cracks, holes or loose clamps and replace or repair defective parts.
☐ Fuel tank breather hose obstructed.
☐ Intake air leak. Check for loose throttle body-to-intake manifold and air duct connections, loose or missing vacuum take-off caps, or loose injectors (Chapter 3).

Compression low

☐ Spark plugs loose. Remove the plugs and inspect their threads. Reinstall and tighten to the specified torque (Chapter 1).
☐ Cylinder head not sufficiently tightened down. If the cylinder head is suspected of being loose, then there's a chance that the gasket and head are damaged if the problem has persisted for any length of time. The head nuts should be tightened to the proper torque in the correct sequence (Chapter 2).
☐ Improper valve clearance. This means that the valve is not closing completely and compression pressure is leaking past the valve. Check and adjust the valve clearances (Chapter 1).
☐ Cylinder and/or piston worn. Excessive wear will cause compression pressure to leak past the rings. This is usually

accompanied by worn rings as well. A top end overhaul is necessary (Chapter 2).
☐ Piston rings worn, weak, broken, or sticking. Broken or sticking piston rings usually indicate a lubrication or mixture problem that causes excess carbon deposits or seizures to form on the pistons and rings. Top-end overhaul is necessary (Chapter 2).
☐ Piston ring-to-groove clearance excessive. This is caused by excessive wear of the piston ring lands. Piston replacement is necessary (Chapter 2).
☐ Cylinder head gasket damaged. If the head is allowed to become loose, or if excessive carbon build-up on the piston crown and combustion chamber causes extremely high compression, the head gasket may leak. Retorquing the head is not always sufficient to restore the seal, so gasket replacement is necessary (Chapter 2).
☐ Cylinder head warped. This is caused by overheating or improperly tightened head nuts. Machine shop resurfacing or head replacement is necessary (Chapter 2).
☐ Valve spring broken or weak. Caused by component failure or wear; the springs must be replaced (Chapter 2).
☐ Valve not seating properly. This is caused by a bent valve (from over-revving or improper valve adjustment), burned valve or seat or an accumulation of carbon deposits on the seat. The valves must be cleaned and/or replaced and the seats serviced (Chapter 2).

Poor acceleration

☐ Timing not advancing. The timing sensor or the Motronic control unit may be defective. Refer to a BMW dealer equipped with the diagnostic tester.
☐ Throttle bodies not synchronised. Adjust them with a vacuum gauge set or manometer (Chapter 1).
☐ Engine oil viscosity too high. Using a heavier oil than that recommended in Chapter 1 can damage the oil pump or lubrication system and cause drag on the engine.
☐ Brakes dragging. Usually caused by debris which has entered the brake piston seals, or from a warped disc or bent axle. Repair as necessary (Chapter 6).

3 Poor running or no power at high speed

Firing incorrect

- [ ] Air filter restricted. Clean or replace filter (Chapter 1).
- [ ] Spark plugs fouled, defective or worn out. See Chapter 1 for spark plug maintenance.
- [ ] Spark plug caps or HT wiring defective. See Chapters 1 and 4 for details of the ignition system.
- [ ] Spark plug caps not in good contact.
- [ ] Incorrect spark plugs. Wrong type, heat range or cap configuration. Check and install correct plugs listed in Chapter 1.
- [ ] Motronic control unit defective. Refer to a BMW dealer equipped with the diagnostic tester.
- [ ] Ignition coils defective. See Chapter 4.

Fuel/air mixture incorrect

- [ ] Injector clogged. Dirt, water or other contaminants can clog the injectors. Renew the fuel filter (Chapter 1).
- [ ] Air filter clogged, poorly sealed, or missing (Chapter 1).
- [ ] Air filter housing poorly sealed. Look for cracks, holes or loose clamps, and replace or repair defective parts.
- [ ] Fuel tank breather hose obstructed.
- [ ] Intake air leak. Check for loose throttle body-to-intake manifold and air duct connections, loose or missing vacuum take-off caps, or loose injectors (Chapter 3).

Compression low

- [ ] Spark plugs loose. Remove the plugs and inspect their threads. Reinstall and tighten to the specified torque (Chapter 1).
- [ ] Cylinder head not sufficiently tightened down. If the cylinder head is suspected of being loose, then there's a chance that the gasket and head are damaged if the problem has persisted for any length of time. The head nuts should be tightened to the proper torque in the correct sequence (Chapter 2).
- [ ] Improper valve clearance. This means that the valve is not closing completely and compression pressure is leaking past the valve. Check and adjust the valve clearances (Chapter 1).
- [ ] Cylinder and/or piston worn. Excessive wear will cause compression pressure to leak past the rings. This is usually accompanied by worn rings as well. A top-end overhaul is necessary (Chapter 2).
- [ ] Piston rings worn, weak, broken, or sticking. Broken or sticking piston rings usually indicate a lubrication or mixture problem that causes excess carbon deposits or seizures to form on the pistons and rings. Top-end overhaul is necessary (Chapter 2).
- [ ] Piston ring-to-groove clearance excessive. This is caused by excessive wear of the piston ring lands. Piston replacement is necessary (Chapter 2).

- [ ] Cylinder head gasket damaged. If the head is allowed to become loose, or if excessive carbon build-up on the piston crown and combustion chamber causes extremely high compression, the head gasket may leak. Retorquing the head is not always sufficient to restore the seal, so gasket replacement is necessary (Chapter 2).
- [ ] Cylinder head warped. This is caused by overheating or improperly tightened head nuts. Machine shop resurfacing or head replacement is necessary (Chapter 2).
- [ ] Valve spring broken or weak. Caused by component failure or wear; the springs must be replaced (Chapter 2).
- [ ] Valve not seating properly. This is caused by a bent valve (from over-revving or improper valve adjustment), burned valve or seat or an accumulation of carbon deposits on the seat. The valves must be cleaned and/or replaced and the seats serviced (Chapter 2).

Knocking or pinking

- [ ] Carbon build-up in combustion chamber. Use of a fuel additive that will dissolve the adhesive bonding the carbon particles to the crown and chamber is the easiest way to remove the build-up. Otherwise, the cylinder head will have to be removed and decarbonised (Chapter 2).
- [ ] Incorrect or poor quality fuel. Old or improper grades of fuel can cause detonation. This causes the piston to rattle, thus the knocking or pinking sound. Drain old fuel and always use the recommended fuel grade.
- [ ] Spark plug heat range incorrect. Uncontrolled detonation indicates the plug heat range is too hot. The plug in effect becomes a glow plug, raising cylinder temperatures. Install the proper heat range plug (Chapter 1).
- [ ] Improper air/fuel mixture. This will cause the cylinder to run hot, which leads to detonation. Refer to a BMW dealer equipped with the diagnostic tester.

Miscellaneous causes

- [ ] Throttle body valve doesn't open fully. Adjust the throttle cable freeplay and throttle body synchronisation (Chapter 1).
- [ ] Clutch slipping. May be caused by loose or worn clutch components. Refer to Chapter 2 for clutch overhaul procedures.
- [ ] Timing not advancing. Faulty Motronic control unit – refer to a BMW dealer equipped with the diagnostic tester.
- [ ] Engine oil viscosity too high. Using a heavier oil than the one recommended in Chapter 1 can damage the oil pump or lubrication system and cause drag on the engine.
- [ ] Brakes dragging. Usually caused by debris which has entered the brake piston seals, or from a warped disc or bent axle. Repair as necessary.

4 Overheating

Engine overheats

- ☐ Oil cooling circuit defective. Check the oil coolers, hoses and the oil pump (see Chapter 2).

Firing incorrect

- ☐ Spark plugs fouled, defective or worn out. See Chapter 1 for spark plug maintenance.
- ☐ Incorrect spark plugs.
- ☐ Motronic control unit defective. Refer to a BMW dealer equipped with the diagnostic tester.
- ☐ Faulty ignition HT coils (Chapter 4).

Fuel/air mixture incorrect

- ☐ Injector clogged. Dirt, water and other contaminants can clog the injectors. Renew the fuel filter (Chapter 1).
- ☐ Air filter clogged, poorly sealed or missing (Chapter 1).
- ☐ Air filter housing poorly sealed. Look for cracks, holes or loose clamps and replace or repair.
- ☐ Fuel tank breather hose obstructed
- ☐ Intake air leak. Check for loose throttle body-to-intake manifold and air duct connections, loose or missing vacuum take-off caps, or loose injectors (Chapter 3).

Compression too high

- ☐ Carbon build-up in combustion chamber. Use of a fuel additive that will dissolve the adhesive bonding the carbon particles to the piston crown and chamber is the easiest way to remove the build-up. Otherwise, the cylinder head will have to be removed and decarbonised (Chapter 2).
- ☐ Improperly machined head surface or installation of incorrect gasket during engine assembly.

Engine load excessive

- ☐ Clutch slipping. Can be caused by damaged, loose or worn clutch components. Refer to Chapter 2 for overhaul procedures.
- ☐ Engine oil level too high. The addition of too much oil will cause pressurisation of the crankcase and inefficient engine operation. Check Specifications and drain to proper level (Chapter 1).
- ☐ Engine oil viscosity too high. Using a heavier oil than the one recommended in Chapter 1 can damage the oil pump or lubrication system as well as cause drag on the engine.
- ☐ Brakes dragging. Usually caused by debris which has entered the brake piston seals, or from a warped disc or bent axle. Repair as necessary.

Lubrication inadequate

- ☐ Engine oil level too low. Friction caused by intermittent lack of lubrication or from oil that is overworked can cause overheating. The oil provides a definite cooling function in the engine. Check the oil level (Chapter 1).
- ☐ Poor quality engine oil or incorrect viscosity or type. Oil is rated not only according to viscosity but also according to type. Some oils are not rated high enough for use in this engine. Check the Specifications section and change to the correct oil (Chapter 1).

Miscellaneous causes

- ☐ Modification to exhaust system. Most aftermarket exhaust systems cause the engine to run leaner, which make them run hotter.

5 Clutch problems

Clutch slipping

- ☐ Clutch plate worn or warped. Overhaul the clutch assembly (Chapter 2).
- ☐ Clutch diaphragm spring broken or weak. Remove and replace (Chapter 2).
- ☐ Clutch release mechanism or cable defective. Replace any defective parts (Chapter 2).
- ☐ Clutch contaminated with oil. Check the crankshaft and gearbox input shaft oil seals (see Chapter 2).
- ☐ Clutch cable freeplay insufficient (Chapter 1).
- ☐ Oil leaking on to clutch plate. Dismantle clutch (Chapter 2), renew clutch plate, wash off all traces of oil and trace source of leak.

Clutch not disengaging completely (drag)

- ☐ Clutch cable freeplay excessive (Chapter 1).
- ☐ Clutch plate warped or damaged. This will cause clutch drag, which in turn will cause the machine to creep. Overhaul the clutch assembly (Chapter 2).
- ☐ Clutch diaphragm spring broken or weak. Remove and replace (Chapter 2).
- ☐ Clutch release mechanism defective. Overhaul the components (Chapter 2).

6 Gearchanging problems

Doesn't go into gear or lever doesn't return

- [ ] Clutch not disengaging.
- [ ] Selector fork(s) bent or seized. Overhaul the transmission (Chapter 2).
- [ ] Gear(s) stuck on shaft. Most often caused by a lack of lubrication or excessive wear in transmission bearings and bushings. Overhaul the transmission (Chapter 2).
- [ ] Selector drum binding. Caused by lubrication failure or excessive wear. Replace the drum and bearing (Chapter 2).
- [ ] Gearchange lever return spring weak or broken (Chapter 2).
- [ ] Gearchange lever broken. Splines stripped out of lever or shaft, caused by allowing the lever to get loose. Replace necessary parts (Chapter 2).
- [ ] Stopper arm broken or worn. Full engagement and rotary movement of shift drum results. Replace the arm (Chapter 2).
- [ ] Stopper arm spring broken. Allows arm to float, causing sporadic shift operation. Replace spring (Chapter 2).

Jumps out of gear

- [ ] Shift fork(s) worn. Overhaul the transmission (Chapter 2).
- [ ] Gear groove(s) worn. Overhaul the transmission (Chapter 2).
- [ ] Gear dogs or dog slots worn or damaged. The gears should be inspected and replaced. No attempt should be made to service the worn parts.

Overshifts

- [ ] Stopper arm spring weak or broken (Chapter 2).
- [ ] Gearshift shaft return spring post broken or distorted (Chapter 2).

7 Abnormal engine noise

Knocking or pinking

- [ ] Carbon build-up in combustion chamber. Use of a fuel additive that will dissolve the adhesive bonding the carbon particles to the piston crown and chamber is the easiest way to remove the build-up. Otherwise, the cylinder head will have to be removed and decarbonised (Chapter 2).
- [ ] Incorrect or poor quality fuel. Old or improper fuel can cause detonation. This causes the pistons to rattle, thus the knocking or pinking sound. Drain the old fuel and always use the recommended grade fuel (Chapter 3).
- [ ] Spark plug heat range incorrect. Uncontrolled detonation indicates that the plug heat range is too hot. The plug in effect becomes a glow plug, raising cylinder temperatures. Install the proper heat range plug (Chapter 1).
- [ ] Improper air/fuel mixture. This will cause the cylinders to run hot and lead to detonation. Clogged injectors or an air leak can cause this imbalance. Refer to a BMW dealer equipped with the diagnostic tester.

Piston slap or rattling

- [ ] Cylinder-to-piston clearance excessive. Caused by improper assembly. Inspect and overhaul top-end parts (Chapter 2).
- [ ] Connecting rod bent. Caused by over-revving, trying to start a badly flooded engine or from ingesting a foreign object into the combustion chamber. Replace the damaged parts (Chapter 2).
- [ ] Piston pin or piston pin bore worn or seized from wear or lack of lubrication. Replace damaged parts (Chapter 2).
- [ ] Piston ring(s) worn, broken or sticking. Overhaul the top-end (Chapter 2).
- [ ] Piston seizure damage. Usually from lack of lubrication or overheating. Replace the pistons and bore the cylinders, as necessary (Chapter 2).
- [ ] Connecting rod upper or lower end clearance excessive. Caused by excessive wear or lack of lubrication. Replace worn parts.

Valve noise

- [ ] Incorrect valve clearances. Adjust the clearances by referring to Chapter 1.
- [ ] Valve spring broken or weak. Check and replace weak valve springs (Chapter 2).
- [ ] Camshafts or auxiliary shaft worn, or their bearing surfaces worn (Chapter 2).

Other noise

- [ ] Cylinder head gasket leaking.
- [ ] Exhaust pipe leaking at cylinder head connection. Caused by improper fit of pipe(s) or loose exhaust flange. All exhaust fasteners should be tightened evenly and carefully. Failure to do this will lead to a leak.
- [ ] Crankshaft runout excessive. Caused by a bent crankshaft (from over-revving) or damage from an upper cylinder component failure.
- [ ] Crankshaft bearings worn (Chapter 2).
- [ ] Cam chain tensioner defective. Replace according to the procedure in Chapter 2.
- [ ] Cam chain, sprockets or guides worn (Chapter 2).

8 Abnormal driveline noise

Clutch noise

- [ ] Loose or damaged clutch components (Chapter 2).

Transmission noise

- [ ] Bearings worn. Also includes the possibility that the shafts are worn. Overhaul the transmission (Chapter 2).
- [ ] Gears worn or chipped (Chapter 2).
- [ ] Metal chips jammed in gear teeth. Probably pieces from a broken gear or shift mechanism that were picked up by the gears. This will cause early bearing failure (Chapter 2).
- [ ] Gearbox oil level too low. Causes a howl from transmission (Chapter 1).

Final drive noise

- [ ] Final drive oil level low (Chapter 1).
- [ ] Final drive gear lash incorrect (Chapter 5).
- [ ] Final drive gears worn or damaged (Chapter 5).
- [ ] Final drive bearings worn (Chapter 5).
- [ ] Driveshaft splines worn and slipping (Chapter 5).

9 Abnormal frame and suspension noise

Front end noise

☐ Telelever ball joint loose or worn. Check and replace if necessary (see Chapter 5).

☐ Telelever mountings loose or bearings worn. Check and tighten or replace as required (see Chapter 5).

☐ Worn fork slider bushes (Chapter 5).

☐ Shock absorber faulty (see below).

☐ Steering head bearing (R, RT and GS) or ball joint (RS) loose or damaged. Clicks when braking. Check and replace as necessary (Chapter 5).

☐ Fork bridges loose. Make sure all bolts are tight (Chapter 5).

☐ Fork tube bent. Good possibility if machine has been dropped. Replace both fork tubes with a new ones (Chapter 5).

☐ Front axle or axle clamp bolt loose. Tighten them to the specified torque (Chapter 6).

Shock absorber noise

☐ Fluid level incorrect. Indicates a leak caused by defective seal. Shock will be covered with oil. Replace shock (Chapter 5).

☐ Defective shock absorber with internal damage. This is in the body of the shock and can't be remedied. The shock must be replaced with a new one (Chapter 6).

☐ Bent or damaged shock body. Replace the shock with a new one (Chapter 6).

Brake noise

☐ Squeal caused by dust on brake pads. Usually found in combination with glazed pads. Clean using brake cleaning solvent (Chapter 6).

☐ Contamination of brake pads. Oil, brake fluid or dirt causing brake to chatter or squeal. Clean or replace pads (Chapter 6).

☐ Pads glazed. Caused by excessive heat from prolonged use or from contamination. Do not use sandpaper, emery cloth, carborundum cloth or any other abrasive to roughen the pad surfaces as abrasives will stay in the pad material and damage the disc. A very fine flat file can be used, but pad replacement is suggested as a cure (Chapter 6).

☐ Disc warped. Can cause a chattering, clicking or intermittent squeal. Usually accompanied by a pulsating lever and uneven braking. Replace the disc (Chapter 7).

☐ Loose or worn wheel bearings (front) or final drive bearings (rear).

10 Oil pressure warning light comes on

Engine lubrication system

☐ Engine oil pump defective, blocked oil strainer gauze or failed relief valve. Carry out oil pressure check (Chapter 2).

☐ Engine oil level low. Inspect for leak or other problem causing low oil level and add recommended oil (Daily (pre-ride) checks).

☐ Engine oil viscosity too low. Very old, thin oil or an improper weight of oil used in the engine. Change to correct oil (Chapter 1).

☐ Camshafts, auxiliary shaft or crankshaft bearings worn. Excessive wear causing drop in oil pressure (Chapter 2).

Electrical system

☐ Oil pressure switch defective. Check the switch according to the procedure in Chapter 8. Replace it if it is defective.

☐ Oil pressure indicator light circuit defective. Check for pinched, shorted, disconnected or damaged wiring (Chapter 8).

11 Excessive exhaust smoke

White smoke

☐ Piston oil ring worn. The ring may be broken or damaged, causing oil from the crankcase to be pulled past the piston into the combustion chamber. Replace the rings with new ones (Chapter 2).

☐ Cylinders worn, cracked, or scored. Caused by overheating or oil starvation. Measure cylinder diameter and renew if worn or damaged.

☐ Valve oil seal damaged or worn. Replace oil seals with new ones (Chapter 2).

☐ Valve guide worn. Perform a complete valve job (Chapter 2).

☐ Engine oil level too high, which causes the oil to be forced past the rings. Drain oil to the proper level (Daily (pre-ride) checks).

☐ Head gasket broken between oil return and cylinder. Causes oil to be pulled into the combustion chamber. Replace the head gasket and check the head for warpage (Chapter 2).

☐ Abnormal crankcase pressurisation, which forces oil past the rings. Clogged breather hose is usually the cause.

Black smoke

☐ Air filter clogged. Clean or replace the element (Chapter 1).

☐ Motronic control unit defective. Refer to a BMW dealer equipped with the diagnostic tester.

Brown smoke

☐ Fuel filter clogged. Renew the fuel filter (Chapter 1).

☐ Fuel flow insufficient. Have a BMW dealer perform a fuel pressure check.

☐ Intake air leak. Check for loose throttle body-to-intake manifold and air duct connections, loose or missing vacuum take-off caps, or loose injectors (Chapter 3).

☐ Air filter poorly sealed or not installed (Chapter 1).

12 Poor handling or stability

Handlebars hard to turn
- [ ] Steering head bearing (R, RT or GS) or ball joint (RS) defective. Check and replace if necessary (Chapter 5).
- [ ] Front tyre air pressure too low (Daily (pre-ride) checks).

Handlebar shakes or vibrates excessively
- [ ] Tyres worn or out of balance. Inspect for wear (Daily (pre-ride) checks). Have a tyre specialist balance the wheels.
- [ ] Swingarm (Paralever) or Telelever bearings worn. Replace worn bearings by referring to Chapter 5.
- [ ] Rim(s) warped or damaged. Inspect wheels for runout (Chapter 6).
- [ ] Wheel bearings worn (Chapter 1).
- [ ] Handlebar clamp bolts loose (Chapter 5).
- [ ] Fork bridge bolts loose. Tighten them to the specified torque (Chapter 5).

Handlebar pulls to one side
- [ ] Wheels out of alignment (Chapter 6).
- [ ] Front or rear suspension components damaged. Caused by impact damage or by dropping the motorcycle. Have the machine checked over thoroughly by a BMW dealer or frame specialist.
- [ ] Fork tube bent. Renew both fork tubes (Chapter 5).

Poor shock absorbing qualities
- [ ] Incorrect adjustment or rear shock absorber (Chapter 5).
- [ ] Tyre pressures incorrect (Daily (pre-ride) checks).
- [ ] Front or rear shock absorber damage. Inspect shocks for leakage or damage.

13 Braking problems

Brakes are spongy, don't hold
- [ ] Air in brake line. Caused by inattention to master cylinder fluid level or by leakage. Locate problem and bleed brakes (Chapter 6).
- [ ] Pad or disc worn (Chapters 1 and 6).
- [ ] Brake fluid leak. See paragraph 1.
- [ ] Contaminated pads. Caused by contamination with oil, grease, brake fluid, etc. Clean or replace pads. Clean disc thoroughly with brake cleaner (Chapter 6).
- [ ] Brake fluid deteriorated. Fluid is old or contaminated. Drain system, replenish with new fluid and bleed the system (Chapter 6).
- [ ] Master cylinder internal parts worn or damaged causing fluid to bypass (Chapter 6).
- [ ] Master cylinder bore scratched by foreign material or broken spring. Repair or replace master cylinder (Chapter 6).
- [ ] Disc warped. Replace disc (Chapter 6).

Brake lever or pedal pulsates
- [ ] Disc warped. Replace disc (Chapter 6).
- [ ] Axle bent. Replace axle (Chapter 6).
- [ ] Brake caliper bolts loose (Chapter 6).

- [ ] Brake caliper sliders damaged or sticking (R, RT and GS rear caliper), causing caliper to bind. Lubricate the sliders or replace them if they are corroded or bent (Chapter 6).
- [ ] Wheel warped or otherwise damaged (Chapter 6).
- [ ] Wheel bearings (front) or final drive bearings (rear) damaged or worn (Chapter 6).
- [ ] ABS system faulty (Chapter 6).

Brakes drag
- [ ] Master cylinder piston seized. Caused by wear or damage to piston or cylinder bore (Chapter 6).
- [ ] Lever balky or stuck. Check pivot and lubricate (Chapter 6).
- [ ] Brake caliper binds. Caused by inadequate lubrication or damage to caliper slider pins – R, RT and GS rear caliper (Chapter 6).
- [ ] Brake caliper piston seized in bore. Caused by wear or ingestion of dirt past deteriorated seal (Chapter 6).
- [ ] Brake pad damaged. Pad material separated from backing plate. Usually caused by faulty manufacturing process or from contact with chemicals. Replace pads (Chapter 6).
- [ ] Pads improperly installed (Chapter 6).

14 Electrical problems

Battery dead or weak

☐ Battery faulty. Caused by sulphated plates which are shorted through sedimentation. Also, broken battery terminal making only occasional contact (Chapter 8).
☐ Battery cables making poor contact (Chapter 8).
☐ Load excessive. Caused by addition of high wattage lights or other electrical accessories.
☐ Ignition (main) switch defective. Switch either earths (grounds) internally or fails to shut off system. Replace the switch (Chapter 8).
☐ Charging system defective (Chapter 8).

☐ Wiring faulty. Wiring earthed (grounded) or connections loose in ignition, charging or lighting circuits (Chapter 8).

Battery overcharged

☐ Regulator/rectifier defective. Overcharging is noticed when battery gets excessively warm (Chapter 8).
☐ Battery defective. Replace battery with a new one (Chapter 8).
☐ Battery amperage too low, wrong type or size. Install manufacturer's specified amp-hour battery to handle charging load (Chapter 8).

Fault Finding Equipment

Checking engine compression

● Low compression will result in exhaust smoke, heavy oil consumption, poor starting and poor performance. A compression test will provide useful information about an engine's condition and if performed regularly, can give warning of trouble before any other symptoms become apparent.
● A compression gauge will be required, along with an adapter to suit the spark plug hole thread size. Note that the screw-in type gauge/adapter set up is preferable to the rubber cone type.

● Before carrying out the test, first check the valve clearances as described in Chapter 1.
1 Run the engine until it reaches normal operating temperature, then stop it and remove the spark plug(s), taking care not to scald your hands on the hot components.
2 Install the gauge adapter and compression gauge in No. 1 cylinder spark plug hole (see illustration 1).
3 On kickstart-equipped motorcycles, make sure the ignition switch is OFF, then open the throttle fully and kick the engine over a couple of times until the gauge reading stabilises.
4 On motorcycles with electric start only, the procedure will differ depending on the nature of the ignition system. Flick the engine kill switch (engine stop switch) to OFF and turn

Screw the compression gauge adapter into the spark plug hole, then screw the gauge into the adapter

the ignition switch ON; open the throttle fully and crank the engine over on the starter motor for a couple of revolutions until the gauge reading stabilises. If the starter will not operate with the kill switch OFF, turn the ignition switch OFF and refer to the next paragraph.

5 Install the spark plugs back into their suppressor caps and arrange the plug electrodes so that their metal bodies are earthed (grounded) against the cylinder head; this is essential to prevent damage to the ignition system as the engine is spun over **(see illustration 2)**. Position the plugs well away from the plug holes otherwise there is a risk of atomised fuel escaping from the combustion chambers and igniting. As a safety precaution, cover the top of the valve cover with rag. Now turn the ignition switch ON and kill switch ON, open the throttle fully and crank the engine over on the starter motor for a couple of revolutions until the gauge reading stabilises.

All spark plugs must be earthed (grounded) against the cylinder head

6 After one or two revolutions the pressure should build up to a maximum figure and then stabilise. Take a note of this reading and on multi-cylinder engines repeat the test on the remaining cylinders.

7 The correct pressures are given in Chapter 2 Specifications. If the results fall within the specified range and on multi-cylinder engines all are relatively equal, the engine is in good condition. If there is a marked difference between the readings, or if the readings are lower than specified, inspection of the top-end components will be required.

8 Low compression pressure may be due to worn cylinder bores, pistons or rings, failure of the cylinder head gasket, worn valve seals, or poor valve seating.

9 To distinguish between cylinder/piston wear and valve leakage, pour a small quantity of oil into the bore to temporarily seal the piston rings, then repeat the compression tests **(see illustration 3)**. If the readings show a noticeable increase in pressure this confirms that the cylinder bore, piston, or rings are worn. If, however, no change is indicated, the cylinder head gasket or valves should be examined.

Bores can be temporarily sealed with a squirt of motor oil

10 High compression pressure indicates excessive carbon build-up in the combustion chamber and on the piston crown. If this is the case the cylinder head should be removed and the deposits removed. Note that excessive carbon build-up is less likely with the used on modern fuels.

Checking battery open-circuit voltage

⚠️ *Warning: The gases produced by the battery are explosive - never smoke or create any sparks in the vicinity of the battery. Never allow the electrolyte to contact your skin or clothing - if it does, wash it off and seek immediate medical attention.*

● Before any electrical fault is investigated the battery should be checked.

● You'll need a dc voltmeter or multimeter to check battery voltage. Check that the leads are inserted in the correct terminals on the meter, red lead to positive (+ve), black lead to negative (-ve). Incorrect connections can damage the meter.

● A sound fully-charged 12 volt battery should produce between 12.3 and 12.6 volts across its terminals (12.8 volts for a maintenance-free battery). On machines with a 6 volt battery, voltage should be between 6.1 and 6.3 volts.

1 Set a multimeter to the 0 to 20 volts dc range and connect its probes across the

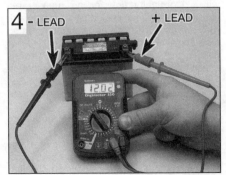

Measuring open-circuit battery voltage

battery terminals. Connect the meter's positive (+ve) probe, usually red, to the battery positive (+ve) terminal, followed by the meter's negative (-ve) probe, usually black, to the battery negative terminal (-ve) **(see illustration 4)**.

2 If battery voltage is low (below 10 volts on a 12 volt battery or below 4 volts on a six volt battery), charge the battery and test the voltage again. If the battery repeatedly goes flat, investigate the motorcycle's charging system.

Checking battery specific gravity (SG)

⚠️ *Warning: The gases produced by the battery are explosive - never smoke or create any sparks in the vicinity of the battery. Never allow the electrolyte to contact your skin or clothing - if it does, wash it off and seek immediate medical attention.*

● The specific gravity check gives an indication of a battery's state of charge.

● A hydrometer is used for measuring specific gravity. Make sure you purchase one which has a small enough hose to insert in the aperture of a motorcycle battery.

● Specific gravity is simply a measure of the electrolyte's density compared with that of water. Water has an SG of 1.000 and fully-charged battery electrolyte is about 26% heavier, at 1.260.

● Specific gravity checks are not possible on maintenance-free batteries. Testing the open-circuit voltage is the only means of determining their state of charge.

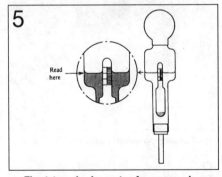

Float-type hydrometer for measuring battery specific gravity

1 To measure SG, remove the battery from the motorcycle and remove the first cell cap. Draw some electrolyte into the hydrometer and note the reading **(see illustration 5)**. Return the electrolyte to the cell and install the cap.

2 The reading should be in the region of 1.260 to 1.280. If SG is below 1.200 the battery needs charging. Note that SG will vary with temperature; it should be measured at 20°C (68°F). Add 0.007 to the reading for every 10°C above 20°C, and subtract 0.007 from the reading for every 10°C below 20°C. Add 0.004 to the reading for every 10°F above 68°F, and subtract 0.004 from the reading for every 10°F below 68°F.

3 When the check is complete, rinse the hydrometer thoroughly with clean water.

Checking for continuity

● The term continuity describes the uninterrupted flow of electricity through an electrical circuit. A continuity check will determine whether an **open-circuit** situation exists.

● Continuity can be checked with an ohmmeter, multimeter, continuity tester or battery and bulb test circuit **(see illustrations 6, 7 and 8)**.

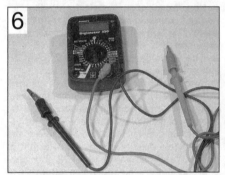

Digital multimeter can be used for all electrical tests

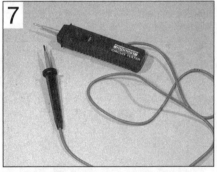

Battery-powered continuity tester

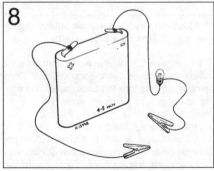

Battery and bulb test circuit

● All of these instruments are self-powered by a battery, therefore the checks are made with the ignition OFF.

● As a safety precaution, always disconnect the battery negative (-ve) lead before making checks, particularly if ignition switch checks are being made.

● If using a meter, select the appropriate ohms scale and check that the meter reads infinity (∞). Touch the meter probes together and check that meter reads zero; where necessary adjust the meter so that it reads zero.

● After using a meter, always switch it OFF to conserve its battery.

Switch checks

1 If a switch is at fault, trace its wiring up to the wiring connectors. Separate the wire connectors and inspect them for security and condition. A build-up of dirt or corrosion here will most likely be the cause of the problem - clean up and apply a water dispersant such as WD40.

Continuity check of front brake light switch using a meter - note split pins used to access connector terminals

2 If using a test meter, set the meter to the ohms x 10 scale and connect its probes across the wires from the switch **(see illustration 9)**. Simple ON/OFF type switches, such as brake light switches, only have two wires whereas combination switches, like the ignition switch, have many internal links. Study the wiring diagram to ensure that you are connecting across the correct pair of wires. Continuity (low or no measurable resistance - 0 ohms) should be indicated with the switch ON and no continuity (high resistance) with it OFF.

3 Note that the polarity of the test probes doesn't matter for continuity checks, although care should be taken to follow specific test procedures if a diode or solid-state component is being checked.

4 A continuity tester or battery and bulb circuit can be used in the same way. Connect its probes as described above **(see illustration 10)**. The light should come on to indicate continuity in the ON switch position, but should extinguish in the OFF position.

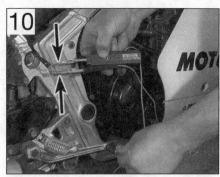

Continuity check of rear brake light switch using a continuity tester

Wiring checks

● Many electrical faults are caused by damaged wiring, often due to incorrect routing or chaffing on frame components.

● Loose, wet or corroded wire connectors can also be the cause of electrical problems, especially in exposed locations.

1 A continuity check can be made on a single length of wire by disconnecting it at each end and connecting a meter or continuity tester across both ends of the wire **(see illustration 11)**.

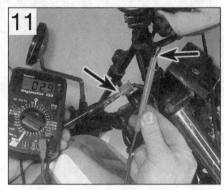

Continuity check of front brake light switch sub-harness

2 Continuity (low or no resistance - 0 ohms) should be indicated if the wire is good. If no continuity (high resistance) is shown, suspect a broken wire.

Checking for voltage

● A voltage check can determine whether current is reaching a component.

● Voltage can be checked with a dc voltmeter, multimeter set on the dc volts scale, test light or buzzer **(see illustrations 12 and 13)**. A meter has the advantage of being able to measure actual voltage.

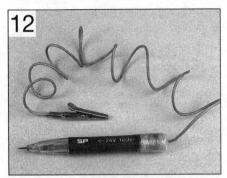

A simple test light can be used for voltage checks

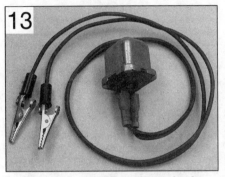

A buzzer is useful for voltage checks

● When using a meter, check that its leads are inserted in the correct terminals on the meter, red to positive (+ve), black to negative (-ve). Incorrect connections can damage the meter.

● A voltmeter (or multimeter set to the dc volts scale) should always be connected in parallel (across the load). Connecting it in series will destroy the meter.

● Voltage checks are made with the ignition ON.

1 First identify the relevant wiring circuit by referring to the wiring diagram at the end of this manual. If other electrical components share the same power supply (ie are fed from the same fuse), take note whether they are working correctly - this is useful information in deciding where to start checking the circuit.

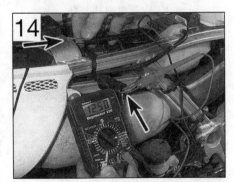

Checking for voltage at the rear brake light power supply wire using a meter . . .

2 If using a meter, check first that the meter leads are plugged into the correct terminals on the meter (see above). Set the meter to the dc volts function, at a range suitable for the battery voltage. Connect the meter red probe (+ve) to the power supply wire and the black probe to a good metal earth (ground) on the motorcycle's frame or directly to the battery negative (-ve) terminal **(see illustration 14)**. Battery voltage should be shown on the meter with the ignition switched ON.

3 If using a test light or buzzer, connect its positive (+ve) probe to the power supply terminal and its negative (-ve) probe to a good earth (ground) on the motorcycle's frame or directly to the battery negative (-ve) terminal **(see illustration 15)**. With the ignition ON, the test light should illuminate or the buzzer sound.

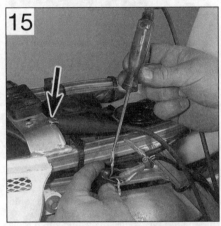

. . . or a test light - note the earth connection to the frame (arrow)

4 If no voltage is indicated, work back towards the fuse continuing to check for voltage. When you reach a point where there is voltage, you know the problem lies between that point and your last check point.

Checking the earth (ground)

● Earth connections are made either directly to the engine or frame (such as sensors, neutral switch etc. which only have a positive feed) or by a separate wire into the earth circuit of the wiring harness. Alternatively a short earth wire is sometimes run directly from the component to the motorcycle's frame.

● Corrosion is often the cause of a poor earth connection.

● If total failure is experienced, check the security of the main earth lead from the negative (-ve) terminal of the battery and also the main earth (ground) point on the wiring harness. If corroded, dismantle the connection and clean all surfaces back to bare metal.

1 To check the earth on a component, use an insulated jumper wire to temporarily bypass its earth connection **(see illustration 16)**. Connect one end of the jumper wire between the earth terminal or metal body of the component and the other end to the motorcycle's frame.

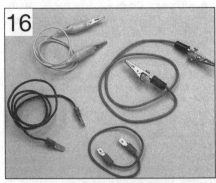

A selection of jumper wires for making earth (ground) checks

2 If the circuit works with the jumper wire installed, the original earth circuit is faulty. Check the wiring for open-circuits or poor connections. Clean up direct earth connections, removing all traces of corrosion and remake the joint. Apply petroleum jelly to the joint to prevent future corrosion.

Tracing a short-circuit

● A short-circuit occurs where current shorts to earth (ground) bypassing the circuit components. This usually results in a blown fuse.

● A short-circuit is most likely to occur where the insulation has worn through due to wiring chafing on a component, allowing a direct path to earth (ground) on the frame.

1 Remove any bodypanels necessary to access the circuit wiring.

2 Check that all electrical switches in the circuit are OFF, then remove the circuit fuse and connect a test light, buzzer or voltmeter (set to the dc scale) across the fuse terminals. No voltage should be shown.

3 Move the wiring from side to side whilst observing the test light or meter. When the test light comes on, buzzer sounds or meter shows voltage, you have found the cause of the short. It will usually shown up as damaged or burned insulation.

4 Note that the same test can be performed on each component in the circuit, even the switch.

A

ABS (Anti-lock braking system) A system, usually electronically controlled, that senses incipient wheel lockup during braking and relieves hydraulic pressure at wheel which is about to skid.

Aftermarket Components suitable for the motorcycle, but not produced by the motorcycle manufacturer.

Allen key A hexagonal wrench which fits into a recessed hexagonal hole.

Alternating current (ac) Current produced by an alternator. Requires converting to direct current by a rectifier for charging purposes.

Alternator Converts mechanical energy from the engine into electrical energy to charge the battery and power the electrical system.

Ampere (amp) A unit of measurement for the flow of electrical current. Current = Volts ˆ Ohms.

Ampere-hour (Ah) Measure of battery capacity.

Angle-tightening A torque expressed in degrees. Often follows a conventional tightening torque for cylinder head or main bearing fasteners **(see illustration)**.

Angle-tightening cylinder head bolts

Antifreeze A substance (usually ethylene glycol) mixed with water, and added to the cooling system, to prevent freezing of the coolant in winter. Antifreeze also contains chemicals to inhibit corrosion and the formation of rust and other deposits that would tend to clog the radiator and coolant passages and reduce cooling efficiency.

Anti-dive System attached to the fork lower leg (slider) to prevent fork dive when braking hard.

Anti-seize compound A coating that reduces the risk of seizing on fasteners that are subjected to high temperatures, such as exhaust clamp bolts and nuts.

API American Petroleum Institute. A quality standard for 4-stroke motor oils.

Asbestos A natural fibrous mineral with great heat resistance, commonly used in the composition of brake friction materials. Asbestos is a health hazard and the dust created by brake systems should never be inhaled or ingested.

ATF Automatic Transmission Fluid. Often used in front forks.

ATU Automatic Timing Unit. Mechanical device for advancing the ignition timing on early engines.

ATV All Terrain Vehicle. Often called a Quad.

Axial play Side-to-side movement.

Axle A shaft on which a wheel revolves. Also known as a spindle.

B

Backlash The amount of movement between meshed components when one component is held still. Usually applies to gear teeth.

Ball bearing A bearing consisting of a hardened inner and outer race with hardened steel balls between the two races.

Bearings Used between two working surfaces to prevent wear of the components and a build-up of heat. Four types of bearing are commonly used on motorcycles: plain shell bearings, ball bearings, tapered roller bearings and needle roller bearings.

Bevel gears Used to turn the drive through 90°. Typical applications are shaft final drive and camshaft drive **(see illustration)**.

Bevel gears are used to turn the drive through 90°

BHP Brake Horsepower. The British measurement for engine power output. Power output is now usually expressed in kilowatts (kW).

Bias-belted tyre Similar construction to radial tyre, but with outer belt running at an angle to the wheel rim.

Big-end bearing The bearing in the end of the connecting rod that's attached to the crankshaft.

Bleeding The process of removing air from an hydraulic system via a bleed nipple or bleed screw.

Bottom-end A description of an engine's crankcase components and all components contained there-in.

BTDC Before Top Dead Centre in terms of piston position. Ignition timing is often expressed in terms of degrees or millimetres BTDC.

Bush A cylindrical metal or rubber component used between two moving parts.

Burr Rough edge left on a component after machining or as a result of excessive wear.

C

Cam chain The chain which takes drive from the crankshaft to the camshaft(s).

Canister The main component in an evaporative emission control system (California market only); contains activated charcoal granules to trap vapours from the fuel system rather than allowing them to vent to the atmosphere.

Castellated Resembling the parapets along the top of a castle wall. For example, a castellated wheel axle or spindle nut.

Catalytic converter A device in the exhaust system of some machines which converts certain pollutants in the exhaust gases into less harmful substances.

Charging system Description of the components which charge the battery, ie the alternator, rectifer and regulator.

Circlip A ring-shaped clip used to prevent endwise movement of cylindrical parts and shafts. An internal circlip is installed in a groove in a housing; an external circlip fits into a groove on the outside of a cylindrical piece such as a shaft. Also known as a snap-ring.

Clearance The amount of space between two parts. For example, between a piston and a cylinder, between a bearing and a journal, etc.

Coil spring A spiral of elastic steel found in various sizes throughout a vehicle, for example as a springing medium in the suspension and in the valve train.

Compression Reduction in volume, and increase in pressure and temperature, of a gas, caused by squeezing it into a smaller space.

Compression damping Controls the speed the suspension compresses when hitting a bump.

Compression ratio The relationship between cylinder volume when the piston is at top dead centre and cylinder volume when the piston is at bottom dead centre.

Continuity The uninterrupted path in the flow of electricity. Little or no measurable resistance.

Continuity tester Self-powered bleeper or test light which indicates continuity.

Cp Candlepower. Bulb rating common found on US motorcycles.

Crossply tyre Tyre plies arranged in a criss-cross pattern. Usually four or six plies used, hence 4PR or 6PR in tyre size codes.

Cush drive Rubber damper segments fitted between the rear wheel and final drive sprocket to absorb transmission shocks **(see illustration)**.

Cush drive rubbers dampen out transmission shocks

D

Degree disc Calibrated disc for measuring piston position. Expressed in degrees.

Dial gauge Clock-type gauge with adapters for measuring runout and piston position. Expressed in mm or inches.

Diaphragm The rubber membrane in a master cylinder or carburettor which seals the upper chamber.

Diaphragm spring A single sprung plate often used in clutches.

Direct current (dc) Current produced by a dc generator.

Decarbonisation The process of removing carbon deposits - typically from the combustion chamber, valves and exhaust port/system.

Detonation Destructive and damaging explosion of fuel/air mixture in combustion chamber instead of controlled burning.

Diode An electrical valve which only allows current to flow in one direction. Commonly used in rectifiers and starter interlock systems.

Disc valve (or rotary valve) A induction system used on some two-stroke engines.

Double-overhead camshaft (DOHC) An engine that uses two overhead camshafts, one for the intake valves and one for the exhaust valves.

Drivebelt A toothed belt used to transmit drive to the rear wheel on some motorcycles. A drivebelt has also been used to drive the camshafts. Drivebelts are usually made of Kevlar.

Driveshaft Any shaft used to transmit motion. Commonly used when referring to the final driveshaft on shaft drive motorcycles.

E

Earth return The return path of an electrical circuit, utilising the motorcycle's frame.

ECU (Electronic Control Unit) A computer which controls (for instance) an ignition system, or an anti-lock braking system.

EGO Exhaust Gas Oxygen sensor. Sometimes called a Lambda sensor.

Electrolyte The fluid in a lead-acid battery.

EMS (Engine Management System) A computer controlled system which manages the fuel injection and the ignition systems in an integrated fashion.

Endfloat The amount of lengthways movement between two parts. As applied to a crankshaft, the distance that the crankshaft can move side-to-side in the crankcase.

Endless chain A chain having no joining link. Common use for cam chains and final drive chains.

EP (Extreme Pressure) Oil type used in locations where high loads are applied, such as between gear teeth.

Evaporative emission control system Describes a charcoal filled canister which stores fuel vapours from the tank rather than allowing them to vent to the atmosphere. Usually only fitted to California models and referred to as an EVAP system.

Expansion chamber Section of two-stroke engine exhaust system so designed to improve engine efficiency and boost power.

F

Feeler blade or gauge A thin strip or blade of hardened steel, ground to an exact thickness, used to check or measure clearances between parts.

Final drive Description of the drive from the transmission to the rear wheel. Usually by chain or shaft, but sometimes by belt.

Firing order The order in which the engine cylinders fire, or deliver their power strokes, beginning with the number one cylinder.

Flooding Term used to describe a high fuel level in the carburettor float chambers, leading to fuel overflow. Also refers to excess fuel in the combustion chamber due to incorrect starting technique.

Free length The no-load state of a component when measured. Clutch, valve and fork spring lengths are measured at rest, without any preload.

Freeplay The amount of travel before any action takes place. The looseness in a linkage, or an assembly of parts, between the initial application of force and actual movement. For example, the distance the rear brake pedal moves before the rear brake is actuated.

Fuel injection The fuel/air mixture is metered electronically and directed into the engine intake ports (indirect injection) or into the cylinders (direct injection). Sensors supply information on engine speed and conditions.

Fuel/air mixture The charge of fuel and air going into the engine. See **Stoichiometric ratio**.

Fuse An electrical device which protects a circuit against accidental overload. The typical fuse contains a soft piece of metal which is calibrated to melt at a predetermined current flow (expressed as amps) and break the circuit.

G

Gap The distance the spark must travel in jumping from the centre electrode to the side electrode in a spark plug. Also refers to the distance between the ignition rotor and the pickup coil in an electronic ignition system.

Gasket Any thin, soft material - usually cork, cardboard, asbestos or soft metal - installed between two metal surfaces to ensure a good seal. For instance, the cylinder head gasket seals the joint between the block and the cylinder head.

Gauge An instrument panel display used to monitor engine conditions. A gauge with a movable pointer on a dial or a fixed scale is an analogue gauge. A gauge with a numerical readout is called a digital gauge.

Gear ratios The drive ratio of a pair of gears in a gearbox, calculated on their number of teeth.

Glaze-busting see **Honing**

Grinding Process for renovating the valve face and valve seat contact area in the cylinder head.

Gudgeon pin The shaft which connects the connecting rod small-end with the piston. Often called a piston pin or wrist pin.

H

Helical gears Gear teeth are slightly curved and produce less gear noise that straight-cut gears. Often used for primary drives.

Installing a Helicoil thread insert in a cylinder head

Helicoil A thread insert repair system. Commonly used as a repair for stripped spark plug threads **(see illustration)**.

Honing A process used to break down the glaze on a cylinder bore (also called glaze-busting). Can also be carried out to roughen a rebored cylinder to aid ring bedding-in.

HT High Tension Description of the electrical circuit from the secondary winding of the ignition coil to the spark plug.

Hydraulic A liquid filled system used to transmit pressure from one component to another. Common uses on motorcycles are brakes and clutches.

Hydrometer An instrument for measuring the specific gravity of a lead-acid battery.

Hygroscopic Water absorbing. In motorcycle applications, braking efficiency will be reduced if DOT 3 or 4 hydraulic fluid absorbs water from the air - care must be taken to keep new brake fluid in tightly sealed containers.

I

lbf ft Pounds-force feet. An imperial unit of torque. Sometimes written as ft-lbs.

lbf in Pound-force inch. An imperial unit of torque, applied to components where a very low torque is required. Sometimes written as in-lbs.

IC Abbreviation for Integrated Circuit.

Ignition advance Means of increasing the timing of the spark at higher engine speeds. Done by mechanical means (ATU) on early engines or electronically by the ignition control unit on later engines.

Ignition timing The moment at which the spark plug fires, expressed in the number of crankshaft degrees before the piston reaches the top of its stroke, or in the number of millimetres before the piston reaches the top of its stroke.

Infinity (∞) Description of an open-circuit electrical state, where no continuity exists.

Inverted forks (upside down forks) The sliders or lower legs are held in the yokes and the fork tubes or stanchions are connected to the wheel axle (spindle). Less unsprung weight and stiffer construction than conventional forks.

J

JASO Quality standard for 2-stroke oils.

Joule The unit of electrical energy.

Journal The bearing surface of a shaft.

K

Kickstart Mechanical means of turning the engine over for starting purposes. Only usually fitted to mopeds, small capacity motorcycles and off-road motorcycles.

Kill switch Handebar-mounted switch for emergency ignition cut-out. Cuts the ignition circuit on all models, and additionally prevent starter motor operation on others.

km Symbol for kilometre.

kph Abbreviation for kilometres per hour.

L

Lambda (λ) sensor A sensor fitted in the exhaust system to measure the exhaust gas oxygen content (excess air factor).

Lapping see **Grinding**.
LCD Abbreviation for Liquid Crystal Display.
LED Abbreviation for Light Emitting Diode.
Liner A steel cylinder liner inserted in a aluminium alloy cylinder block.
Locknut A nut used to lock an adjustment nut, or other threaded component, in place.
Lockstops The lugs on the lower triple clamp (yoke) which abut those on the frame, preventing handlebar-to-fuel tank contact.
Lockwasher A form of washer designed to prevent an attaching nut from working loose.
LT Low Tension Description of the electrical circuit from the power supply to the primary winding of the ignition coil.

M

Main bearings The bearings between the crankshaft and crankcase.
Maintenance-free (MF) battery A sealed battery which cannot be topped up.
Manometer Mercury-filled calibrated tubes used to measure intake tract vacuum. Used to synchronise carburettors on multi-cylinder engines.
Micrometer A precision measuring instrument that measures component outside diameters **(see illustration)**.

Tappet shims are measured with a micrometer

MON (Motor Octane Number) A measure of a fuel's resistance to knock.
Monograde oil An oil with a single viscosity, eg SAE80W.
Monoshock A single suspension unit linking the swingarm or suspension linkage to the frame.
mph Abbreviation for miles per hour.
Multigrade oil Having a wide viscosity range (eg 10W40). The W stands for Winter, thus the viscosity ranges from SAE10 when cold to SAE40 when hot.
Multimeter An electrical test instrument with the capability to measure voltage, current and resistance. Some meters also incorporate a continuity tester and buzzer.

N

Needle roller bearing Inner race of caged needle rollers and hardened outer race. Examples of uncaged needle rollers can be found on some engines. Commonly used in rear suspension applications and in two-stroke engines.
Nm Newton metres.
NOx Oxides of Nitrogen. A common toxic pollutant emitted by petrol engines at higher temperatures.

O

Octane The measure of a fuel's resistance to knock.
OE (Original Equipment) Relates to components fitted to a motorcycle as standard or replacement parts supplied by the motorcycle manufacturer.
Ohm The unit of electrical resistance. Ohms = Volts ÷ Current.
Ohmmeter An instrument for measuring electrical resistance.
Oil cooler System for diverting engine oil outside of the engine to a radiator for cooling purposes.
Oil injection A system of two-stroke engine lubrication where oil is pump-fed to the engine in accordance with throttle position.
Open-circuit An electrical condition where there is a break in the flow of electricity - no continuity (high resistance).
O-ring A type of sealing ring made of a special rubber-like material; in use, the O-ring is compressed into a groove to provide the
Oversize (OS) Term used for piston and ring size options fitted to a rebored cylinder.
Overhead cam (sohc) engine An engine with single camshaft located on top of the cylinder head.
Overhead valve (ohv) engine An engine with the valves located in the cylinder head, but with the camshaft located in the engine block or crankcase.
Oxygen sensor A device installed in the exhaust system which senses the oxygen content in the exhaust and converts this information into an electric current. Also called a Lambda sensor.

P

Plastigauge A thin strip of plastic thread, available in different sizes, used for measuring clearances. For example, a strip of Plastigauge is laid across a bearing journal. The parts are assembled and dismantled; the width of the crushed strip indicates the clearance between journal and bearing.
Polarity Either negative or positive earth (ground), determined by which battery lead is connected to the frame (earth return). Modern motorcycles are usually negative earth.
Pre-ignition A situation where the fuel/air mixture ignites before the spark plug fires. Often due to a hot spot in the combustion chamber caused by carbon build-up. Engine has a tendency to 'run-on'.
Pre-load (suspension) The amount a spring is compressed when in the unloaded state. Preload can be applied by gas, spacer or mechanical adjuster.
Premix The method of engine lubrication on older two-stroke engines. Engine oil is mixed with the petrol in the fuel tank in a specific ratio. The fuel/oil mix is sometimes referred to as "petroil".
Primary drive Description of the drive from the crankshaft to the clutch. Usually by gear or chain.
PS Pfedestärke - a German interpretation of BHP.
PSI Pounds-force per square inch. Imperial measurement of tyre pressure and cylinder pressure measurement.
PTFE Polytetrafluroethylene. A low friction substance.

Pulse secondary air injection system A process of promoting the burning of excess fuel present in the exhaust gases by routing fresh air into the exhaust ports.

Q

Quartz halogen bulb Tungsten filament surrounded by a halogen gas. Typically used for the headlight **(see illustration)**.

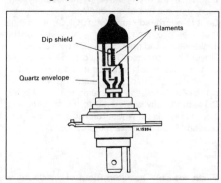

Quartz halogen headlight bulb construction

R

Rack-and-pinion A pinion gear on the end of a shaft that mates with a rack (think of a geared wheel opened up and laid flat). Sometimes used in clutch operating systems.
Radial play Up and down movement about a shaft.
Radial ply tyres Tyre plies run across the tyre (from bead to bead) and around the circumference of the tyre. Less resistant to tread distortion than other tyre types.
Radiator A liquid-to-air heat transfer device designed to reduce the temperature of the coolant in a liquid cooled engine.
Rake A feature of steering geometry - the angle of the steering head in relation to the vertical **(see illustration)**.

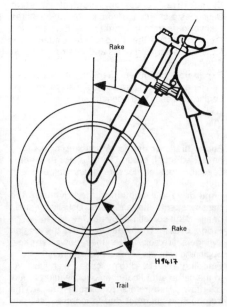

Steering geometry

Rebore Providing a new working surface to the cylinder bore by boring out the old surface. Necessitates the use of oversize piston and rings.

Rebound damping A means of controlling the oscillation of a suspension unit spring after it has been compressed. Resists the spring's natural tendency to bounce back after being compressed.

Rectifier Device for converting the ac output of an alternator into dc for battery charging.

Reed valve An induction system commonly used on two-stroke engines.

Regulator Device for maintaining the charging voltage from the generator or alternator within a specified range.

Relay A electrical device used to switch heavy current on and off by using a low current auxiliary circuit.

Resistance Measured in ohms. An electrical component's ability to pass electrical current.

RON (Research Octane Number) A measure of a fuel's resistance to knock.

rpm revolutions per minute.

Runout The amount of wobble (in-and-out movement) of a wheel or shaft as it's rotated. The amount a shaft rotates 'out-of-true'. The out-of-round condition of a rotating part.

S

SAE (Society of Automotive Engineers) A standard for the viscosity of a fluid.

Sealant A liquid or paste used to prevent leakage at a joint. Sometimes used in conjunction with a gasket.

Service limit Term for the point where a component is no longer useable and must be renewed.

Shaft drive A method of transmitting drive from the transmission to the rear wheel.

Shell bearings Plain bearings consisting of two shell halves. Most often used as big-end and main bearings in a four-stroke engine. Often called bearing inserts.

Shim Thin spacer, commonly used to adjust the clearance or relative positions between two parts. For example, shims inserted into or under tappets or followers to control valve clearances. Clearance is adjusted by changing the thickness of the shim.

Short-circuit An electrical condition where current shorts to earth (ground) bypassing the circuit components.

Skimming Process to correct warpage or repair a damaged surface, eg on brake discs or drums.

Slide-hammer A special puller that screws into or hooks onto a component such as a shaft or bearing; a heavy sliding handle on the shaft bottoms against the end of the shaft to knock the component free.

Small-end bearing The bearing in the upper end of the connecting rod at its joint with the gudgeon pin.

Spalling Damage to camshaft lobes or bearing journals shown as pitting of the working surface.

Specific gravity (SG) The state of charge of the electrolyte in a lead-acid battery. A measure of the electrolyte's density compared with water.

Straight-cut gears Common type gear used on gearbox shafts and for oil pump and water pump drives.

Stanchion The inner sliding part of the front forks, held by the yokes. Often called a fork tube.

Stoichiometric ratio The optimum chemical air/fuel ratio for a petrol engine, said to be 14.7 parts of air to 1 part of fuel.

Sulphuric acid The liquid (electrolyte) used in a lead-acid battery. Poisonous and extremely corrosive.

Surface grinding (lapping) Process to correct a warped gasket face, commonly used on cylinder heads.

T

Tapered-roller bearing Tapered inner race of caged needle rollers and separate tapered outer race. Examples of taper roller bearings can be found on steering heads.

Tappet A cylindrical component which transmits motion from the cam to the valve stem, either directly or via a pushrod and rocker arm. Also called a cam follower.

TCS Traction Control System. An electronically-controlled system which senses wheel spin and reduces engine speed accordingly.

TDC Top Dead Centre denotes that the piston is at its highest point in the cylinder.

Thread-locking compound Solution applied to fastener threads to prevent slackening. Select type to suit application.

Thrust washer A washer positioned between two moving components on a shaft. For example, between gear pinions on gearshaft.

Timing chain See **Cam Chain.**

Timing light Stroboscopic lamp for carrying out ignition timing checks with the engine running.

Top-end A description of an engine's cylinder block, head and valve gear components.

Torque Turning or twisting force about a shaft.

Torque setting A prescribed tightness specified by the motorcycle manufacturer to ensure that the bolt or nut is secured correctly. Undertightening can result in the bolt or nut coming loose or a surface not being sealed. Overtightening can result in stripped threads, distortion or damage to the component being retained.

Torx key A six-point wrench.

Tracer A stripe of a second colour applied to a wire insulator to distinguish that wire from another one with the same colour insulator. For example, Br/W is often used to denote a brown insulator with a white tracer.

Trail A feature of steering geometry. Distance from the steering head axis to the tyre's central contact point.

Triple clamps The cast components which extend from the steering head and support the fork stanchions or tubes. Often called fork yokes.

Turbocharger A centrifugal device, driven by exhaust gases, that pressurises the intake air. Normally used to increase the power output from a given engine displacement.

TWI Abbreviation for Tyre Wear Indicator. Indicates the location of the tread depth indicator bars on tyres.

U

Universal joint or U-joint (UJ) A double-pivoted connection for transmitting power from a driving to a driven shaft through an angle. Typically found in shaft drive assemblies.

Unsprung weight Anything not supported by the bike's suspension (ie the wheel, tyres, brakes, final drive and bottom (moving) part of the suspension).

V

Vacuum gauges Clock-type gauges for measuring intake tract vacuum. Used for carburettor synchronisation on multi-cylinder engines.

Valve A device through which the flow of liquid, gas or vacuum may be stopped, started or regulated by a moveable part that opens, shuts or partially obstructs one or more ports or passageways. The intake and exhaust valves in the cylinder head are of the poppet type.

Valve clearance The clearance between the valve tip (the end of the valve stem) and the rocker arm or tappet/follower. The valve clearance is measured when the valve is closed. The correct clearance is important - if too small the valve won't close fully and will burn out, whereas if too large noisy operation will result.

Valve lift The amount a valve is lifted off its seat by the camshaft lobe.

Valve timing The exact setting for the opening and closing of the valves in relation to piston position.

Vernier caliper A precision measuring instrument that measures inside and outside dimensions. Not quite as accurate as a micrometer, but more convenient.

VIN Vehicle Identification Number. Term for the bike's engine and frame numbers.

Viscosity The thickness of a liquid or its resistance to flow.

Volt A unit for expressing electrical "pressure" in a circuit. Volts = current x ohms.

W

Water pump A mechanically-driven device for moving coolant around the engine.

Watt A unit for expressing electrical power. Watts = volts x current.

Wear limit see **Service limit**

Wet liner A liquid-cooled engine design where the pistons run in liners which are directly surrounded by coolant **(see illustration).**

Wet liner arrangement

Wheelbase Distance from the centre of the front wheel to the centre of the rear wheel.

Wiring harness or loom Describes the electrical wires running the length of the motorcycle and enclosed in tape or plastic sheathing. Wiring coming off the main harness is usually referred to as a sub harness.

Woodruff key A key of semi-circular or square section used to locate a gear to a shaft. Often used to locate the alternator rotor on the crankshaft.

Wrist pin Another name for gudgeon or piston pin.

Note: References throughout this index are in the form - *"Chapter number"•"Page number"*

Haynes Motorcycle Manuals – The Complete List

Title	Book No.
BMW	
BMW 2-valve Twins (70 - 96)	0249
BMW R850 & R1100 4-valve Twins (93 - 97)	3466
BMW K100 & 75 2-valve Models (83 - 96)	1373
BSA	
BSA Bantam (48 - 71)	0117
BSA Unit Singles (58 - 72)	0127
BSA Pre-unit Singles (54 - 61)	0326
BSA A7 & A10 Twins (47 - 62)	0121
BSA A50 & A65 Twins (62 - 73)	0155
BULTACO	
Bultaco Competition Bikes (72 - 75)	0219
CZ	
CZ 125 & 175 Singles (69 - 90)	◊ 0185
DUCATI	
Ducati 600, 750 & 900 2-valve V-Twins (91 - 96)	3290
HARLEY-DAVIDSON	
Harley-Davidson Sportsters (70 - 93)	0702
Harley-Davidson Big Twins (70 - 93)	0703
HONDA	
Honda SH50 City Express (84 - 89)	◊ 1597
Honda NB, ND, NP & NS50 Melody (81 - 85)	◊ 0622
Honda NE/NB50 Vision & SA50 Vision Met-in (85 - 95)	◊ 1278
Honda MB, MBX, MT & MTX50 (80 - 93)	0731
Honda C50, C70 & C90 (67 - 95)	0324
Honda ATC70, 90, 110, 185 & 200 (71 - 85)	0565
Honda CR80R & CR125R (86 - 97)	2220
Honda XR80R & XR100R (85 - 96)	2218
Honda XL/XR 80, 100, 125, 185 & 200 2-valve Models (78 - 87)	0566
Honda CB100N & CB125N (78 - 86)	◊ 0569
Honda H100 & H100S Singles (80 - 92)	◊ 0734
Honda CB/CD125T & CM125C Twins (77 - 88)	◊ 0571
Honda CG125 (76 - 94)	◊ 0433
Honda NS125 (86 - 93)	◊ 3056
Honda CB125, 160, 175, 200 & CD175 Twins (64 - 78)	0067
Honda MBX/MTX125 & MTX200 (83 - 93)	◊ 1132
Honda CD/CM185 200T & CM250C 2-valve Twins (77 - 85)	0572
Honda XL/XR 250 & 500 (78 - 84)	0567
Honda XR250L, XR250R & XR400R (86 - 97)	2219
Honda CB250RS Singles (80 - 84)	◊ 0732
Honda CB250 & CB400N Super Dreams (78 - 84)	◊ 0540
Honda CR250R & CR500R (86 - 97)	2222
Honda Elsinore 250 (73 - 75)	0217
Honda TRX300 Shaft Drive ATVs (88 - 95)	2125
Honda CB400 & CB550 Fours (73 - 77)	0262
Honda CX/GL500 & 650 V-Twins (78 - 86)	0442
Honda CBX550 Four (82 - 86)	◊ 0940
Honda XL600R & XR600R (83 - 96)	2183
Honda CBR600F1 & 1000F Fours (87 - 96)	1730
Honda CBR600F2 Fours (91 - 94)	2070
Honda CB650 sohc Fours (78 - 84)	0665
Honda NTV600 & 650 V-Twins (88 - 96)	3243
Honda CB750 sohc Four (69 - 79)	0131
Honda V45/65 Sabre & Magna (82 - 88)	0820
Honda VFR750 & 700 V-Fours (86 - 97)	2101
Honda CB750 & CB900 dohc Fours (78 - 84)	0535
Honda CBR900RR FireBlade (92 - 97)	2161
Honda GL1000 Gold Wing (75 - 79)	0309
Honda GL1100 Gold Wing (79 - 81)	0669
Honda ST1100 Pan European V-Fours (90 - 97)	3384
KAWASAKI	
Kawasaki AE/AR 50 & 80 (81 - 95)	1007
Kawasaki KC, KE & KH100 (75 - 93)	1371
Kawasaki AR125 (82 - 94)	◊ 1006
Kawasaki KMX125 & 200 (86 - 96)	◊ 3046
Kawasaki 250, 350 & 400 Triples (72 - 79)	0134
Kawasaki 400 & 440 Twins (74 - 81)	0281
Kawasaki 400, 500 & 550 Fours (79 - 91)	0910
Kawasaki EN450 & 500 Twins (Ltd/Vulcan) (85 - 93)	2053
Kawasaki EX500 (GPZ500S) Twins (87 - 93)	2052
Kawasaki ZX600 (Ninja ZX-6, ZZ-R600) Fours (90 - 97)	2146
Kawasaki ZX600 (GPZ600R, GPX600R, Ninja 600R & RX) & ZX750 (GPX750R, Ninja 750R) Fours (85 - 97)	1780
Kawasaki 650 Four (76 - 78)	0373
Kawasaki 750 Air-cooled Fours (80 - 91)	0574
Kawasaki ZR550 & 750 Zephyr Fours (90 - 97)	3382
Kawasaki ZX750 (Ninja ZX-7 & ZXR750) Fours (89 - 95)	2054
Kawasaki 900 & 1000 Fours (73 - 77)	0222
Kawasaki ZX900, 1000 & 1100 Liquid-cooled Fours (83 - 97)	1681
MOTO GUZZI	
Moto Guzzi 750, 850 & 1000 V-Twins (74 - 78)	0339
MZ	
MZ TS125 (76 - 86)	◊ 1270
MZ ETZ Models (81 - 95)	◊ 1680

Title	Book No.
NORTON	
Norton 500, 600, 650 & 750 Twins (57 - 70)	0187
Norton Commando (68 - 77)	0125
SUZUKI	
Suzuki FR50, 70 & 80 (74 - 87)	◊ 0801
Suzuki GT, ZR & TS50 (77 - 90)	◊ 0799
Suzuki TS50X (84 - 95)	◊ 1599
Suzuki 100, 125, 185 & 250 Air-cooled Trail bikes (79 - 89)	0797
Suzuki GP100 & 125 Singles (78 - 93)	◊ 0576
Suzuki GS & DR125 Singles (82 - 94)	◊ 0888
Suzuki 250 & 350 Twins (68 - 78)	0120
Suzuki GT250X7, GT200X5 & SB200 Twins (78 - 83)	◊ 0469
Suzuki GS/GSX250, 400 & 450 Twins (79 - 85)	0736
Suzuki GS500E Twin (89 - 97)	3238
Suzuki GS550 (77 - 82) & GS750 Fours (76 - 79)	0363
Suzuki GS/GSX550 4-valve Fours (83 - 88)	1133
Suzuki GSF600 & 1200 Bandit Fours (95 - 97)	3367
Suzuki GS850 Fours (78 - 88)	0536
Suzuki GS1000 Four (77 - 79)	0484
Suzuki GSX-R750, GSX-R1100, GSX600F, GSX750F, GSX1100F (Katana) Fours (85 - 96)	2055

Title	Book No.
Suzuki GS/GSX1000, 1100 & 1150 4-valve Fours (79 - 88)	0737
TOMOS	
Tomos A3K, A3M, A3MS & A3ML Mopeds (82 - 91)	◊ 1062
TRIUMPH	
Triumph Tiger Cub & Terrier (52 - 68)	0414
Triumph 350 & 500 Unit Twins (58 - 73)	0137
Triumph Pre-Unit Twins (47 - 62)	0251
Triumph 650 & 750 2-valve Unit Twins (63 - 83)	0122
Triumph Trident & BSA Rocket 3 (69 - 75)	0136
Triumph Triples & Fours (91 - 95)	2162
VESPA	
Vespa P/PX125, 150 & 200 Scooters (78 - 95)	0707
Vespa Scooters (59 - 78)	0126
YAMAHA	
Yamaha FS1E, FS1 & FS1M (72 - 90)	◊ 0166
Yamaha RD50 & 80 (78 - 89)	◊ 1255
Yamaha DT50 & 80 Trail Bikes (78 - 95)	◊ 0800
Yamaha T50 & 80 Townmate (83 - 95)	◊ 1247
Yamaha YT, YFM, YTM & YTZ ATVs (80 - 85)	1154
Yamaha YB100 Singles (73 - 91)	◊ 0474
Yamaha 100, 125 & 175 Trail bikes (71 - 85)	0210
Yamaha RS/RXS100 & 125 Singles (74 - 95)	0331
Yamaha RD & DT125LC (82 - 87)	◊ 0887
Yamaha TZR125 (87 - 93) & DT125R (88 - 95)	◊ 1655
Yamaha TY50, 80, 125 & 175 (74 - 84)	◊ 0464
Yamaha XT & SR125 (82 - 96)	1021
Yamaha 250 & 350 Twins (70 - 79)	0040
Yamaha XS250, 360 & 400 sohc Twins (75 - 84)	0378
Yamaha YBF250 Timberwolf ATV (92 - 96)	2217
Yamaha YFM350 Big Bear and ER ATVs (87 - 95)	2126
Yamaha RD250 & 350LC Twins (80 - 82)	0803
Yamaha RD350 YPVS Twins (83 - 95)	1158
Yamaha RD400 Twin (75 - 79)	0333
Yamaha XT, TT & SR500 Singles (75 - 83)	0342
Yamaha XZ550 Vision V-Twins (82 - 85)	0821
Yamaha FJ, FZ, XJ & YX600 Radian (84 - 92)	2100
Yamaha XJ600S (Seca II, Diversion) & XJ600N Fours (92 - 95 UK) (92 - 96 USA)	2145
Yamaha 650 Twins (70 - 83)	0341
Yamaha XJ650 & 750 Fours (80 - 84)	0738
Yamaha XS750 & 850 Triples (76 - 85)	0340
Yamaha FZR600, 750 & 1000 Fours (87 - 96)	2056
Yamaha XV V-Twins (81 - 96)	0802
Yamaha XJ900F Fours (83 - 94)	3239
Yamaha FJ1100 & 1200 Fours (84 - 96)	2057
PRACTICAL MANUALS	
ATV Basics	10450
Motorcycle Basics Manual	1083
Motorcycle Carburettor Manual	0603
Motorcycle Electrical Manual (2nd Edition)	0446
Motorcycle Workshop Practice Manual	1454

◊ denotes manual not available in the USA.

The manuals featured on this page are available through good motorcycle dealers and accessory shops. In case of difficulty, contact:
Haynes Publishing (UK) +44 1963 440635
(USA) +1 805 4986703
(France) +33 1 47 03 61 80
(Sweden) +46 18 124016
(Australia/New Zealand) +61 3 9763 8100

MCL05.01/98

Preserving Our Motoring Heritage

< The Model J Duesenberg Derham Tourster. Only eight of these magnificent cars were ever built – this is the only example to be found outside the United States of America

Almost every car you've ever loved, loathed or desired is gathered under one roof at the Haynes Motor Museum. Over 300 immaculately presented cars and motorbikes represent every aspect of our motoring heritage, from elegant reminders of bygone days, such as the superb Model J Duesenberg to curiosities like the bug-eyed BMW Isetta. There are also many old friends and flames. Perhaps you remember the 1959 Ford Popular that you did your courting in? The magnificent 'Red Collection' is a spectacle of classic sports cars including AC, Alfa Romeo, Austin Healey, Ferrari, Lamborghini, Maserati, MG, Riley, Porsche and Triumph.

A Perfect Day Out

Each and every vehicle at the Haynes Motor Museum has played its part in the history and culture of Motoring. Today, they make a wonderful spectacle and a great day out for all the family. Bring the kids, bring Mum and Dad, but above all bring your camera to capture those golden memories for ever. You will also find an impressive array of motoring memorabilia, a comfortable 70 seat video cinema and one of the most extensive transport book shops in Britain. The Pit Stop Cafe serves everything from a cup of tea to wholesome, home-made meals or, if you prefer, you can enjoy the large picnic area nestled in the beautiful rural surroundings of Somerset.

> John Haynes O.B.E., Founder and Chairman of the museum at the wheel of a Haynes Light 12.

< The 1936 490cc sohc-engined International Norton – well known for its racing success

The Museum is situated on the A359 Yeovil to Frome road at Sparkford, just off the A303 in Somerset. It is about 40 miles south of Bristol, and 25 minutes drive from the M5 intersection at Taunton.

Open 9.30am - 5.30pm (10.00am - 4.00pm Winter) 7 days a week, *except Christmas Day, Boxing Day and New Years Day*

Special rates available for schools, coach parties and outings Charitable Trust No. 292048